Renewable Energy: Geographic Information Systems and Remote Sensing Techniques

Renewable Energy: Geographic Information Systems and Remote Sensing Techniques

Editor: Daxton Owen

R CALLISTO REFERENCE

www.callistoreference.com

Callisto Reference,
118-35 Queens Blvd., Suite 400,
Forest Hills, NY 11375, USA

Visit us on the World Wide Web at:
www.callistoreference.com

ISBN: 978-1-64116-830-4 (Hardback)

Cataloging-in-Publication Data

Renewable energy : geographic information systems and remote sensing
techniques / edited by Daxton Owen.
        p. cm.
Includes bibliographical references and index.
ISBN 978-1-64116-830-4
1. Renewable energy sources--Remote sensing. 2. Renewable energy sources--Geographic
information systems. 3. Environmental monitoring--Remote sensing. I. Owen, Daxton.
TJ808 .R46 2023
333.794--dc23

Table of Contents

Permissions

List of Contributors

Index

Preface

This book aims to highlight the current researches and provides a platform to further the scope of innovations in this area. This book is a product of the combined efforts of many researchers and scientists, after going through thorough studies and analysis from different parts of the world. The objective of this book is to provide the readers with the latest information of the field.

Renewable energy is the type of energy that is derived from renewable sources of energy. Hydro energy, geothermal energy, solar energy, biomass energy, wind energy, and tidal energy are some of the most popular renewable energy sources. Remote sensing and geographic information systems (GIS) technologies are widely used in the field of renewable energy. One of the most important steps in developing high-penetration renewable energy systems is evaluating the potential of renewable energy sources. A variety of tools and techniques can be utilized to accomplish this, including on-site remote sensing tools, in situ observations, reanalysis datasets, and satellite image data. The techniques based on remote sensing and GIS are beneficial in finding out optimum locations of renewable energy resources for developing renewable energy infrastructure. This book is a compilation of chapters that discuss the most vital concepts and emerging trends in the use of geographic information systems and remote sensing techniques for renewable energy. It is appropriate for students seeking detailed information in this area of study as well as for experts.

I would like to express my sincere thanks to the authors for their dedicated efforts in the completion of this book. I acknowledge the efforts of the publisher for providing constant support. Lastly, I would like to thank my family for their support in all academic endeavors.

Editor

Real-Time Automatic Cloud Detection using a Low-Cost Sky Camera

Joaquín Alonso-Montesinos [1,2]

[1] Department of Chemistry and Physics, University of Almería, 04120 Almería, Spain; joaquin.alonso@ual.es
[2] CIESOL, Joint Centre of the University of Almería-CIEMAT, 04120 Almería, Spain

Abstract: Characterizing the atmosphere is one of the most complex studies one can undertake due to the non-linearity and phenomenological variability. Clouds are also among the most variable atmospheric constituents, changing their size and shape over a short period of time. There are several sectors in which the study of cloudiness is of vital importance. In the renewable field, the increasing development of solar technology and the emerging trend for constructing and operating solar plants across the earth's surface requires very precise control systems that provide optimal energy production management. Similarly, airports are hubs where cloud coverage is required to provide high-precision periodic observations that inform airport operators about the state of the atmosphere. This work presents an autonomous cloud detection system, in real time, based on the digital image processing of a low-cost sky camera. An algorithm was developed to identify the clouds in the whole image using the relationships established between the channels of the RGB and Hue, Saturation, Value (HSV) color spaces. The system's overall success rate is approximately 94% for all types of sky conditions; this is a novel development which makes it possible to identify clouds from a ground perspective without the use of radiometric parameters.

Keywords: cloud detection; cloud coverage; sky camera; image processing; remote sensing; CSP plants; solar energy; solar irradiance forecasting; total sky imagery

1. Introduction

With the expansion in solar plant development [1,2], comprehensive knowledge of the events that might affect plant production quality is required. Solar technologies involving energy transformation generally have inherent issues that must be overcome. Knowing when clouds will appear in the solar field is essential information for solar plant operators [3]. With this knowledge, operators can perform a range of actions to optimize solar plant operation.

At airports, a daily meteorological report is provided showing the state of the cloud cover; this is usually carried out by a human observer with experience in cloud visualization. It is, therefore, necessary to develop a real-time meteorological cloud detection system capable of repeatedly providing concise information on the state of the atmosphere.

Clouds have been detected using a wide variety of tools, one of which is satellite imagery. A simple classification method was developed based on the split-window technique. This system provided a detection accuracy of 44%, with an underestimation error of 56%, correctly classifying the areas in 88% of cases [4]. Another work presented two machine-learning methods to determine cloud masking using Spinning Enhanced Visible and Infrared Imager (SEVIRI) images that measured the reflectance values obtained from the IR3.9 channel.

In general, the determination coefficient (r^2) presented higher results than in 75% of the cases analyzed with MODIS (Moderate Resolution Imaging Spectroradiometer) and CLM (the Cloud Mask product from EUMETSAT) images [5]. Furthermore, several authors have used Landsat 8 data to deal with the problem of detecting clouds in visible and multispectral imagery from high-resolution satellite cameras [6]. Focusing on the importance of monitoring cloud cover over solar power plant areas, satellites have been used for cloud estimation, thus making it possible to track clouds and forecast their future position to predict when the sun will be blocked [7]. Other authors developed the function of mask (Fmask) 4.0 algorithm to automatically detect cloud and cloud shadow in Landsats 4-8 and Sentinel-2 images, where normally the computational time needed to process a single Landsat or Sentinel-2 image takes between 1.5 and 10 min (depending on the percentage of cloud cover and the land surface characteristics) [8]. However, space satellites have certain drawbacks, such as spatial and temporal resolution. Geostationary satellites provide images at a frequency of about 15 min and a spatial resolution of several square kilometers. Polar satellites have a higher spatial resolution, in the order of meters, but usually take only one image per day. Added to this, the process for performing certain tasks with these images (matrix calculations) involves more time and may not be very efficient, depending on how the obtained information is applied.

Sky cameras are a way of providing a view of the sky that complements the satellites, where clouds can be identified more accurately and at a higher temporal resolution. In some works, dust episodes have also been studied with this technology, as in the case of the Saharan dust intrusion over southern Spain in 2017. This appeared as though it were an overcast image that consequently affected the Direct Normal Irradiance (DNI) [9]. The study of aerosol optical properties is another important field that uses sky cameras [10]. Some authors have created their own sky cameras for cloud cover assessment without using conventional solar occulting devices [11]. Furthermore, to counteract the temporal resolution limitations of satellite images, a digital camera was used for night-time cloud detection, detailing the percentage of cloud cover at 5-min intervals over the Manila Observatory [12]. In addition, sky cameras have been used to record cloud detection from solar radiation data [13] and then to predict the solar resource over the short term using digital levels and maximum cross correlation [14,15]. Other authors have compared cloud detection data using satellite imagery and sky cameras [16,17]; in [16], the comparison was made in south-eastern Brazil over a period of approximately three months. Good agreement was obtained for clear sky and overcast conditions, with detection probabilities of 92.8% and 80.7%, respectively. For partially cloudy skies, the agreement was around 40%. In that article, the authors cited problems with the sky camera, for example, very bright areas around the sun, which were sometimes identified as clouds, leading to cloud cover overestimation. A similar observation was cited in [18]. In [17] the observations took place in Germany and New Zealand, over time frames of 3 and 2 months, respectively. For a clear sky, the authors found detection probabilities of between 72 and 76%. They recommended that more automated ground-based instruments (in the form of cloud cameras) should be installed as they cover larger areas of the sky than less automated ground-based versions. These cameras could be an invaluable supplement to SYNOP observation as they cover the same spectral wavelengths as the human eye. It is also common to use the R/B, R − B or (R − B)/(R + B) ratios to obtain better cloud characterization, where R is the red channel pixels and B the blue channel pixels [19]. Nevertheless, these ratios lack precision when the image is processed, especially in the solar area where problems with pixel brightness tend to overestimate the presence of clouds.

This work presents an automatic and autonomous cloud detection system using a low-cost sky camera (Mobotix Q24). The system mainly uses digital image levels and the solar height angle calculated at minute intervals. For each image, the system generates a processed image representing the original image; this is accomplished by identifying the clouds in white and the sky in blue. In general, the novelty of this approach lies in the high overall cloud detection accuracy in the sun area without needing to use radiation data to detect the clouds; this is made possible by optimally defining the parameters related to pixel brightness. In other works, only the RGB ratios are used.

The article is structured as follows: Section 2 presents the properties of the image acquisition system. Section 3 gives a step-by-step description of the methodology used to identify clouds (this has been divided into subsections due to its size). Section 4 presents the most important results, both visual and numerical, after comparing this system with another consolidated cloud detection system. Finally, Section 5 sets out the main conclusions.

2. Sky Camera Acquisition and Optical Properties

This work used images from a hemispheric low-cost sky camera, (model Mobotix Q24), placed on the rooftop of the Solar Energy Research Center (CIESOL) at the University of Almería, Spain ($36.8°N$, $2.4°W$; at sea level). Figure 1 shows the camera installation on the CIESOL building.

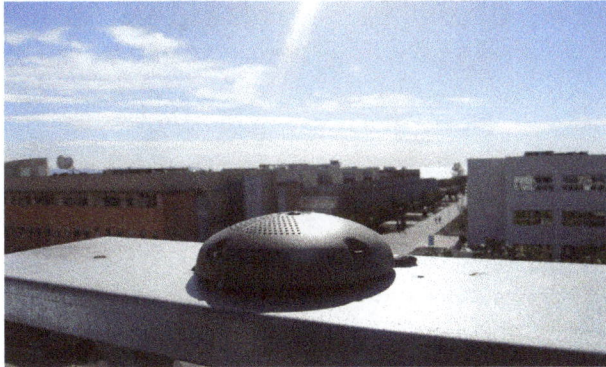

Figure 1. Mobotix Q24 camera installed in the University of Almería.

The facility's location has a Mediterranean climate with a high maritime aerosol presence. The images produced are high resolution from a fully digital color CMOS (2048×1536 pixels). One image is recorded every minute in JPEG format, the optimal time for identifying clouds in the sky. The three distinct channels represent the red, green and blue levels. Each image pixel is made up of 8 bits, resulting in values of between 0 and 255.

For this work, images were selected from all possible sky types, spanning the earliest times of the day to just before sunset, and at different times of the year. The period studied was from 2013 to 2019.

For this work, images were selected from all possible sky types, spanning the earliest times of the day to the latest, just before sunset, and at different times of the year. The period studied was from 2013 to 2019.

3. Methodology

This section describes the particular tasks undertaken to find the different clouds that can appear in a sky capture, defining the methodology and the steps involved to process an image. All developments have been carried out in the MATLAB environment as it is the optimal platform for operating with matrices in record time. Its efficiency is the main reason this software was used for the methodology. As the starting point, a raw image is presented in Figure 2, where the sky is represented as a common image.

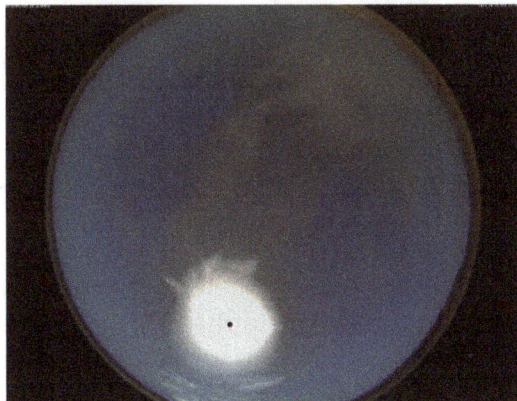

Figure 2. Sky image acquired from the low-cost sky camera on 19 October 2014 at 11:18 (Universal Time Coordinated - UTC).

Here, we see the circular representation of the sky appearing over a black background. The first step is to determine the area of interest from the raw image by applying a white mask, thus obtaining the image seen in Figure 3, where the sky is perfectly identified in the circular area.

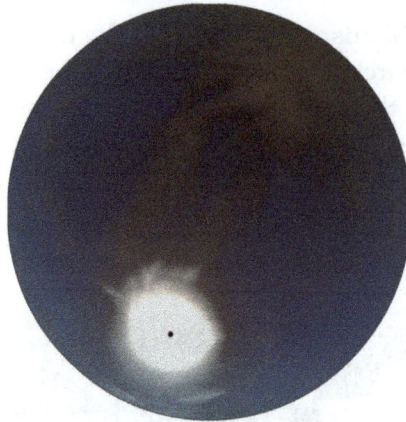

Figure 3. Application of a white mask to the raw sky image on 19 October 2014 at 11:18 (UTC).

After applying this customized mask, the image is ready for use in the developed algorithm to detect clouds. The MATLAB environment allows one to make changes in the image's color space so that specific properties can be studied. This is the case with HSV (Hue, Saturation, Value), NTSC and yCbCr color spaces. The first, HSV, gives information about the gray color scale, where pixels vary between 0 and 1. The second, NTSC, is really an RGB color space, where the first component, luminance, represents grayscale information and the latter two components make up the chrominance (color information). The last, yCbCr, is used to digitally encode the color information in the computing systems: y represents the brightness of the color, Cb is the blue component relative to the green component and Cr is the red component relative to the green component. Figure 4 shows the image presented beforehand in the three color spaces.

Figure 4. Representation of different color spaces for the image acquired from the low-cost sky camera.

The different colors of the three images represent the main image characteristics. Focusing solely on the inside of the circle, the blue color identifies the more saturated areas of the HSV image. These areas are represented by red and yellow in the NTSC image, and by orange, rose and red in the yCbCr image. Normally, in the NTSC and yCbCr color spaces, the pixel values acquire inflexible static values in each color space channel whereas in the HSV color space, the pixels (represented by the three bands/channels) provide values with better precision for the purpose of cloud detection. Moreover, the clouds are not perfectly represented in the three color spaces so it is important to define the most significant color space to work with. Given that the HSV color space represents cloudy pixels better (including those clouds that are more difficult to classify), and more clearly, it has been used together with the RGB color space to identify clouds in the image processing procedure [13,14,20]. For the complete image processing procedure, the developed algorithm was structured into different parts, as described in the following subsections. The different tables define specific criteria for image processing. Basically, each table has been created for the purpose of each subsection and to set the intensity levels of the channels, so as to precisely detect the zones (clouds or sky). To obtain a special classification of the pixels in each zone, different labels are defined and attributed to the pixels, as shown in Table 1.

Table 1. Criteria for labeling pixels in the whole sky cam image.

Pixel Label	Color Assigned	Pixel Condition
256	Green	Initial saturated pixels
257	White	Cloudy pixel in the solar area
258	White	Cloudy pixel in general (excluding solar area)
259	Blue	Cloudless pixel in general (excluding solar area)
260	Red	Cloudless pixel in the solar area
261	Yellow	Pixels around solar area with intermediate bright

To accomplish this, different images have been analyzed at different times of the day and at different times of the year. Following this, a general initial state has been assumed to precisely adjust the intensity values. Normally, the general state of each table starts by analyzing the R value and the comparison with the G and B channels. If the intensities of these channels are enclosed in particular values, the HSV channels are also analyzed. The uncertainty of each channel is the sensitivity of each one, i.e., in the RGB color space, ± 1, it is ± 0.001 for the H channel, ± 0.0001 for the S channel and ± 1 for the V channel. Therefore, this algorithm has been gradually formed in parts so that in the end, it remains connected and sequential, in the same order as it appears in the manuscript.

3.1. Recognition of the Solar Area: Classification of Pixels

The first step for detecting clouds in the whole sky image is to determine the solar area. Being able to recognize this area is fundamental for establishing the sun's position in the image. To track the angle of solar altitude (α) at each minute, the Cartesian coordinates are obtained, with the south represented by the bottom-center pixels and the east by the centeR − right pixels. Subsequently, the original image (in JPG format), defined by the RGB color space, is also converted into the HSV color space. As seen in the previous images, the sun appears as bright pixels, so one needs to consider the position of the pixels to determine the bright solar pixels. To do this, after locating the sun pixel (according to the solar altitude and azimuthal angles), a matrix is created to determine the distance of the other pixels from it (Dis). This operation allows one to classify whether a bright pixel is a 'solar pixel' or not (based on its position). As a general rule, when the value of the red, blue and green pixels is greater than 160, the pixels are identified as being in the sun area. Figure 5 shows the general detection of the sun pixels.

Figure 5. Identification of sun pixels applying a green mask.

The main step consists of applying a green mask to pixels that are placed in the sun area. After that, the idea is to detect if these pixels are cloudless or overcast. Table 2 shows the rules for determining cloudless pixels in the solar area. In general, the table collects the relationships between the pixel levels (according to the corresponding channel) that satisfy the criteria for determining cloudless pixels in the solar area. The parameters and thresholds have been defined based on the cases studied for the proposed model.

Table 2. Criteria for selecting cloudless pixels in the solar area.

Red (R)	Green (G)	Blue (B)	Hue (H)	Saturation (S)	Special Condition
[171,180]	-	-	0	0	R = G & G = B
[225,235]	-	R − 2	(0.166, 0.168)	(0.0070, 0.0100)	R = G
≥200	-	R − 2	(0.166, 0.168)	(0.0070, 0.0100)	R = G
228	-	R − 2	(0.166, 0.168)	(0.008, 0.009)	R = G
≥230	-	-	>0	-	R = G & G = B
≥210	-	-	≥0.416	≤0.005	-
[173,218]	-	-	[0.416, 0.418]	(0.0112, 0.1130)	Dis < 135
[173,178]	≥178	-	(0.166, 0.168)	(0.0110, 0.1120)	Dis ≥ 135
[172,177]	-	-	(0.166, 0.168)	(0.0100, 0.1113]	R = G
≥176	-	R − 2	≠0.167	-	Dis < 135 & R = G
≥176	-	R − 2	(0.166, 0.168)	(0.0103, 0.0107)	Dis < 135 & R = G
≥172	-	-	(0.416, 0.418)	(0.0109, 0.0113)	R = (G − 2) & R = (B − 1)
≥180	-	-	(0.166, 0.168)	(0.0100, 0.0110)	R = G & R = (B + 2)
≥172	-	-	(0.416, 0.418)	(0.0105, 0.0107)	R = (G − 2) & R = (B − 1)
[172,185]	-	-	(0.166, 0.168)	(0.0112, 0.0113)	R = G & R = (B + 2)
[170,174]	R + 2	R + 1	(0.416, 0.418)	(0.0107, 0.0109)	-
[190,194]	R + 2	R + 1	(0.166, 0.168)	(0.0112, 0.0113)	R = G & R = (B + 2)
194	R + 2	R − 1	(0.277, 0.279)	(0.0152, 0.154)	-
≥225	≥225	≥225	0	0	α ≤ 14
≥200	≥200	≥200	[0.416, 0.418]	≥0.0080	α ≤ 14

After the different strategies have been carried out to determine the cloudless pixels in the sun area, according to the pixel intensity in each image channel (Table 2), Figure 6 shows the general detection of the cloudless pixels in the sun area (represented in red) after this filter has been applied.

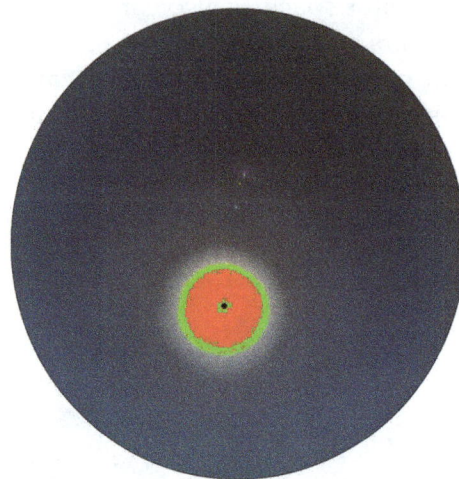

Figure 6. Determination of cloudless pixels in the solar area of the sky cam image.

Subsequently, the algorithm looks for cloudy pixels in the same area to detect if any clouds are present. Table 3 shows the condition for classifying the pixels in the solar area as cloudy.

Table 3. Criteria for selecting cloudy pixels in the solar area.

Red (R)	Green (G)	Blue (B)	Hue (H)	Saturation (S)	Special Condition
≥172	-	-	<0.168	≥0.013	R = G

Only one sentence (criterion) is applied for detecting cloudy pixels in the solar area. In these situations, when a cloud is identified by means of a pixel, the mask applied is also green. When the solar area has been fully treated, the algorithm focuses on the rest of the image, starting with the solar area periphery.

3.2. Detection of Bright Zones around the Solar Area

The pixels located around the solar area have an intermediate bright characteristic. In other words, the pixels present values lower than the solar area pixels but higher than those in the rest of the image. The size of this area varies according to the day and the atmospheric conditions at each moment. Table 4 shows the adjusted criteria for determining these pixels.

Table 4. Criteria for detecting bright pixels around the solar area.

Red (R)	Green (G)	Blue (B)	Hue (H)	Saturation (S)	Special Condition
>125	-	-	-	-	(B − R > 8) & (B − 4 > 4)
130	≥140	≥150	-	≥0.1900	Dis < 650
(130,140)	≥150	≥159	-	≥0.1900	Dis < 650
[140,150)	≥160	≥169	-	≥0.1900	Dis < 650
[150,165]	≥165	≥175	-	≥0.1900	Dis < 650
>130	-	-	-	-	(Dis < 650) & (B − R > 30)
>130	≠R	-	<0.1	≥0.1000	(Dis < 650) & (R − G ≥ 8)
-	>110	-	-	-	(Dis < 650) & (G > R) & (G > B) & (G − R < 8)
>140	-	-	-	-	(Dis < 650) & (R − G ≤ 10) & (R − B ≥ 25)
(≠ 257) & (≠ 260)	≥130	-	[0.400, 0.580]	≤0.0800	(Dis < 650)
(≠ 257) & (≠ 260)	-	-	≤0.090	≤0.0900	(Dis < 650)
(≠ 257) & (≠ 260)	-	-	≤0.090	≤0.0900	(Dis < 650) & (G < R) & (R − G < 8) & (|G − B| < 3)
(≥100) & (≠ 257) & (≠ 260)	-	-	-	-	(Dis < 650) & (G − R > 20) & (B − R > 35)
(≥100) & (≠ 257) & (≠ 260)	-	-	-	-	(Dis < 650) & (G − R > 20) & (B − R > 35)

One of the most important tasks is to locate each pixel. The *Dis* variable almost always appears because the pixel emplacement is very important in this process. Therefore, to distinguish previously classified areas in subsequent processes, the yellow color is used to mark the new area (Figure 7).

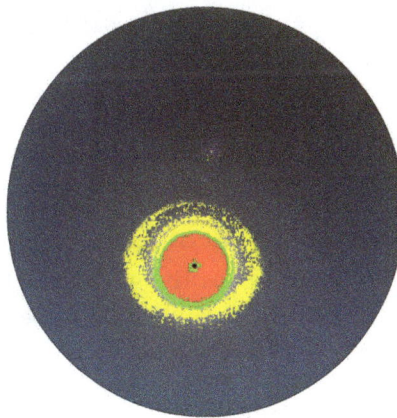

Figure 7. Detection of the pixels with an intermediate value of bright around the solar area.

The new pixels classified as yellow do not represent a homogeneous area; they are dispersed across the image but at a distance of less than 650 units from the central solar pixel. In this new preprocessing, there are gaps between the yellow and green pixels that need to be classified beforehand. With the solar and surrounding area processed, the algorithm looks for cloudy pixels in the rest of the image.

3.3. Detection of Cloudy Pixels in the Rest of the Image

In general, clouds present several characteristics that allow us to identify the most common cloud types (white or extremely dark clouds). Table 5 shows the general pattern for detecting the clouds in the complete image by characterizing the digital levels of these common cloud types.

Table 5. Criteria for cloud detection excluding the solar area.

Red (R)	Green (G)	Blue (B)	Hue (H)	Saturation (S)	Special Condition
B	B	-	-	-	-
$\neq[259,261]$	-	-	-	>0.2500	(V ≠ 256) & (R ≥ B)
$\neq[259,261]$	-	≥100	-	≤0.1800	(V ≠ 256) & (R/B ≤ 0.90)
$\neq[259,261]$	-	-	≤0.200	≤0.2000	(V ≠ 256)
$\neq45$	-	≥80	<0.600	≥0.3500	-

If a cloudy pixel is detected, it is marked in white. There are many cases in which some pixels are identified as cloudy although no clouds are present in the sky. This is caused by the similarity in the range of channel values, whereby dark skies can be confused for dark clouds. However, this mistake can be remedied during the algorithm's following steps. An example is presented in Figure 8, where certain pixels are classified as cloudy in color white.

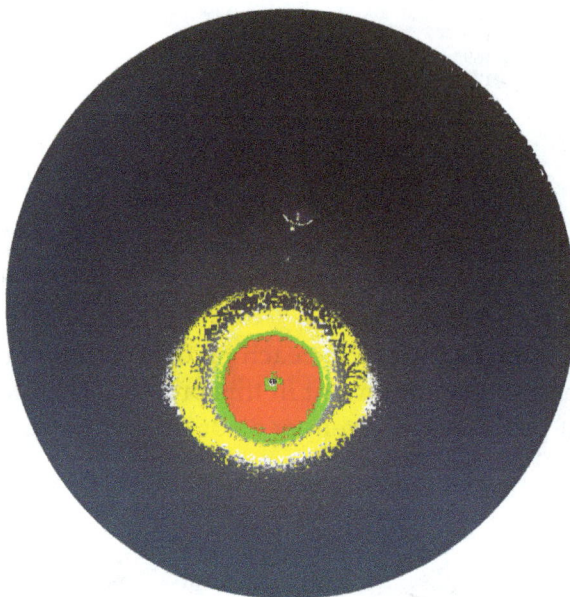

Figure 8. Detection of cloudy pixels in the total area of the sky cam.

Only a few pixels are classified as cloudy near the sun area. The first picture for this day showed no clouds appearing in the image, thus no cloudy pixels could be generated. Despite this, a few pixels are interpreted as cloud. When solar area pixels and cloudy pixels are evaluated, the process continues to detect the pixels as unclassified.

3.4. Detection of Cloudless Pixels in the Image Excluding the Solar Area

After the solar area has been classified, the rest of the image is analyzed to identify if a pixel represents a cloud or not. Table 6 represents the set of sentences implemented to detect the cloud-free pixels in the parts of the image that do not include the solar area.

Table 6. Criteria for detecting cloudless pixels in the image excepting in the solar area.

Red (R)	Green (G)	Blue (B)	Hue (H)	Saturation (S)	Special Condition
-	-	≤83	≥0.250	≥0.3000	-
≥55	-	≤83	≥0.580	≥0.5700	-
≤52	-	≥100	≥0.570	≥0.5700	-
≤50	≤60	≥83	≥0.610	-	-
-	-	-	≥0.500	≥0.4500	-
[50,60]	[57,63]	[66,75]	[0.610, 0.650]	[0.2000, 0.3000]	-
≤70	-	≥75	≤0.700	≤0.3100	-
≥55	-	-	≥0.600	-	$R/B \leq 0.80$
[50,68]	-	-	-	≥0.4600	$R/B \leq 0.80$
>G	-	-	≥0.660	-	$(R/B \geq 0.94)$ & $(G/B \geq 0.88)$
-	-	≥90	≥0.620	-	-
-	-	≥115	≥0.550	≥0.3000	-
-	-	-	≥0.800	-	-
[65,72]	[80,90]	≥108	-	-	-
[49,54]	[60,69]	[86,93]	-	-	-
G	-	-	-	-	$B > G$
≥100	≥110	≥130	-	-	-
-	B	>70	-	-	$R < B$
[70,75]	-	[84,89]	-	-	$G - R < 4$
-	-	[90,100]	-	-	$(G - R \in [6,7])$ & $(G - R \in [9,12]))$
-	-	-	-	-	$(\lvert R - B \rvert \leq 3)$ & $(G{>}B)$ & $(R/G \in [0.90, 0.92])$
[80,85]	[82,90]	[88,94]	-	-	$(G - R \leq 3)$ & $(B - R \leq 7)$
[59,64]	[69,75]	-	-	-	$(G - R \geq 6)$ & $(B - R \geq 12)$

One can see that the *Dis* variable was not used even though we have presented the criteria to identify the pixels outside the solar area have been presented. This is because, for cloudy pixels, the digital pixel levels never appear in the range levels shown in the table. For this reason, it was not necessary to include the aforementioned variable in the sentences used (Figure 9).

Figure 9. A representation of the sky cam image processing where the sun area and a part of the sky have been processed.

In the image, the sun area and the surrounding area are processed, along with a small part of the remaining image. Therefore, at this point in the algorithm, it is still possible that a large part of the image has not been processed. Consequently, a further step is needed to conclude the algorithm and classify all the pixels.

3.5. Determination of Non-Classified Pixels

The final steps for classifying the pixels in a complete image establish a statistical criterion based on the pixels that have already been classified. Knowing the number of pixels for each color, we determine those pixels that do not have a label. There are different strategies for establishing the classification criteria of these, as yet unclassified pixels. Table 7 shows the steps to determine whether the pixels should be classified as cloudless; if not, they will be classified as cloudy.

Table 7. Criteria for the determination of cloudless pixels for non-labeled pixels.

Main Condition	Other Conditions
$\alpha < 5$	(SkyPixels > 1,000,000) & (RedPixels $\geq$ 7000)
$\alpha \geq 5$	CloudPixels $\leq$ 25,000
$\alpha \geq 5$	(CloudPixels > 25,000) & (GreenPixels $\leq$ 1.6 RedPixels)
$\alpha \geq 5$	(CloudPixels > 25,000) & (RedPixels $\geq$ 0.7 (GreenPixels + YellowPixles))
$\alpha \geq 5$	(CloudPixels > 25,000) & (NonClass $\geq$ 1,100,000 & (RedPixels $\geq$ 2 GreenPixels) & (SkyPixels > CloudPixels))
$\alpha \geq 5$	(CloudPixels > 25,000) & (NonClass $\geq$ 800,000 & (SkyPixels $\geq$ 400,000) & (CloudPixels $\leq$ 140,000))
$\alpha \geq 5$	(CloudPixels > 25,000) & (NonClass $\geq$ 600,000 & (SkyPixels $\geq$ 700,000) & (CloudPixels $\leq$ 120,000))

In the table, different expressions appear. *SkyPixels* are pixels that have been classified as cloudless, whereas *CloudPixels* are those that have been labeled as cloudy. Red, green and yellow pixels have been obtained in the previous processes and *NonClass* is used to refer to the pixels that remain unclassified. For these operations, the MATLAB environment allows one to perform matrix operations in an efficient way. This part of the algorithm results in a matrix where all the pixels have been labeled, as shown in Figure 10.

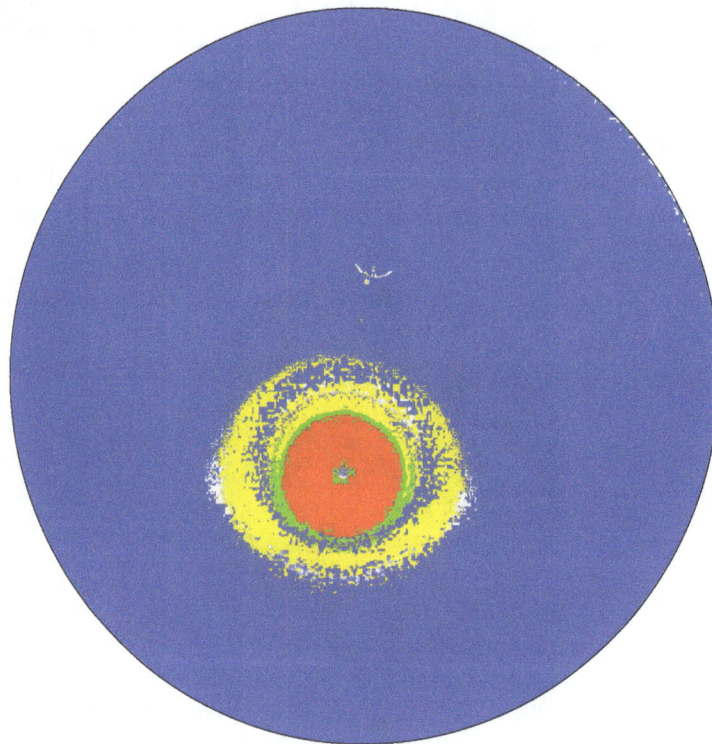

Figure 10. Graphical representation of the identification of the pixels in the sky cam image.

As one can see, all the pixels have been assigned a color: blue, yellow, red or green. Now, the aim is to finish the classification process according to a common criterion.

3.6. Final Step in the Sky Cam Image Classification

To finish the sky cam image processing, a final step is needed in which the differently colored pixels are converted so one can determine whether they are cloudless or cloudy. This process has been developed from experience gathered working with a great number of images and scenarios. Table 8 shows the specific criterion for assigning the final pixel classification.

Table 8. Criteria for the last classification of pixels in the sky cam image processing.

Label	Classification Conditions	Final Classification
Green	(Dis $\geq$ 700)	Cloud
Yellow	(Dis $\geq$ 700)	Cloud
Green/Yellow	($\alpha \geq 5$) & (RedPixels $\geq$ 0.5 GreenPixels) & (GreenPixels > 25,000)	Cloudless
Green/Yellow	($\alpha \geq 5$) & (RedPixels $\geq$ 0.3 GreenPixels) & (SkyPixels $\geq$ 1,000,000)	Cloudless
Green/Yellow	($\alpha \geq 5$) & (RedPixels $\geq$ 0.3 GreenPixels) & (RedPixels $\leq$ 15,000)	Cloud
Green/Yellow	($\alpha \geq 5$) & (RedPixels $\geq$ 0.3 GreenPixels) & (CloudPixels $\geq$ 500,000) & (RedPixels $\leq$ 0.5 GreenPixels)	Cloud
Green/Yellow	($\alpha \geq 5$) & (RedPixels $\geq$ 0.3 GreenPixels) & (CloudPixels $\geq$ 500,000) & (GreenPixels $\leq$ 25,000)	Cloud
Green/Yellow	($\alpha \geq 5$) & (RedPixels $\geq$ 0.3 GreenPixels) & (CloudPixels $\geq$ 500,000) & (RedPixels ϵ [15,000,30,000])	Cloud
Green/Yellow	($\alpha \geq 5$) & (RedPixels $\geq$ 0.5 GreenPixels) & (GreenPixels > 25,000)	Cloudless
Green/Yellow	($\alpha \geq 5$) & (RedPixels $\geq$ 0.3 GreenPixels) & (SkyPixels $\geq$ 1,000,000)	Cloudless
Green/Yellow	($\alpha \geq 5$) & (SkyPixels $\geq$ 1,000,000) & (RedPixels $\leq$ 15,000)	Cloud
Green/Yellow	($\alpha \geq 5$) & (SkyPixels $\geq$ 1,000,000) & (CloudPixels $\geq$ 500,000) & (RedPixels $\leq$ 0.5 GreenPixels)	Cloud
Green/Yellow	($\alpha \geq 5$) & (SkyPixels $\geq$ 1,000,000) & (CloudPixels $\geq$ 500,000) & (GreenPixels $\leq$ 25,000)	Cloud
Green/Yellow	($\alpha \geq 5$) & (SkyPixels $\geq$ 1,000,000) & (CloudPixels $\geq$ 500,000) & (RedPixels ϵ [15,000,30,000])	Cloud
Green/Yellow	($\alpha < 5$) & (RedPixels $\geq$ 7000)	Cloudless
Green/Yellow	($\alpha < 5$) & (RedPixels $\geq$ 3 GreenPixels)	Cloudless
Green/Yellow	($\alpha < 5$) & (RedPixels $\geq$ 10,000) & (SkyPixels $\geq$ 1,000,000)	Cloudless
Green/Yellow	OtheR $-$ case	Cloud
Red	(Dis $\geq$ 700)	Cloud
Red	-	Cloudless

Following the assignations in the table, the final image can be generated, the result of which is shown in Figure 11.

Figure 11. Result of the sky cam image processing.

The criteria presented in the above tables have been carefully defined, thus changes in the correlations represent alterations in the final processed image. Each criterion has an associated sensitivity according to the number of pixels involved (the pixels that fulfill the determined criteria). Therefore, the sensitivity associated with each criterion affects the pixels that fulfill the condition and, consequently, the final processed image. An error in one of the criteria presented in the tables would mean a cloud detection error, and therefore the image would not have been processed in a valid way and would be identified as wrong.

4. Results

In this section, we present the results of the cloud detection algorithm. To analyze the behavior of the software developed under different sky conditions, this section presents several pictures from various sky scenarios. A total of 850 images were taken from 2013 to 2019 at different times (from sunrise to sunset). The images were processed with the analytical objective of obtaining an accurate identification of clear sky and clouds. Therefore, this section is divided into subsections.

4.1. Sky Images Processed under all Sky Conditions

To analyze the quality of the developed model, several images have been processed and studied. In general, the processed image should show the most important clouds appearing in the original image (when visually inspected). The important clouds are those that can be identified clearly (not only by a few pixels). To view several examples, the following figures represent the image processing procedures carried out for the algorithm that was presented in the previous section. Figure 12 shows two examples chosen randomly, in which a clear and cloudless sky appear.

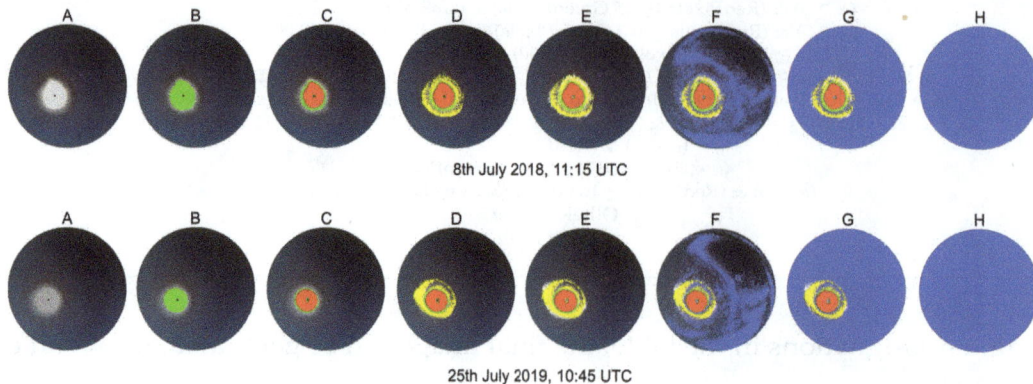

Figure 12. Result of the Sky Cam Image Processing Under Cloudless Sky Conditions.

In the case of cloudless skies, the sun can vary its form and size depending on the solar altitude angle. The algorithm contemplates this angle to identify clouds based on the variability in pixel intensity according to the sun position. Specifically, the sun is the key intensity point of the pixel value and the solar position determines a pathway for performing the cloud recognition. Despite this, for the two cloudless days represented in the image, the sky is free of clouds, as represented by images with the letter H. As one can observe, for each original image (marked with the letter A), a sequence of images appears showing the steps the algorithm takes to finally obtain the image (by identifying the sky and cloudy pixels). In these cases, no cloud was detected and, therefore, the final images are completely blue.

When the sky is not completely free of clouds, it can be either partially or completely covered. In the first case, varying portions of the sky and cloud may appear when observing a sky camera image. Therefore, it is important to determine the boundaries between the cloud and the sky as effectively as possible. Figure 13 shows two different partially cloudy situations.

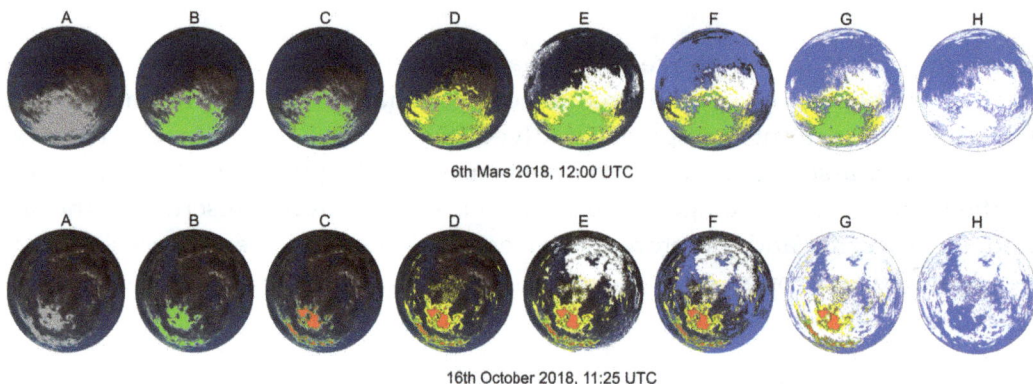

Figure 13. Result of the sky cam image processing under partially cloudy sky conditions.

Two scenes have been represented for cloud identification, each differing in various ways. In the first sequence of images (the top line), one can appreciate how most of the clouds are around the solar area. In images B and C, the algorithm recognizes the large green and yellow areas. Subsequently, clouds are detected in the adjoining areas (image E) before finally classifying those green and yellow pixels as a cloud. In the other image sequence (the bottom line), the area detected in green turns red where the sun appears. This is because the algorithm's established criteria have not been met for identifying the pixels as a cloud; therefore, they are marked in red. Following this, the clouds are optimally detected in image E and the blue sky is detected in image F. Finally, the image is resolved, classifying the pixels in red, green and yellow as cloudless. It is curious how the clouds have generated significant brightness in the solar area, mistakenly classifying the solar area pixels as clouds.

Figure 14 shows two cases in which virtually the entire image is covered by clouds.

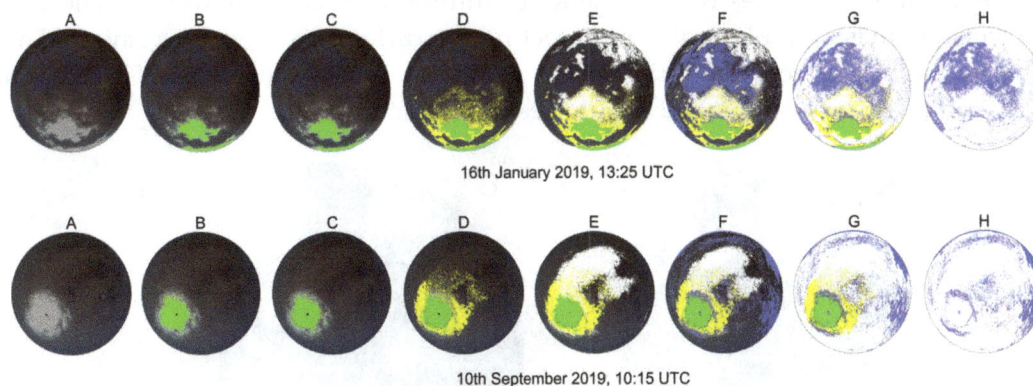

Figure 14. Result of the sky cam image processing under overcast sky conditions.

The top sequence of images shows a day when there was a lot of cloudiness and only small portions of sky. As the algorithm is executed, it is interesting to see how no red zone is detected (attributing the solar area as cloudless) because the clouds in this case have a profile that is perfectly identified by the set of sentences presented in the previous tables. Here, the breaking clouds have been correctly classified in image F. To conclude, image H shows the result of the process with the identification and classification of all the processed pixels being virtually identical to the original image. The bottom image sequence shows another day when there were more clouds in the image, and again one can observe that no red pixels have appeared. By following the steps described in the previous sections, the image processing very precisely determines the areas of blue sky and clouds (image H).

4.2. Statistical Results and Comparison with TSI-880

To make a statistical evaluation of the developed model's efficacy, we used a model that is already established and published [13,21]. This model works with images from a sky camera (installed on the CIESOL building) that has a rotational shadow band (the TSI-880 model) providing a hemispheric view of the sky (fish-eye vision). In short, the TSI-880 camera model is based on a sky classification that uses direct, diffuse and global radiation data. The sky is classified as clear, covered or partially covered. For each sky type, digital image processing is performed, and an image is obtained with the cloud identification. Figure 15 shows an example of a partially cloudy day at a time shortly after sunrise.

Looking at the TSI-880 camera image, it is significant how the pixel intensities are very different from those represented by the Mobotix camera. The blue sky generally appears lighter because of the camera's optics. For this reason, it is not possible to extrapolate the development of the TSI-880 camera to that of the Mobotix Q24 camera.

Taking this model as a reference, a total of 845 images were taken between the months of August and December 2013 (by that time, the images from both cameras were available). To make the comparison, a routine was established whereby an image was taken approximately once an hour (whenever possible) and the images from both cameras were processed. In total, 419 images corresponded to clear skies, 202 to partially covered skies and 224 to overcast skies. A probability function was used to determine the model's success rate in percentage values. It was considered a success if the processed image represented the original image; that is to say, it differentiated the areas of clear sky and clouds correctly. Equation (1) shows the probability function (*PF*):

$$PF(\%) = 100 \cdot \frac{Successes}{Total\ cases}, \tag{1}$$

In Equation (1), two variables are defined: Successes and Total cases. The first is based on obtaining a processed image with practically perfect cloud identification. Almost perfect means that the processed image adequately represents what appears in the original image (either with the TSI-880 camera or with the Mobotix camera). Therefore, a hit will be a final processed image that is like the original. The total number of cases will be the total number of images analyzed. The evaluation is done visually since there is no tool capable of detecting the difference between a raw and a processed image (that is why a cloud detection algorithm is so valuable). In this sense, the reference is always the original image, and any processed image should resemble the original.

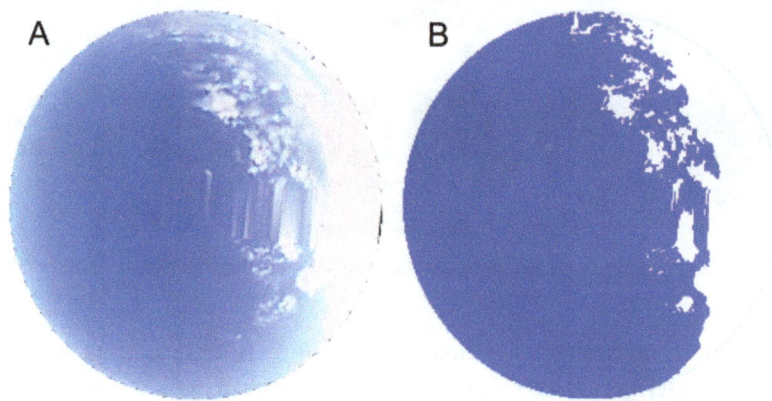

Figure 15. Image from TSI-880 on 31st March 2012, 7:15 UTC, where (**A**) is the original and (**B**) is the processed image.

Figure 16 shows the image processing efficacy once the function was defined by comparing the Mobotix Q24 camera to the TSI-880 camera in terms of the sky classification.

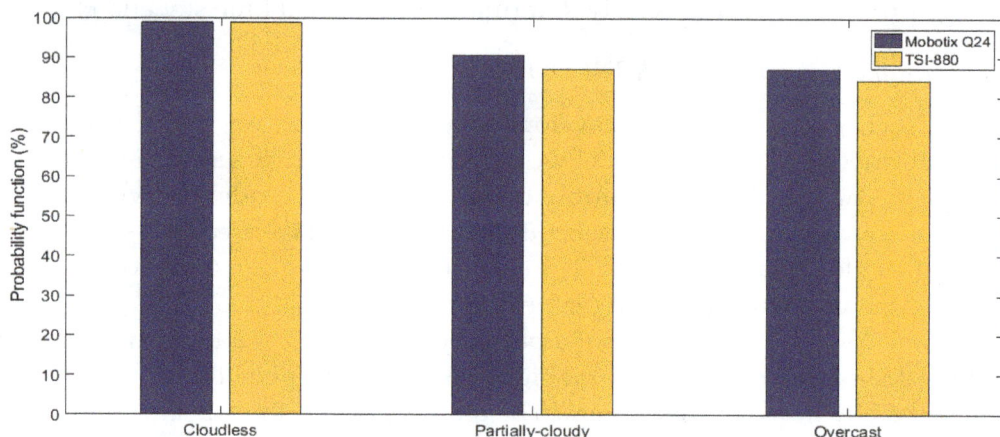

Figure 16. Graphic success representation of image processing for the cloud detection by using different sky cams and algorithms, according to sky conditions.

The results have been divided into three basic groups: one representing the probabilistic results for clear skies, another for partially covered skies and the third for covered skies. In each group, a bar represents the success rate in (%). As one can observe, all the presented results are above 80%. The best results were obtained for clear skies, where the two cameras had the same success rate, 98.8%.

For partially covered skies, the Q24 camera provided better results than the TSI-880, with a success rate of 90.6% (the TSI-880 success rate was 87.1%). In the case of overcast skies, the Mobotix camera again had a higher success rate, at 87.1%, while the TSI-880 camera had a value of 84.4%. In general, one can say that the cameras had a very similar success rate despite the slight differences found on days with clouds. Figure 17 shows a comparative graph of the overall hits in cloud detection.

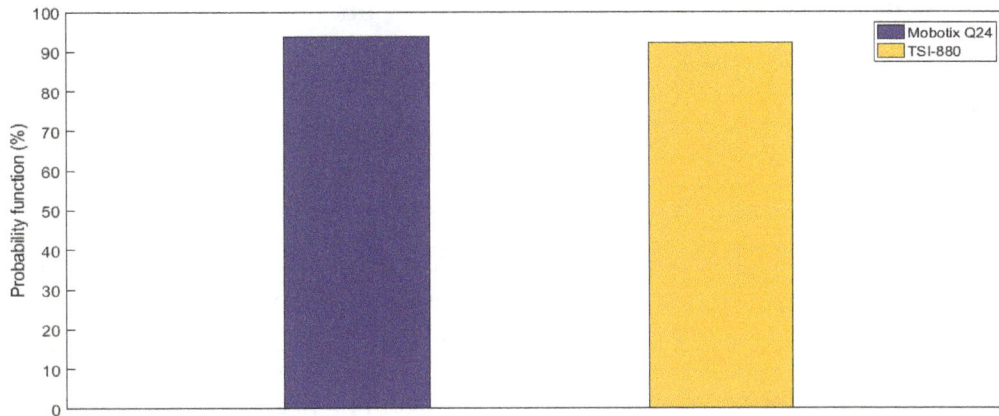

Figure 17. Graphic success representation of image processing for the cloud detection by using different sky cams and algorithms, for all cases.

As can be seen in the graph, the two cameras had very similar values overall; the cloud detection image processing for the Mobotix camera had a success rate of 93.7% while the TSI-880 camera had a value of 92.3%.

In addition, it should be noted that the TSI-880 camera requires a high level of maintenance to ensure the optimal quality of the images taken; this is due to its special design, in which the glass is a rotating dome that must be cleaned periodically, taking care not to scratch it. Moreover, the glass is rotated by a motor that needs to be checked regularly to ensure proper operation. In contrast, the MobotixQ24 camera is similar in dimension to a surveillance camera, with only a small glass panel protecting the lens; this means that its maintenance requirements are significantly less. For this work, one can state that it was not necessary to clean the glass for several months. The device produced sharp, appropriate images allowing the algorithm to correctly identify the clouds present. Consequently, this article demonstrates that a new algorithm has been developed which is capable of offering the same performance as the TSI-880 camera, without the need for radiation measurements to perform the digital image processing and requiring only minimal maintenance to acquire quality images for the cloud detection process.

5. Conclusions

This work presents a model for detecting the cloudiness present in real time images. It uses a low-cost Mobotix Q24 sky camera which only requires the digital image levels.

To detect clouds in the camera images, different areas of the image are differentiated. First, pixels are tagged in the solar area and the surrounding areas, assigning them with a red, green or yellow color. Subsequently, the algorithm detects cloudy pixels in the rest of the image and then clear sky pixels. Finally, the tagged pixels (such as cloud or sky) are classified, obtaining a final image that resembles the original.

The cloud detection system developed has been compared to a published and referenced system that is also based on digital image levels but uses a TSI-880 camera. In general, the results are very similar for both models. Under all sky conditions, the system developed with the Mobotix Q24 camera presented a higher success rate (93%) than the TSI-880 camera (which was around 92%). Under clear sky conditions, the processing from both cameras gave the same result (a 98% success rate). Under partially covered skies, the Mobotix camera performed better with a success rate higher than 90% (the TSI-880 success rate was 87%). Under overcast skies, the Mobotix camera had a success rate of 87% while the TSI camera's success rate was 3% less.

One of the main advantages of the new system is that there is no need for direct, diffuse and global radiation data to perform the image processing (as is the case of the TSI-880); this greatly reduces costs as it makes cloud detection possible using sky cameras. Another major advantage is the minimal maintenance required to clean the camera, meaning the system is almost autonomous and can automatically obtain high quality images in which the clouds can be defined optimally.

With this method, a new system is presented which combines the digital channels from a very low-cost sky camera. It can be installed in the control panel of any solar plant or airport. The system represents a new development in predicting cloud cover and solar radiation over the short term.

Moreover, this new development makes it possible to extrapolate the algorithm to other cameras. This task is probably not easy or direct since each camera has its own optics, which makes it more difficult to adapt a custom-made algorithm. However, a novel idea has appeared to adapt this new system to other cameras (with very slight modifications), to see if it is possible to obtain hits in the same range as for the Mobotix camera. The modifications will be necessary because each lens has its own properties in terms of saturation levels and exposure, etc. Saturation is an important factor due to the correlation between the final image and the exposure time of the camera capture. Therefore, the saturation can be adapted to simulate the Mobotix camera, depending on the camera's technology. Perhaps, it will be possible to determine a correlation between the intensity of the pixel channels for different technologies.

Acknowledgments: The author would like to thank the PRESOL Project (references ENE2014-59454-C3-1, 2 and 3) and the PVCastSOIL Project (references ENE2017-83790-C3-1, 2 and 3), which were funded by the *Ministerio de Economía, Industria y Competitividad and co-financed by the European Regional Development Fund.*

References

1.   Chapman, A.J.; McLellan, B.C.; Tezuka, T. Prioritizing mitigation efforts considering co-benefits, equity and energy justice: Fossil fuel to renewable energy transition pathways. *Appl. Energy* **2018**, *219*, 187–198. [CrossRef]

2.   Li, H.; Edwards, D.; Hosseini, M.; Costin, G. A review on renewable energy transition in Australia: An updated depiction. *J. Clean. Prod.* **2020**, *242*. [CrossRef]

3.   Alonso, J.; Batlles, F. Short and medium-term cloudiness forecasting using remote sensing techniques and sky camera imagery. *Energy* **2014**, *73*, 890–897. [CrossRef]

4.   Paszkuta, M.; Zapadka, T.; Krężel, A. Assessment of cloudiness for use in environmental marine research. *Int. J. Remote Sens.* **2019**, *40*, 9439–9459. [CrossRef]

5.   Hadizadeh, M.; Rahnama, M.; Hesari, B. Verification of two machine learning approaches for cloud masking based on reflectance of channel IR3.9 using Meteosat Second Generation over Middle East maritime. *Int. J. Remote Sens.* **2019**, *40*, 8899–8913. [CrossRef]

6.   Francis, A.; Sidiropoulos, P.; Muller, J.P. CloudFCN: Accurate and robust cloud detection for satellite imagery with deep learning. *Remote Sens.* **2019**, *11*, 2312. [CrossRef]

7.   Escrig, H.; Batlles, F.; Alonso, J.; Baena, F.; Bosch, J.; Salbidegoitia, I.; Burgaleta, J. Cloud detection, classification and motion estimation using geostationary satellite imagery for cloud cover forecast. *Energy* **2013**, *55*, 853–859. [CrossRef]

8.   Qiu, S.; Zhu, Z.; He, B. Fmask 4.0: Improved cloud and cloud shadow detection in Landsats 4–8 and Sentinel-2 imagery. *Remote Sens. Environ.* **2019**, *231*. [CrossRef]

9.   Alonso-Montesinos, J.; Barbero, J.; Polo, J.; López, G.; Ballestrín, J.; Batlles, F. Impact of a Saharan dust intrusion over southern Spain on DNI estimation with sky cameras. *Atmos. Environ.* **2017**, *170*, 279–289. [CrossRef]

10.   Román, R.; Torres, B.; Fuertes, D.; Cachorro, V.; Dubovik, O.; Toledano, C.; Cazorla, A.; Barreto, A.; Bosch, J.;

Lapyonok, T.; et al. Remote sensing of lunar aureole with a sky camera: Adding information in the nocturnal retrieval of aerosol properties with GRASP code. *Remote Sens. Environ.* **2017**, *196*, 238–252. [CrossRef]

11. Fa, T.; Xie, W.; Wang, Y.; Xia, Y. Development of an all-sky imaging system for cloud cover assessment. *Appl. Opt.* **2019**, *58*, 5516–5524. [CrossRef] [PubMed]

12. Gacal, G.; Antioquia, C.; Lagrosas, N. Trends of night-time hourly cloud-cover values over Manila Observatory: Ground-based remote-sensing observations using a digital camera for 13 months. *Int. J. Remote Sens.* **2018**, *39*, 7628–7642. [CrossRef]

13. Alonso, J.; Batlles, F.; López, G.; Ternero, A. Sky camera imagery processing based on a sky classification using radiometric data. *Energy* **2014**, *68*, 599–608. [CrossRef]

14. Alonso-Montesinos, J.; Batlles, F. The use of a sky camera for solar radiation estimation based on digital image processing. *Energy* **2015**, *90*, 377–386. [CrossRef]

15. Alonso-Montesinos, J.; Batlles, F.; Portillo, C. Solar irradiance forecasting at one-minute intervals for different sky conditions using sky camera images. *Energy Convers. Manag.* **2015**, *105*, 1166–1177. [CrossRef]

16. Luiz, E.; Martins, F.; Costa, R.; Pereira, E. Comparison of methodologies for cloud cover estimation in Brazil—A case study. *Energy Sustain. Dev.* **2018**, *43*, 15–22. [CrossRef]

17. Werkmeister, A.; Lockhoff, M.; Schrempf, M.; Tohsing, K.; Liley, B.; Seckmeyer, G. Comparing satellite- to ground-based automated and manual cloud coverage observations - A case study. *Atmos. Meas. Tech.* **2015**, *8*, 2001–2015. [CrossRef]

18. Román, R.; Cazorla, A.; Toledano, C.; Olmo, F.; Cachorro, V.; de Frutos, A.; Alados-Arboledas, L. Cloud cover detection combining high dynamic range sky images and ceilometer measurements. *Atmos. Res.* **2017**, *196*, 224–236. [CrossRef]

19. Yang, J.; Min, Q.; Lu, W.; Yao, W.; Ma, Y.; Du, J.; Lu, T.; Liu, G. An automated cloud detection method based on the green channel of total-sky visible images. *Atmos. Meas. Tech.* **2015**, *8*, 4671–4679. [CrossRef]

20. Jayadevan, V.; Rodriguez, J.; Cronin, A. A new contrast-enhancing feature for cloud detection in ground-based sky images. *J. Atmos. Ocean. Technol.* **2015**, *32*, 209–219. [CrossRef]

21. Alonso, J.; Batlles, F.; Villarroel, C.; Ayala, R.; Burgaleta, J. Determination of the sun area in sky camera images using radiometric data. *Energy Convers. Manag.* **2014**, *78*, 24–31. [CrossRef]

2

Modified Search Strategies Assisted Crossover Whale Optimization Algorithm with Selection Operator for Parameter Extraction of Solar Photovoltaic Models

Guojiang Xiong [1,*]**, Jing Zhang** [1]**, Dongyuan Shi** [2]**, Lin Zhu** [3]**, Xufeng Yuan** [1] **and Gang Yao** [4]

[1] Guizhou Key Laboratory of Intelligent Technology in Power System, College of Electrical Engineering, Guizhou University, Guiyang 550025, China; zhangjing@gzu.edu.cn (J.Z.); xfyuan@gzu.edu.cn (X.Y.)
[2] State Key Laboratory of Advanced Electromagnetic Engineering and Technology, Huazhong University of Science and Technology, Wuhan 430074, China; dongyuanshi@hust.edu.cn
[3] Department of Electrical Engineering and Computer Science, University of Tennessee, Knoxville, TN 37996, USA; lzhu12@utk.edu
[4] Guizhou Electric Power Grid Dispatching and Control Center, Guiyang 550002, China; yaogang0319@gz.csg.cn
* Correspondence: gjxiong1@gzu.edu.cn

Abstract: Extracting accurate values for involved unknown parameters of solar photovoltaic (PV) models is very important for modeling PV systems. In recent years, the use of metaheuristic algorithms for this problem tends to be more popular and vibrant due to their efficacy in solving highly nonlinear multimodal optimization problems. The whale optimization algorithm (WOA) is a relatively new and competitive metaheuristic algorithm. In this paper, an improved variant of WOA referred to as MCSWOA, is proposed to the parameter extraction of PV models. In MCSWOA, three improved components are integrated together: (i) Two modified search strategies named WOA/rand/1 and WOA/current-to-best/1 inspired by differential evolution are designed to balance the exploration and exploitation; (ii) a crossover operator based on the above modified search strategies is introduced to meet the search-oriented requirements of different dimensions; and (iii) a selection operator instead of the "generate-and-go" operator used in the original WOA is employed to prevent the population quality getting worse and thus to guarantee the consistency of evolutionary direction. The proposed MCSWOA is applied to five PV types. Both single diode and double diode models are used to model these five PV types. The good performance of MCSWOA is verified by various algorithms.

Keywords: metaheuristic; parameter extraction; solar photovoltaic; whale optimization algorithm

1. Introduction

Solar energy is an inexhaustible and carbon emission-free energy source to promote sustainable development. Solar photovoltaic (PV) is becoming the preferred choice for meeting the rapidly growing power demands globally [1,2]. It is a clean energy according to the principle of sustainability. Take China as an example, according to the latest data from the National Energy Administration, PV added 5.20GW capacity, which was more than that of wind (added 4.78GW) in the first quarter of 2019 [3]. In addition, by the end of the first quarter of 2019, the total installed PV capacity had reached 180GW, accounting for 24.3% of renewable energy, only 0.09GW below that of wind, and the gap is narrowing. Along with the increasing installed capacity of PV, its impact on the connected power system is growing, and thereby, analyzing PV systems' dynamic conversion behavior is quite important and necessary. Thereinto, accurate modeling of the PV system's basic device, i.e., the PV cell or module, is the premise and crux. The most widely used modeling tool is the single diode (SDM) and double diode (DDM) equivalent circuit models [4]. The SDM and DDM have five and seven unknown model parameters, respectively, and extracting accurate values for these parameters is just the purpose of this study.

Many methods have been proposed to solve the parameter extraction problem of PV models. They can be categorized into analytical methods and optimization methods approximately. Analytical methods mainly use some special data points such as short-circuit point, open-circuit point, and maximum power point of the current-voltage (I–V) characteristic curve under standard test conditions (STC) to formulate a few mathematical equations for the unknown model parameters. They have the features of simplicity, rapidity, and convenience. Their extraction accuracy is directly subject to the selected special data points provided by the manufacturers. In this context, the incorrectly specified values for these data points will degrade the extraction accuracy considerably due to the extraction strategy of "taking a part for the whole" [5,6]. In addition, those employed special data points are factory measured under the STC, while the PV degradation makes the model parameters change over time [7], which further influences the extraction accuracy of the "taking a part for the whole" methods.

Different from the analytical methods, the optimization methods abandon the heavy dependence on several special data points and use a number of actual measured data points to extract the unknown model parameters. First, an optimization objective function is constructed to reflect the difference between the measured data and the calculated data based on the idea of curve fitting. Then, solution optimization methods, including deterministic methods and metaheuristic methods, are designed to minimize the objective function and thereby to obtain the values for the unknown model parameters. These solution methods can overcome the shortcomings of the analytical methods thanks to "taking all actual measured data" rather than "taking a part of factory measured data" for the whole. The deterministic methods such as the Newton method, Newton–Raphson method and Levenberg–Marquardt algorithm are gradient-based methods. They are likely to get stuck in local optima especially for complicated multimodal problems such as the one considered in this work. Additionally, simplification and linearization are frequently performed to ease the optimization procedure. Consequently, they may result in poor approximate and unreliable solutions [8].

Metaheuristic methods, alternatively, do not use the gradient information and make no simplification or linearization to the optimization procedure. Therefore, they can hedge the problems of deterministic methods and have attracted growing attention recently. Many metaheuristic methods concluding particle swarm optimization (PSO) [9–11], differential evolution (DE) [12], teaching-learning-based optimization (TLBO) [13,14], supply-demand-based optimization (SDO) [15], symbiotic organisms search algorithm (SOS) [16], JAYA algorithm [17], artificial bee colony (ABC) [18], imperialist competitive algorithm [19], flower pollination algorithm (FPA) [20], hybrid algorithms [21–24], etc., have been applied to the parameter extraction problem of PV models.

The whale optimization algorithm (WOA) [25] is a new and versatile metaheuristic method inspired by the special spiral bubble-net hunting behavior of humpback whales. It performs effectively, competitively, and has been applied to various engineering optimization problems, including the parameter extraction problem of PV models. For example, Oliva et al. [26] utilized the chaotic maps to improve the performance of WOA and then applied the modified WOA to the concerned problem here. Abd Elaziz and Oliva [27] employed the opposition-based learning to enhance the exploration of WOA and applied the resultant WOA variant to both benchmark optimization functions and the problem considered here. Xiong et al. [28] developed two improved search strategies to balance WOA's local exploitation and global exploration, and then applied the improved WOA to different PV models. In reference [29], Xiong et al. used DE to enhance the exploration of WOA and then employed the hybrid algorithm to both benchmark optimization functions and different PV models.

From our previous works [28,29], we know that the original WOA performs well in local exploitation but badly in global exploration, which easily leads to premature convergence. They also reveal that the use of both improved search strategies and DE can enhance the performance of WOA significantly. Having noticed this, in this paper, we propose two modified search strategies named

WOA/rand/1 and WOA/current-to-best/1 inspired by DE. The former uses one random weighted difference vector to perturb a randomly selected individual and thus to improve the exploration; while the latter simultaneously adopts one current-to-best weighted difference vector and one random weighted difference vector to perturb the current individual and thereby to maintain the exploitation. In addition, in the original WOA, the values of all dimensions of each offspring completely come from a vector generated by one search strategy, which cannot meet the exploration and exploitation performance requirements of different dimensions. In this case, a crossover operator based on the modified search strategies is designed. It adopts two different search strategies to generate each offspring simultaneously, which can further promote the balance between exploration and exploitation. Moreover, the original WOA preserves the generated vector regardless of its quality. This "generate-and-go" strategy may result in retrogression or oscillation in evolutionary process. To prevent this phenomenon from occurring, a selection operator instead of the "generate-and-go" operator is implemented to guarantee the consistency of evolutionary direction. The resultant improved variant of WOA, referred to as MCSWOA, is applied to five PV types modeled by both SDM and DDM.

The main contributions of this paper are the following:

(1)   An improved variant of WOA, i.e., MCSWOA, is presented to parameter extraction of PV models. In MCSWOA, three improved components, including two modified search strategies, a crossover operator, and a selection operator are developed and integrated well to enhance its performance.

(2)   MCSWOA is applied to five PV types, including RTC France cell, Photowatt-PWP201 module, STM6-40/36 module, STP6-120/36 module, and Sharp ND-R250A5 module. Both SDM and DDM are used to model these five PV types.

(3)   The good performance of MCSWOA in extracting accurate parameters of PV models is fully verified through comparison with other 31 algorithms in terms of the parameter accuracy, convergence speed, robustness, and statistics.

The rest of this paper is organized as follows. In Section 2, the mathematical formulation of the parameter extraction problem is described. Section 3 introduces the original WOA. Section 4 gives the proposed MCSWOA. Section 5 presents the experimental results and comparisons. The discussions are provided in Section 6. Finally, the paper is concluded in Section 7.

2. Problem Formulation

2.1. Single Diode Model (SDM)

The equivalent circuit of SDM is presented in Figure 1.

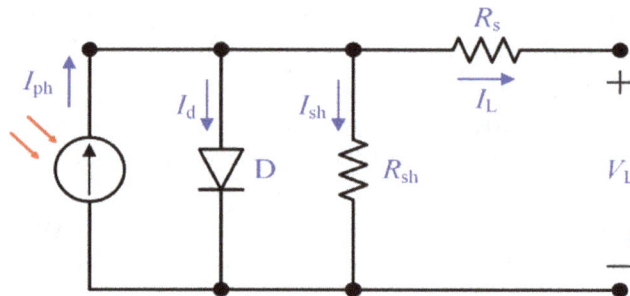

Figure 1. Equivalent circuit of a single diode model (SDM).

The output current I_L can be achieved according to the Kirchhoff's current law:

$$I_L = I_{ph} - I_d - I_{sh} \tag{1}$$

where I_{ph}, I_{sh} and I_d are the photogenerated current, shunt resistor current, and diode current, respectively. I_d and I_{sh} are calculated as follows [4,6]:

$$I_d = I_{sd} \cdot [\exp(\frac{V_L + R_s \cdot I_L}{nV_t}) - 1] \tag{2}$$

$$V_t = \frac{kT}{q} \tag{3}$$

$$I_{sh} = \frac{V_L + R_s \cdot I_L}{R_{sh}} \tag{4}$$

where I_{sd} is the saturation current, V_L is the output voltage, R_s and R_{sh} are the series and shunt resistances, respectively, n is the diode ideal factor, k is the Boltzmann constant ($1.3806503 \times 10^{-23}$ J/K), q is the electron charge ($1.60217646 \times 10^{-19}$ C), and T is the cell temperature (K).

The output current I_L can be obtained by substituting Equations (2) and (4) into (1):

$$I_L = I_{ph} - I_{sd} \cdot [\exp(\frac{V_L + R_s \cdot I_L}{nV_t}) - 1] - \frac{V_L + R_s \cdot I_L}{R_{sh}} \tag{5}$$

From Equation (5), it can be seen that the SDM has 5 unknown parameters (i.e., $I_{ph}, I_{sd}, R_s, R_{sh}$, and n) that need to be extracted.

2.2. Double Diode Model (DDM)

When considering the effect of the recombination current loss in the depletion region, we can get the equivalent circuit of DDM, as shown in Figure 2. It performs well in some applications [4].

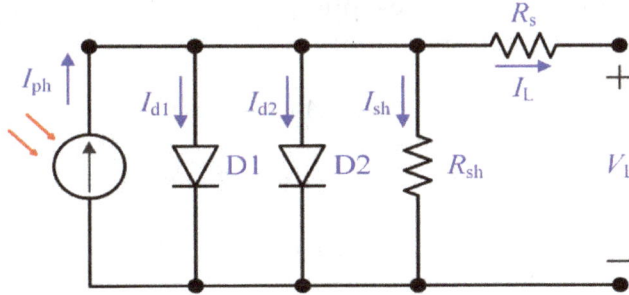

Figure 2. Equivalent circuit of a double diode model (DDM).

The output current I_L is calculated as follows:

$$\begin{aligned} I_L &= I_{ph} - I_{d1} - I_{d2} - I_{sh} \\ &= I_{ph} - I_{sd1} \cdot [\exp(\frac{V_L + R_s \cdot I_L}{n_1 V_t}) - 1] \\ &\quad - I_{sd2} \cdot [\exp(\frac{V_L + R_s \cdot I_L}{n_2 V_t}) - 1] - \frac{V_L + R_s \cdot I_L}{R_{sh}} \end{aligned} \tag{6}$$

where I_{sd1} and I_{sd2} are diode currents, n_1 and n_2 are diode ideal factors. The DDM has 7 unknown parameters (i.e., $I_{ph}, I_{sd1}, I_{sd2}, R_s, R_{sh}, n_1$ and n_2) that need to be extracted.

2.3. PV Module Model

For a PV module with $N_s \times N_p$ solar cells in series and/or in parallel, its output current I_L can be formulated as follows:

For the SDM based PV module:

$$I_L = N_p \left\{ I_{ph} - I_{sd} \cdot [\exp(\frac{V_L/N_s + R_s I_L/N_p}{nV_t}) - 1] - \frac{V_L/N_s + R_s I_L/N_p}{R_{sh}} \right\} \tag{7}$$

For the DDM based PV module:

$$I_{\mathrm{L}} = N_{\mathrm{P}}\left\{ \begin{array}{l} I_{\mathrm{ph}} - I_{\mathrm{sd1}}\cdot[\exp(\frac{V_{\mathrm{L}}/N_{\mathrm{s}}+R_{\mathrm{s}}I_{\mathrm{L}}/N_{\mathrm{P}}}{n_1 V_{\mathrm{t}}})-1] \\ -I_{\mathrm{sd2}}\cdot[\exp(\frac{V_{\mathrm{L}}/N_{\mathrm{s}}+R_{\mathrm{s}}I_{\mathrm{L}}/N_{\mathrm{P}}}{n_2 V_{\mathrm{t}}})-1] - \frac{V_{\mathrm{L}}/N_{\mathrm{s}}+R_{\mathrm{s}}I_{\mathrm{L}}/N_{\mathrm{P}}}{R_{\mathrm{sh}}} \end{array}\right\} \tag{8}$$

2.4. Objective Function

One way to extract the unknown parameters of PV models is to construct an objective function to reflect the difference between the measured data and the calculated data. Commonly, the root mean square error (RMSE) between the measured current $I_{L,\mathrm{measured}}$ and the calculated current $I_{L,\mathrm{calculated}}$ as shown in Equation (9) is recommended [6,8,9,30,31].

$$\min f(x) = \mathrm{RMSE}(x) = \sqrt{\frac{1}{N}\sum_{k=1}^{N}[I_{L,\mathrm{calculated}}^{k}(x) - I_{L,\mathrm{measured}}^{k}]^2} \tag{9}$$

where N is the number of measured data, x is the vector of unknown parameters.

3. Whale Optimization Algorithm

WOA [25] is an effective metaheuristic inspired by the special spiral bubble-net hunting behavior of humpback whales. In WOA, the position of each whale (i.e., population individual) is represented as $x_i^t = [x_{i,1}^t, x_{i,2}^t, \ldots, x_{i,D}^t]$, where $i = 1, 2, \ldots, ps$, $t = 1, 2, \ldots, t_{\max}$, ps is the population size, $t_{\max}$ is the maximum number of iterations, and D is the dimension of one individual. WOA contains the following three parts:

(1)   Encircling prey

WOA defines the position of a current best humpback whale as the target prey, and other whales encircle the prey using the following formulation:

$$x_i^{t+1} = x_g^t - A\cdot|C\cdot x_g^t - x_i^t| \tag{10}$$

where x_g^t is the best position found so far. A and C are coefficient parameters and calculated for each individual using the following method:

$$A = 2\cdot a\cdot r - a \tag{11}$$

$$C = 2\cdot r \tag{12}$$

where a linearly decreases from 2 to 0 with the increasing of iterations. r is a random real number in (0,1).

(2)   Bubble-net attacking method

WOA employs both shrinking encircling and spiraling to spin around the prey with the same probability as follows:

$$x_i^{t+1} = x_g^t - A\cdot|C\cdot x_g^t - x_i^t| \quad \text{if } p < 0.5 \tag{13}$$

$$x_i^{t+1} = x_g^t + \exp(bl)\cdot\cos(2\pi l)\cdot|x_g^t - x_i^t| \quad \text{if } p \geq 0.5 \tag{14}$$

where b is a constant for defining the shape of the logarithmic spiral, l and p are random real numbers in (0,1).

(3)   Searching for prey

Before finding the prey, humpback whales swim around and select a random whale to search for prey. This behavior is formulated as follows and continues if $|A| \geq 1$.

$$x_i^{t+1} = x_r^t - A\cdot|C\cdot x_r^t - x_i^t| \tag{15}$$

where $r \in \{1, 2, \ldots, ps\}$ is different from i.

4. The Proposed MCSWOA

4.1. Modified Search Strategies

It is well-known that balancing exploration and exploitation is very important for a metaheuristic algorithm. For the original WOA, it emphasizes the exploitation excessively and thus easily suffers from premature convergence [28]. In order to solve this issue, one active method is to modify its search strategy.

Differential evolution (DE) [32] has proved its efficiency in solving different real-world problems. The efficiency of DE comes largely from its versatile mutation strategies. The following are 2 popular mutation strategies widely used in the literature:

$$\text{DE/rand/1}: v_i^t = x_{r1}^t + F \cdot (x_{r2}^t - x_{r3}^t) \tag{16}$$

$$\text{DE/current} - \text{to} - \text{best/1}: v_i^t = x_i^t + F \cdot (x_g^t - x_i^t) + F \cdot (x_{r1}^t - x_{r2}^t) \tag{17}$$

where $r1$, $r2$ and $r3$ are random distinct integers selected from $\{1, 2, \cdots, ps\}$ and are also different from i, the parameter F is the scaling factor. The former, i.e., DE/rand/1 strategy, usually presents good exploration while the latter, i.e., DE/current-to-best/1 strategy exhibits good exploitation.

Inspired by the mutation strategies of DE, in this paper, two modified search strategies are proposed to generate new donor individuals as follows:

$$\text{WOA/rand/1}: v_i^t = x_{r1}^t - A \cdot |x_{r2}^t - x_{r3}^t| \tag{18}$$

$$\text{WOA/current} - \text{to} - \text{best/1}: v_i^t = x_i^t - A \cdot |x_g^t - x_i^t| - A \cdot |x_{r1}^t - x_{r2}^t| \tag{19}$$

The above-modified search strategies are employed to replace Equations (15) and (13), respectively.

4.2. Modified Search Strategies Assisted Crossover Operator

In the original WOA, the random parameter p is generated for each individual, indicating that all dimensions would perform the same search strategy. For example, on the premise of $|A| \geq 1$, if $p < 0.5$, then the current individual would perform Equation (15). According to Equation (15), WOA updates the current individual around a random individual x_r^t, which is beneficial for the exploration but harmful to the exploitation. In fact, different dimensions of an individual have different performance requirements for exploration and exploitation. For one dimension, it is wise to perform the exploration-oriented search strategy if the population diversity associated with this dimension is high; otherwise, it is wise to perform the exploitation-oriented search strategy. In order to meet the performance requirements of different dimensions, a crossover operator based on the abovementioned modified search strategies is proposed and shown in Figure 3. In the crossover operator, for each dimension of each individual, the random parameter p is regenerated, and thereby the target dimension of the donor individual has the same chance of deriving from 2 search strategies, which is able to promote the balance between the exploration and exploitation. This crossover operator can be formulated as follows:

$$v_{i,d}^t = \begin{cases} \begin{cases} x_{r1,d}^t - A \cdot |x_{r2,d}^t - x_{r3,d}^t| & \text{if } p < 0.5 \\ x_{g,d}^t + \exp(bl) \cdot \cos(2\pi l) \cdot |x_{g,d}^t - x_{i,d}^t| & \text{if } p \geq 0.5 \end{cases} & \text{if } |A| \geq 1 \\ \begin{cases} x_{i,d}^t - A \cdot |x_{g,d}^t - x_{i,d}^t| - A \cdot |x_{r1,d}^t - x_{r2,d}^t| & \text{if } p < 0.5 \\ x_{g,d}^t + \exp(bl) \cdot \cos(2\pi l) \cdot |x_{g,d}^t - x_{i,d}^t| & \text{if } p \geq 0.5 \end{cases} & \text{if } |A| < 1 \end{cases} \tag{20}$$

WOA   MCSWOA

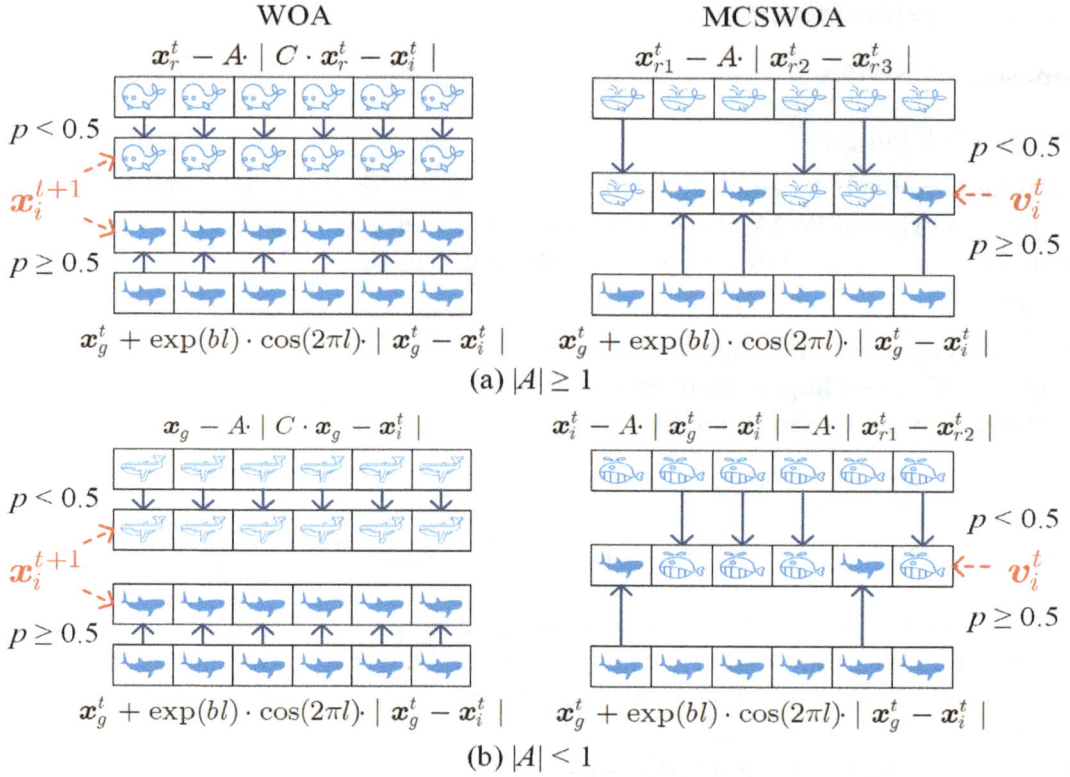

Figure 3. Modified search strategies assisted crossover operator sketch.

4.3. Selection Operator

In the original WOA, the target individual is directly replaced by the newly generated vector regardless of its quality. This "generate-and-go" operator is not very effective because the newly generated vector may be worse than the target individual. In order to guarantee the consistency of evolutionary direction, a selection operator is employed to determine whether the target individual or the donor individual survives to the next iteration. This selection operator is formulated as follows:

$$x_i^{t+1} = \begin{cases} v_i^t & \text{if } f(v_i^t) \leq f(x_i^t) \\ x_i^t & \text{if } f(v_i^t) > f(x_i^t) \end{cases} \tag{21}$$

Hence, the prerequisite of using the donor individual to replace the target individual is that the donor individual achieves an equal or better fitness value; otherwise, the donor individual is abandoned, and the target individual is retained and passed on to the next iteration. Consequently, the population either gains quality improvement or maintains the current quality level, but never gets worse.

4.4. The Main Procedure of MCSWOA

By combining the abovementioned 3 improved components into WOA, the MCSWOA is developed and presented in Algorithm 1. Compared with the original WOA, it can be seen that: (1) MCSWOA needs only a small extra computational cost in comparing the fitness values of current individuals with those of donor individuals. (2) The structure of MCSWOA also remains very simple, and no new parameter that needed to be adjusted is introduced. (3) The use of the selection operator makes MCSWOA an elitist method that is able to preserve best individuals in the population.

Algorithm 1: The main procedure of MCSWOA

1: Generate a random initial population
2: Evaluate the fitness for each individual
3: Select the best individual x_{best}^0 and set it as x_g^0
4: Initialize the iteration counter $t = 1$
5: **While** the stopping condition is not satisfied **do**
6:    **for** $i = 1$ to ps **do**
7:       Update a, A, and l
8:       **for** $d = 1$ to D **do**
9:          Update p
10:         **if** $p < 0.5$ **then**
11:            Select three random individuals $x_{r1}^t \neq x_{r2}^t \neq x_{r3}^t \neq x_i^t$
12:            **if** $|A| \geq 1$ **then**
13:               $v_{i,d}^t = x_{r1,d}^t - A \cdot |x_{r2,d}^t - x_{r3,d}^t|$
14:            **else**
15:               $v_{i,d}^t = x_{i,d}^t - A \cdot |x_{g,d}^t - x_{i,d}^t| - A \cdot |x_{r1,d}^t - x_{r2,d}^t|$
16:            **end if**
17:         **else**
18:            $v_{i,d}^{t+1} = x_{g,d}^t + \exp(bl) \cdot \cos(2\pi l) \cdot |x_{g,d}^t - x_{i,d}^t|$
19:         **end if**
20:       **end for**
21:    **end for**
22:    Evaluate the fitness for each donor individual
23:    **for** $i = 1$ to ps **do**
24:       **if** $f(v_i^t) \leq f(x_i^t)$ **then**
25:          $x_i^{t+1} = v_i^t$
26:       **else**
27:          $x_i^{t+1} = x_i^t$
28:       **end if**
29:    **end for**
30:    Select the best individual x_{best}^t of the updated population
31:    **if** $f(x_{\text{best}}^t) \leq f(x_g^t)$ **then**
32:       $x_g^{t+1} = x_{\text{best}}^t$
33:    **else**
34:       $x_g^{t+1} = x_g^t$
35:    **end if**
36:    $t = t + 1$
37: End while

---

5. Experimental Results

5.1. Test Cases

In this work, the proposed MCSWOA was applied to five PV types, including RTC France cell, Photowatt-PWP201 module, STM6-40/36 module, STP6-120/36 module, and Sharp ND-R250A5 module. Both the SDM and DDM were adopted to model them, and thus we could get 10 test cases. The detailed information about these 10 test cases is tabulated in Table 1. The search ranges of involved parameters are presented in Table 2. They are kept the same as those used in [6,9,10].

Table 1. Test photovoltaic (PV) models in this work.

Case	PV Type	Number of Cells $(N_s \times N_p)$	Irradiance (W/m^2)	Temperature $(^\circ C)$	PV model
1/2	RTC France cell	1×1	1000	33	SDM/DDM
3/4	Photowatt-PWP201 module	36×1	1000	45	SDM/DDM
5/6	STM6-40/36 module	36×1	NA	51	SDM/DDM
7/8	STP6-120/36 module	36×1	NA	55	SDM/DDM
9/10	Sharp ND-R250A5 module	60×1	1040	59	SDM/DDM

NA denotes the value is not available in the literature.

Table 2. Ranges of parameters of PV models.

Parameter	RTC France Cell		Photowatt-PWP201 Module		STM6-40/36 Module		STP6-120/36 Module		Sharp ND-R250A5 Module	
	LB	UB	LB	UB	LB	UB	LB	UB	LB	UB
I_{ph} (A)	0	1	0	2	0	2	0	8	0	10
I_{sd} (μA)	0	1	0	50	0	50	0	50	0	10
R_s (Ω)	0	0.5	0	2	0	0.36	0	0.36	0	2
R_{sh} (Ω)	0	100	0	2000	0	1000	0	1500	0	5000
n, n_1, n_2	1	2	1	50	1	60	1	50	1	50

5.2. Experimental Settings

In this work, the maximum number of fitness evaluations (Max_FEs) setting as 50,000 [15,17,24,33] was employed as the stopping condition. All involved algorithms used the same population size with the value $ps = 50$ [14,24]. With regard to other parameters associated with the compared algorithms, the same values in their original literature were used for a fair comparison. In addition, 50 independent runs for each algorithm on each test case were performed in MATLAB 2017a.

5.3. Experimental Results

5.3.1. Comparison of MCSWOA with WOA

In this subsection, the proposed MCSWOA was compared with the original WOA to demonstrate its effectiveness. The experimental results tabulated in Table 3 contain the minimum (Min), maximum (Max), mean, and standard deviation (Std Dev) values of the RMSE values over 50 independent runs. The best results on each case are highlighted in **boldface**. It can be seen that MCSWOA was significantly better than WOA in all terms of RMSE values in all cases, indicating that the proposed modified components could improve the performance of WOA considerably.

The extracted values corresponding to the minimum RMSE given by MCSWOA for the involved unknown parameters are presented in Table 4. By using these extracted parameters, the output current could be easily calculated and given in Tables 5–9, respectively. Two error metrics, i.e., individual absolute error (IAE) and the sum of individual absolute error (SIAE) were used to evaluate the fitting results between the calculated current and the measured current. Tables 5–9 only provide the detailed calculated current of MCSWOA due to the space limitation, while for WOA only the SIAE values were listed. It is obvious that MCSWOA achieved smaller SIAE values than WOA on all cases. Namely, the calculated current obtained by MCSWOA fitted the measured current better than that of WOA, meaning that the parameters extracted by MCSWOA were more accurate. In addition, it can be observed that the DDM obtained slightly smaller SIAE values on the RTC France solar cell and Photowatt-PWP201 module, while the SDM yielded somewhat better results on the STM6-40/36, STP6-120/36 and Sharp ND-R250A5 modules. However, the differences were very small, which could be confirmed by some representative reconstructed I-V and P-V characteristic curves illustrated in Figure 4. Figure 4 also shows that the calculated data given by MCSWOA with both SDM and DDM were highly in agreement with the measured data throughout the entire voltage range.

Table 3. Experimental results of the whale optimization algorithm (WOA) and MCSWOA.

Case	Algorithm	Min	Max	Mean	Std. Dev.
1	WOA	1.0395×10^{-3}	1.1528×10^{-2}	3.3118×10^{-3}	2.5700×10^{-3}
	MCSWOA	$\mathbf{9.8602 \times 10^{-4}}$	$\mathbf{9.8603 \times 10^{-4}}$	$\mathbf{9.8602 \times 10^{-4}}$	$\mathbf{4.8373 \times 10^{-10}}$
2	WOA	1.0381×10^{-3}	1.3797×10^{-2}	3.6217×10^{-3}	2.7791×10^{-3}
	MCSWOA	$\mathbf{9.8250 \times 10^{-4}}$	$\mathbf{1.1903 \times 10^{-3}}$	$\mathbf{1.0078 \times 10^{-3}}$	$\mathbf{3.7264 \times 10^{-5}}$
3	WOA	2.4991×10^{-3}	4.9837×10^{-2}	9.6733×10^{-3}	1.1794×10^{-2}
	MCSWOA	$\mathbf{2.4251 \times 10^{-3}}$	$\mathbf{2.4270 \times 10^{-3}}$	$\mathbf{2.4252 \times 10^{-3}}$	$\mathbf{3.2927 \times 10^{-7}}$
4	WOA	2.4270×10^{-3}	7.5526×10^{-2}	2.4505×10^{-2}	2.2337×10^{-2}
	MCSWOA	$\mathbf{2.4251 \times 10^{-3}}$	$\mathbf{2.4881 \times 10^{-3}}$	$\mathbf{2.4377 \times 10^{-3}}$	$\mathbf{1.3424 \times 10^{-5}}$
5	WOA	2.9904×10^{-3}	3.1090×10^{-1}	2.8343×10^{-2}	6.0554×10^{-2}
	MCSWOA	$\mathbf{1.7298 \times 10^{-3}}$	$\mathbf{1.7364 \times 10^{-3}}$	$\mathbf{1.7311 \times 10^{-3}}$	$\mathbf{1.0774 \times 10^{-6}}$
6	WOA	3.3265×10^{-3}	4.8619×10^{-2}	1.2171×10^{-2}	8.5449×10^{-3}
	MCSWOA	$\mathbf{1.7061 \times 10^{-3}}$	$\mathbf{1.7358 \times 10^{-3}}$	$\mathbf{1.7296 \times 10^{-3}}$	$\mathbf{5.4724 \times 10^{-6}}$
7	WOA	1.6759×10^{-2}	1.4164	1.3390×10^{-1}	3.3374×10^{-1}
	MCSWOA	$\mathbf{1.6601 \times 10^{-2}}$	$\mathbf{1.6741 \times 10^{-2}}$	$\mathbf{1.6632 \times 10^{-2}}$	$\mathbf{2.6486 \times 10^{-5}}$
8	WOA	1.7345×10^{-2}	5.6762×10^{-2}	3.8581×10^{-2}	1.1413×10^{-2}
	MCSWOA	$\mathbf{1.6601 \times 10^{-2}}$	$\mathbf{1.6732 \times 10^{-2}}$	$\mathbf{1.6640 \times 10^{-2}}$	$\mathbf{2.8956 \times 10^{-5}}$
9	WOA	1.1206×10^{-2}	2.1439	1.9117×10^{-1}	5.2271×10^{-1}
	MCSWOA	$\mathbf{1.1183 \times 10^{-2}}$	$\mathbf{1.1244 \times 10^{-2}}$	$\mathbf{1.1187 \times 10^{-2}}$	$\mathbf{9.1358 \times 10^{-6}}$
10	WOA	1.1233×10^{-2}	5.1709×10^{-2}	3.4638×10^{-2}	1.2972×10^{-2}
	MCSWOA	$\mathbf{1.1183 \times 10^{-2}}$	$\mathbf{1.1220 \times 10^{-2}}$	$\mathbf{1.1190 \times 10^{-2}}$	$\mathbf{8.4623 \times 10^{-6}}$

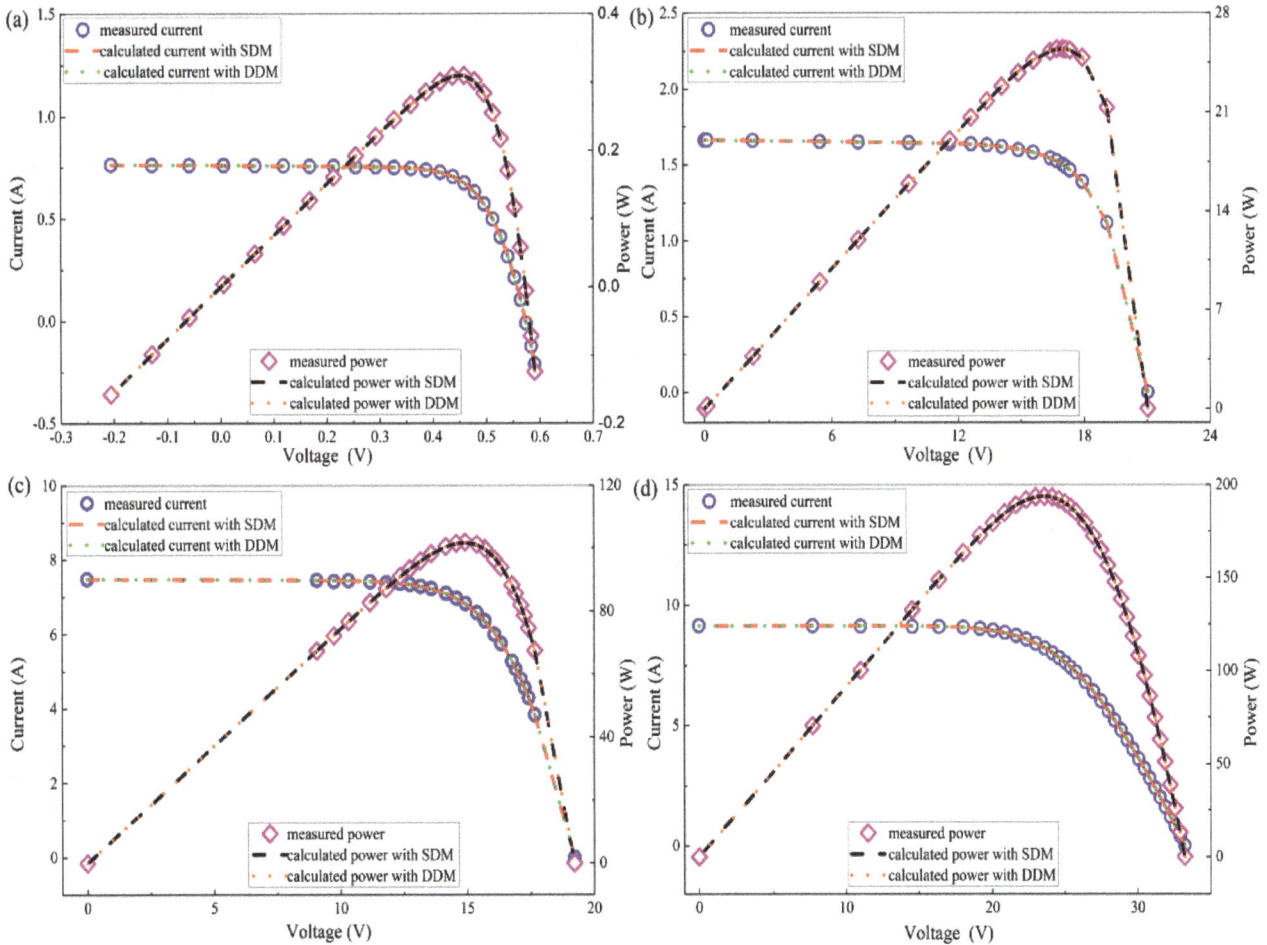

Figure 4. Comparison between the measured and calculated data achieved by MCSWOA. (**a**) RTC France cell; (**b**) STM6-40/36 module; (**c**) STP6-120/36 module; (**d**) Sharp ND-R250A5 module.

Table 4. Extracted value for involved parameters by MCSWOA.

Case	I_{ph} (A)	I_{sd1} (µA)	R_s (Ω)	R_{sh} (Ω)	n_1	I_{sd2} (µA)	n_2	RMSE
1	0.7608	0.3230	0.0364	53.7185	1.4812	—	—	9.8602×10^{-4}
2	0.7608	0.2206	0.0368	53.6255	1.4490	0.7974	2.0000	9.8250×10^{-4}
3	1.0305	3.4822	1.2013	981.9585	48.6428	—	—	2.4251×10^{-3}
4	1.0305	0.3648	1.2017	976.2658	48.6426	3.1036	48.6377	2.4251×10^{-3}
5	1.6639	1.7390	0.0043	15.9294	1.5203	—	—	1.7298×10^{-3}
6	1.6639	0.6103	0.0054	16.9519	1.4224	11.7629	2.1992	1.7061×10^{-3}
7	7.4727	2.3300	0.0046	21.9831	1.2599	—	—	1.6601×10^{-2}
8	7.4722	2.3466	0.0046	22.9095	1.2605	4.8598	49.5302	1.6601×10^{-2}
9	9.1431	1.1142	0.0098	5000	1.2150	—	—	1.1183×10^{-2}
10	9.1431	1.1142	0.0098	5000	1.2150	5.3615×10^{-9}	45.2483	1.1183×10^{-2}

Table 5. Calculated results of MCSWOA for the RTC France solar cell.

Item	V_L (V)	I_L Measured (A)	SDM (Case 1)		DDM (Case 2)	
			I_L Calculated (A)	IAE (A)	I_L Calculated (A)	IAE (A)
1	−0.2057	0.7640	0.76408765	0.00008765	0.76397504	0.00002496
2	−0.1291	0.7620	0.76266264	0.00066264	0.76259878	0.00059878
3	−0.0588	0.7605	0.76135473	0.00085473	0.76133540	0.00083540
4	0.0057	0.7605	0.76015424	0.00034576	0.76017516	0.00032484
5	0.0646	0.7600	0.75905594	0.00094406	0.75911205	0.00088795
6	0.1185	0.7590	0.75804334	0.00095666	0.75812819	0.00087181
7	0.1678	0.7570	0.75709159	0.00009159	0.75719567	0.00019567
8	0.2132	0.7570	0.75614207	0.00085793	0.75625201	0.00074799
9	0.2545	0.7555	0.75508732	0.00041268	0.75518481	0.00031519
10	0.2924	0.7540	0.75366447	0.00033553	0.75372792	0.00027208
11	0.3269	0.7505	0.75138806	0.00088806	0.75139769	0.00089769
12	0.3585	0.7465	0.74734834	0.00084834	0.74729341	0.00079341
13	0.3873	0.7385	0.74009688	0.00159688	0.73998455	0.00148455
14	0.4137	0.7280	0.72739678	0.00060322	0.72725566	0.00074434
15	0.4373	0.7065	0.70695328	0.00045328	0.70682698	0.00032698
16	0.4590	0.6755	0.67529492	0.00020508	0.67522445	0.00027555
17	0.4784	0.6320	0.63088433	0.00111567	0.63088651	0.00111349
18	0.4960	0.5730	0.57208208	0.00091792	0.57214313	0.00085687
19	0.5119	0.4990	0.49949167	0.00049167	0.49957540	0.00057540
20	0.5265	0.4130	0.41349364	0.00049364	0.41356073	0.00056073
21	0.5398	0.3165	0.31721950	0.00071950	0.31724418	0.00074418
22	0.5521	0.2120	0.21210317	0.00010317	0.21208087	0.00008087
23	0.5633	0.1035	0.10272136	0.00077864	0.10266905	0.00083095
24	0.5736	−0.0100	−0.00924878	0.00075122	−0.00929990	0.00070010
25	0.5833	−0.1230	−0.12438136	0.00138136	−0.12439111	0.00139111
26	0.5900	−0.2100	−0.20919308	0.00080692	−0.20914456	0.00085544
	SIAE of MCSWOA			**0.01770381**		**0.01730633**
	SIAE of WOA			0.01928659		0.01876701

Table 6. Calculated results of MCSWOA for the Photowatt-PWP201 module.

Item	V_L (V)	I_L Measured (A)	SDM (Case 3)		DDM (Case 4)	
			I_L Calculated (A)	IAE (A)	I_L Calculated (A)	IAE (A)
1	0.1248	1.0315	1.02912301	0.00237699	1.02914976	0.00235024
2	1.8093	1.0300	1.02738443	0.00261557	1.02740128	0.00259872
3	3.3511	1.0260	1.02574218	0.00025782	1.02575007	0.00024993
4	4.7622	1.0220	1.02410400	0.00210400	1.02410397	0.00210397
5	6.0538	1.0180	1.02228339	0.00428339	1.02227663	0.00427663
6	7.2364	1.0155	1.01991736	0.00441736	1.01990537	0.00440537
7	8.3189	1.0140	1.01635076	0.00235076	1.01633550	0.00233550
8	9.3097	1.0100	1.01049137	0.00049137	1.01047529	0.00047529
9	10.2163	1.0035	1.00067872	0.00282128	1.00066456	0.00283544
10	11.0449	0.9880	0.98465339	0.00334661	0.98464377	0.00335623
11	11.8018	0.9630	0.95969770	0.00330230	0.95969440	0.00330560
12	12.4929	0.9255	0.92304878	0.00245122	0.92305206	0.00244794
13	13.1231	0.8725	0.87258820	0.00008820	0.87259659	0.00009659
14	13.6983	0.8075	0.80731017	0.00018983	0.80732090	0.00017910
15	14.2221	0.7265	0.72795786	0.00145786	0.72796791	0.00146791
16	14.6995	0.6345	0.63646667	0.00196667	0.63647370	0.00197370
17	15.1346	0.5345	0.53569608	0.00119608	0.53569897	0.00119897
18	15.5311	0.4275	0.42881624	0.00131624	0.42881506	0.00131506
19	15.8929	0.3185	0.31866863	0.00016863	0.31866436	0.00016436
20	16.2229	0.2085	0.20785708	0.00064292	0.20785117	0.00064883
21	16.5241	0.1010	0.09835419	0.00264581	0.09834825	0.00265175
22	16.7987	−0.0080	−0.00816923	0.00016923	−0.00817364	0.00017364
23	17.0499	−0.1110	−0.11096847	0.00003153	−0.11096996	0.00003004
24	17.2793	−0.2090	−0.20911761	0.00011761	−0.20911505	0.00011505
25	17.4885	−0.3030	−0.30202234	0.00097766	−0.30201487	0.00098513
		SIAE of MCSWOA		**0.04178694**		**0.04174098**
		SIAE of WOA		0.04521107		0.04308364

Table 7. Calculated results of MCSWOA for the STM6-40/36 module.

Item	V_L (V)	I_L Measured (A)	SDM (Case 5)		DDM (Case 6)	
			I_L Calculated (A)	IAE (A)	I_L Calculated (A)	IAE (A)
1	0.0000	1.6630	1.66345754	0.00045754	1.66335653	0.00035653
2	0.1180	1.6630	1.66325166	0.00025166	1.66316242	0.00016242
3	2.2370	1.6610	1.65955087	0.00144913	1.65966539	0.00133461
4	5.4340	1.6530	1.65391451	0.00091451	1.65427645	0.00127645
5	7.2600	1.6500	1.65056604	0.00056604	1.65099325	0.00099325
6	9.6800	1.6450	1.64543105	0.00043105	1.64576715	0.00076715
7	11.5900	1.6400	1.63923502	0.00076498	1.63929611	0.00070389
8	12.6000	1.6360	1.63371634	0.00228366	1.63357235	0.00242765
9	13.3700	1.6290	1.62728896	0.00171104	1.62699263	0.00200737
10	14.0900	1.6190	1.61831553	0.00068447	1.61791078	0.00108922
11	14.8800	1.5970	1.60306755	0.00606755	1.60262830	0.00562830
12	15.5900	1.5810	1.58158496	0.00058496	1.58123166	0.00023166
13	16.4000	1.5420	1.54232802	0.00032802	1.54223011	0.00023011
14	16.7100	1.5240	1.52122491	0.00277509	1.52126131	0.00273869
15	16.9800	1.5000	1.49920537	0.00079463	1.49936328	0.00063672
16	17.1300	1.4850	1.48527079	0.00027079	1.48549479	0.00049479
17	17.3200	1.4650	1.46564287	0.00064287	1.46594489	0.00094489
18	17.9100	1.3880	1.38759918	0.00040082	1.38804424	0.00004424
19	19.0800	1.1180	1.11837322	0.00037322	1.11798671	0.00001329
20	21.0200	0.0000	−0.00002144	0.00002144	0.00002509	0.00002509
		SIAE of MCSWOA		**0.02177346**		**0.02210631**
		SIAE of WOA		0.04187370		0.04245192

Table 8. Calculated results of MCSWOA for the STP6-120/36 module.

Item	V_L (V)	I_L Measured (A)	SDM (Case 7)		DDM (Case 8)	
			I_L Calculated (A)	IAE (A)	I_L Calculated (A)	IAE (A)
1	19.2100	0.0000	0.00117621	0.00117621	0.00114264	0.00114264
2	17.6500	3.8300	3.83225520	0.00225520	3.83236037	0.00236037
3	17.4100	4.2900	4.27391075	0.01608925	4.27398800	0.01601200
4	17.2500	4.5600	4.54627802	0.01372198	4.54633438	0.01366562
5	17.1000	4.7900	4.78582746	0.00417254	4.78586359	0.00413641
6	16.9000	5.0700	5.08193661	0.01193661	5.08194603	0.01194603
7	16.7600	5.2700	5.27377339	0.00377339	5.27376501	0.00376501
8	16.3400	5.7500	5.77683588	0.02683588	5.77678272	0.02678272
9	16.0800	6.0000	6.03752035	0.03752035	6.03744819	0.03744819
10	15.7100	6.3600	6.34875976	0.01124024	6.34867349	0.01132651
11	15.3900	6.5800	6.56796191	0.01203809	6.56787501	0.01212499
12	14.9300	6.8300	6.81488832	0.01511168	6.81481542	0.01518458
13	14.5800	6.9700	6.95847149	0.01152851	6.95841712	0.01158288
14	14.1700	7.1000	7.08815167	0.01184833	7.08812304	0.01187696
15	13.5900	7.2300	7.21776382	0.01223618	7.21777158	0.01222842
16	13.1600	7.2900	7.28412533	0.00587467	7.28415609	0.00584391
17	12.7400	7.3400	7.33147260	0.00852740	7.33152077	0.00847923
18	12.3600	7.3700	7.36325038	0.00674962	7.36330957	0.00669043
19	11.8100	7.3800	7.39585537	0.01585537	7.39592269	0.01592269
20	11.1700	7.4100	7.42024640	0.01024640	7.42031281	0.01031281
21	10.3200	7.4400	7.43907657	0.00092343	7.43912820	0.00087180
22	9.7400	7.4200	7.44670325	0.02670325	7.44673825	0.02673825
23	9.0600	7.4500	7.45253188	0.00253188	7.45254265	0.00254265
24	0.0000	7.4800	7.47109229	0.00890771	7.47066044	0.00933956
	SIAE of MCSWOA			**0.27780418**		**0.27832466**
	SIAE of WOA			0.28272891		0.28498596

Table 9. Calculated results of MCSWOA for the Sharp ND-R250A5 module.

Item	V_L (V)	I_L Measured (A)	SDM (Case 9)		DDM (Case 10)	
			I_L Calculated (A)	IAE (A)	I_L Calculated (A)	IAE (A)
1	0.0000	9.1500	9.14302743	0.00697257	9.14302768	0.00697232
2	7.7100	9.1400	9.14242378	0.00242378	9.14242403	0.00242403
3	10.9800	9.1200	9.14016661	0.02016661	9.14016685	0.02016685
4	14.5500	9.1100	9.12733899	0.01733899	9.12733920	0.01733920
5	16.3600	9.1000	9.10594093	0.00594093	9.10594110	0.00594110
6	18.0000	9.0700	9.06266719	0.00733281	9.06266730	0.00733270
7	19.1500	9.0200	9.00583091	0.01416909	9.00583095	0.01416905
8	20.0400	8.9500	8.93692097	0.01307903	8.93692095	0.01307905
9	20.8700	8.8600	8.84418281	0.01581719	8.84418274	0.01581726
10	21.6700	8.7300	8.71970414	0.01029586	8.71970401	0.01029599
11	22.3600	8.5800	8.57706890	0.00293110	8.57706873	0.00293127
12	23.0200	8.4000	8.40362835	0.00362835	8.40362815	0.00362815
13	23.6200	8.2000	8.20979996	0.00979996	8.20979975	0.00979975
14	24.1500	8.0000	8.00692218	0.00692218	8.00692197	0.00692197
15	24.6100	7.8000	7.80514823	0.00514823	7.80514802	0.00514802
16	25.0200	7.6000	7.60439716	0.00439716	7.60439697	0.00439697
17	25.3900	7.4000	7.40597697	0.00597697	7.40597679	0.00597679
18	25.7500	7.2000	7.19709834	0.00290166	7.19709818	0.00290182
19	26.3800	6.8000	6.79421478	0.00578522	6.79421466	0.00578534
20	26.9400	6.4000	6.39703240	0.00296760	6.39703233	0.00296767
21	27.4600	6.0000	5.99656297	0.00343703	5.99656293	0.00343707
22	27.9400	5.6000	5.60112090	0.00112090	5.60112090	0.00112090
23	28.4000	5.2000	5.20016085	0.00016085	5.20016088	0.00016088
24	28.8400	4.8000	4.79761966	0.00238034	4.79761971	0.00238029
25	29.2500	4.4000	4.40675456	0.00675456	4.40675462	0.00675462
26	29.6600	4.0000	4.00156633	0.00156633	4.00156640	0.00156640
27	30.0500	3.6000	3.60362789	0.00362789	3.60362796	0.00362796

Table 9. *Cont.*

Item	V_L (V)	I_L Measured (A)	SDM (Case 9)		DDM (Case 10)	
			I_L Calculated (A)	IAE (A)	I_L Calculated (A)	IAE (A)
28	30.4400	3.2000	3.19420724	0.00579276	3.19420732	0.00579268
29	30.8100	2.8000	2.79578571	0.00421429	2.79578579	0.00421421
30	31.1700	2.4000	2.39932544	0.00067456	2.39932550	0.00067450
31	31.5200	2.0000	2.00601421	0.00601421	2.00601426	0.00601426
32	31.8800	1.6000	1.59382496	0.00617504	1.59382498	0.00617502
33	32.2200	1.2000	1.19780705	0.00219295	1.19780706	0.00219294
34	32.5500	0.8000	0.80751916	0.00751916	0.80751914	0.00751914
35	32.8900	0.4000	0.39962407	0.00037593	0.39962402	0.00037598
36	33.2200	0.0000	−0.00159760	0.00159760	−0.00159769	0.00159769
	SIAE of MCSWOA			**0.21759970**		**0.21759985**
	SIAE of WOA			0.24899579		0.26906430

5.3.2. The Benefit of MCSWOA Components

It can be seen from Section 4 that the proposed MCSWOA has three improved components, i.e., modified search strategies, crossover operator, and selection operator. In this subsection, the influence of these three components on MCSWOA was assessed. Six variants of MCSWOA were considered here: (1) WOAwM: The original WOA with modified search strategies; (2) WOAwC: The original WOA with crossover operator; (3) WOAwS: The original WOA with selection operator; (4) MCSWOAwoM: The proposed MCSWOA without modified search strategies; (5) MCSWOAwoC: The proposed MCSWOA without crossover operator; and (6) MCSWOAwoS: The proposed MCSWOA without selection operator.

The mean and standard deviation values of the RMSE values over 50 independent runs are summarized in Table 10. The Wilcoxon's rank sum test was employed to compare the significance between MCSWOA and other algorithms. It is clear that MCSWOA performed significantly better than all of the other algorithms on all cases. Comparing WOAwM, WOAwC, and WOAwS with the original WOA, they won on 7, 10 and 5 cases while lost on 3, 0, and 5 cases, respectively. Additionally, comparison with WOAwM, WOAwC, and WOAwS, MCSWOAwoM beat them on all cases; MCSWOAwoC was better on 9, 4, and 9 cases, respectively; and MCSWOAwoS outperformed WOAwM and WOAwS on all cases, while just lost on cases 9 and 10 when compared with WOAwC. The comparison result indicated that the crossover operator contributed the most to MCSWOA, followed by the selection operator and modified search strategies. Besides, the absence of any improved component would deteriorate the performance of MCSWOA.

5.3.3. Comparison with Advanced WOA Variants

In this subsection, some advanced WOA variants were employed to verify the proposed MCSWOA. These advanced WOA variants included CWOA [34], IWOA [28], Lion_Whale [35], LWOA [36], MWOA [37], OBWOA [27], PSO_WOA [38], RWOA [39], SAWOA [40], WOA−CM [41], and WOABHC [42]. The experimental results are summarized in Table 11. It can be seen that MCSWOA was consistently significantly better than all of the other 11 algorithms on all cases, according to the statistical result of Wilcoxon's rank sum test. In addition, the standard deviation values of RMSE achieved by MCSWOA were also the smallest, meaning that the proposed algorithm was the most robust one among these 12 advanced WOA variants. Furthermore, the Friedman test result presented in Figure 5 manifests that MCSWOA yielded the first ranking, followed by IWOA, WOA−CM, Lion_Whale, MWOA, WOABHC, RWOA, LWOA, SAWOA, PSO_WOA, OBWOA, and CWOA. Some representative convergence curves given in Figure 6 indicate that MCSWOA had the fastest convergence speed overall, while other algorithms converged relatively slowly and had the possibility of being plunged into local optima. IWOA was slightly faster than MCSWOA at the initial stage on Case 2, but it was overtaken and surpassed quickly by MCSWOA.

Table 10. Influence of components on MCSWOA (Mean ± Std. dev.).

Algorithm	Case 1	Case 2	Case 3	Case 4	Case 5
WOA	$3.3118 \times 10^{-3} \pm 2.5700\times10^{-3}$ †	$3.6217 \times 10^{-3} \pm 2.7791\times10^{-3}$ †	$9.6733 \times 10^{-3} \pm 1.1794\times10^{-2}$ †	$2.4505 \times 10^{-2} \pm 2.2337\times10^{-2}$ †	$2.8343 \times 10^{-2} \pm 6.0554\times10^{-2}$ †
WOAwM	$1.9296 \times 10^{-3} \pm 8.6309\times10^{-4}$ †	$2.3822 \times 10^{-3} \pm 1.0539\times10^{-3}$ †	$3.9812 \times 10^{-3} \pm 2.0408\times10^{-3}$ †	$1.9714 \times 10^{-2} \pm 2.6652\times10^{-2}$ †	$9.2314 \times 10^{-3} \pm 8.1319\times10^{-3}$ †
WOAwC	$1.7668 \times 10^{-3} \pm 5.3337\times10^{-4}$ †	$2.2107 \times 10^{-3} \pm 6.2234\times10^{-4}$ †	$3.0516 \times 10^{-3} \pm 9.4430\times10^{-4}$ †	$3.9044 \times 10^{-3} \pm 1.4276\times10^{-3}$ †	$3.4664 \times 10^{-3} \pm 1.0919\times10^{-3}$ †
WOAwS	$1.5324 \times 10^{-3} \pm 6.0777\times10^{-4}$ †	$1.6445 \times 10^{-3} \pm 5.2751\times10^{-4}$ †	$4.0862 \times 10^{-3} \pm 4.2387\times10^{-3}$ †	$9.1345 \times 10^{-3} \pm 1.4631\times10^{-2}$ †	$3.6344 \times 10^{-3} \pm 9.2696\times10^{-3}$ †
MCSWOAwoM	$1.3311 \times 10^{-3} \pm 4.2360\times10^{-4}$ †	$1.5667 \times 10^{-3} \pm 5.5562\times10^{-4}$ †	$2.9301 \times 10^{-3} \pm 1.0113\times10^{-3}$ †	$2.8068 \times 10^{-3} \pm 6.8783\times10^{-4}$ †	$2.6430 \times 10^{-3} \pm 4.3694\times10^{-4}$ †
MCSWOAwoC	$1.3425 \times 10^{-3} \pm 3.5746\times10^{-4}$ †	$1.3755 \times 10^{-3} \pm 3.7302\times10^{-4}$ †	$2.8885 \times 10^{-3} \pm 9.4059\times10^{-4}$ †	$7.6989 \times 10^{-3} \pm 1.2735\times10^{-2}$ †	$3.5022 \times 10^{-2} \pm 9.4205\times10^{-2}$ †
MCSWOAwoS	$1.5019 \times 10^{-3} \pm 4.7729\times10^{-4}$ †	$1.5784 \times 10^{-3} \pm 4.6080\times10^{-4}$ †	$2.7574 \times 10^{-3} \pm 5.5040\times10^{-4}$ †	$2.9587 \times 10^{-3} \pm 9.4949\times10^{-4}$ †	$2.9479 \times 10^{-3} \pm 6.2242\times10^{-4}$ †
MCSWOA	$\mathbf{9.8602 \times 10^{-4}} \pm 4.8373\times10^{-10}$	$\mathbf{1.0078 \times 10^{-3}} \pm 3.7224\times10^{-5}$	$\mathbf{2.4252 \times 10^{-3}} \pm 3.2927\times10^{-7}$	$\mathbf{2.4377 \times 10^{-3}} \pm 1.3424\times10^{-5}$	$\mathbf{1.7311 \times 10^{-3}} \pm 1.0774\times10^{-6}$

Algorithm	Case 6	Case 7	Case 8	Case 9	Case 10
WOA	$1.2171 \times 10^{-2} \pm 8.5449\times10^{-3}$ †	$1.3390 \times 10^{-1} \pm 3.3374\times10^{-1}$ †	$3.8581 \times 10^{-2} \pm 1.1413\times10^{-2}$ †	$1.9117 \times 10^{-1} \pm 5.2271\times10^{-1}$ †	$3.4638 \times 10^{-2} \pm 1.2972\times10^{-2}$ †
WOAwM	$1.1827 \times 10^{-2} \pm 1.0218\times10^{-2}$ †	$1.4783 \times 10^{-1} \pm 4.0155\times10^{-1}$ †	$3.2690 \times 10^{-2} \pm 1.2753\times10^{-2}$ †	$9.2149 \times 10^{-1} \pm 1.1998$ †	$6.2633 \times 10^{-1} \pm 1.1112$ †
WOAwC	$3.8571 \times 10^{-2} \pm 1.3413\times10^{-3}$ †	$3.3596 \times 10^{-2} \pm 1.2829\times10^{-2}$ †	$3.5628 \times 10^{-2} \pm 1.1472\times10^{-2}$ †	$4.5812 \times 10^{-2} \pm 1.0846\times10^{-1}$ †	$2.8767 \times 10^{-2} \pm 1.3194\times10^{-2}$ †
WOAwS	$1.0220 \times 10^{-2} \pm 4.3428\times10^{-2}$ †	$3.0149 \times 10^{-1} \pm 5.3225\times10^{-1}$ †	$1.3955 \times 10^{-1} \pm 3.3301\times10^{-1}$ †	$8.6839 \times 10^{-1} \pm 9.1094\times10^{-1}$ †	$3.0201 \times 10^{-1} \pm 5.5643\times10^{-1}$ †
MCSWOAwoM	$2.7088 \times 10^{-3} \pm 4.7905\times10^{-3}$ †	$2.1776 \times 10^{-2} \pm 3.9977\times10^{-2}$ †	$2.2506 \times 10^{-2} \pm 3.9073\times10^{-3}$ †	$2.0330 \times 10^{-2} \pm 7.6106\times10^{-3}$ †	$2.4356 \times 10^{-2} \pm 8.2859\times10^{-3}$ †
MCSWOAwoC	$1.0494 \times 10^{-2} \pm 4.3369\times10^{-2}$ †	$8.8438 \times 10^{-2} \pm 2.0395\times10^{-2}$ †	$2.9647 \times 10^{-2} \pm 1.4360\times10^{-2}$ †	$3.0895 \times 10^{-1} \pm 6.8799\times10^{-1}$ †	$1.7716 \times 10^{-1} \pm 3.7440\times10^{-1}$ †
MCSWOAwoS	$2.9186 \times 10^{-3} \pm 6.0366\times10^{-4}$ †	$3.3319 \times 10^{-2} \pm 8.6499\times10^{-3}$ †	$3.0825 \times 10^{-2} \pm 9.3727\times10^{-3}$ †	$4.8567 \times 10^{-2} \pm 1.1040\times10^{-1}$ †	$3.2615 \times 10^{-2} \pm 1.0742\times10^{-2}$ †
MCSWOA	$\mathbf{1.7296 \times 10^{-3}} \pm 5.4724\times10^{-6}$	$\mathbf{1.6632 \times 10^{-2}} \pm 2.6486\times10^{-5}$	$\mathbf{1.6640 \times 10^{-2}} \pm 2.8956\times10^{-5}$	$\mathbf{1.1187 \times 10^{-2}} \pm 9.1358\times10^{-6}$	$\mathbf{1.1190 \times 10^{-2}} \pm 8.4623\times10^{-6}$

† denotes MCSWOA is significantly better than the compared algorithm according to the Wilcoxon's rank sum test at 5% significance difference.

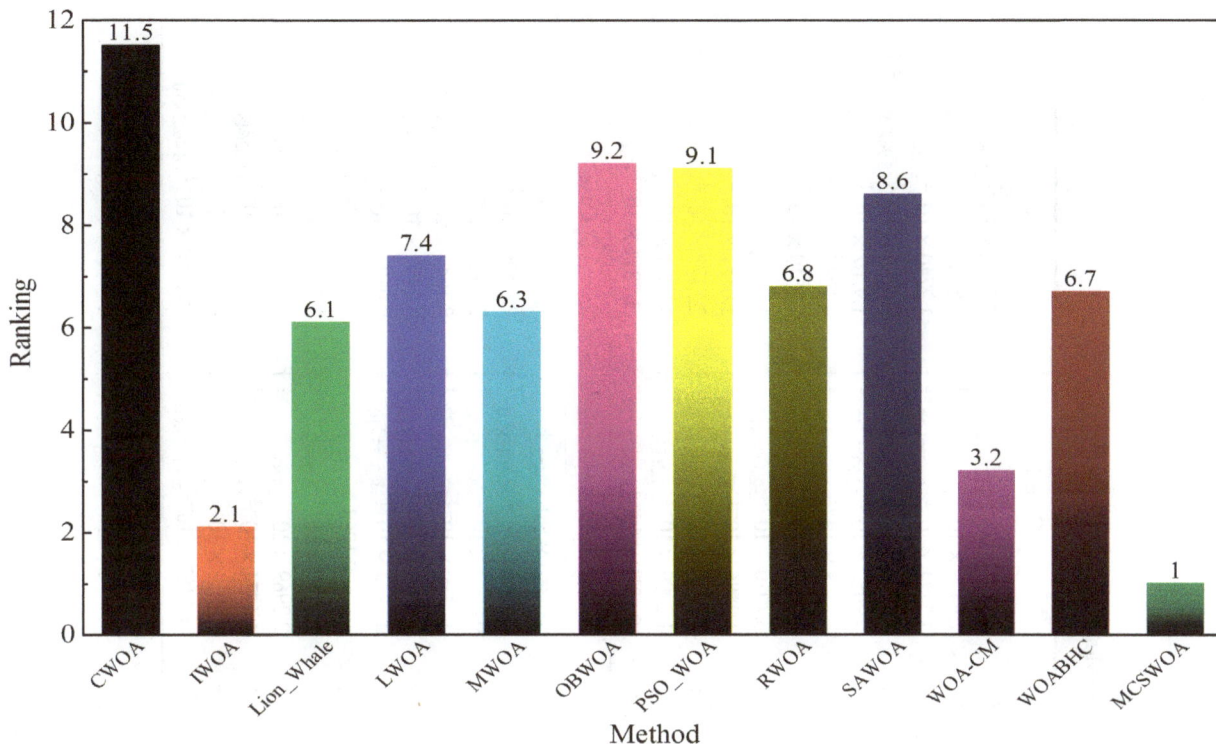

Figure 5. Friedman test result of MCSWOA with advanced WOA variants.

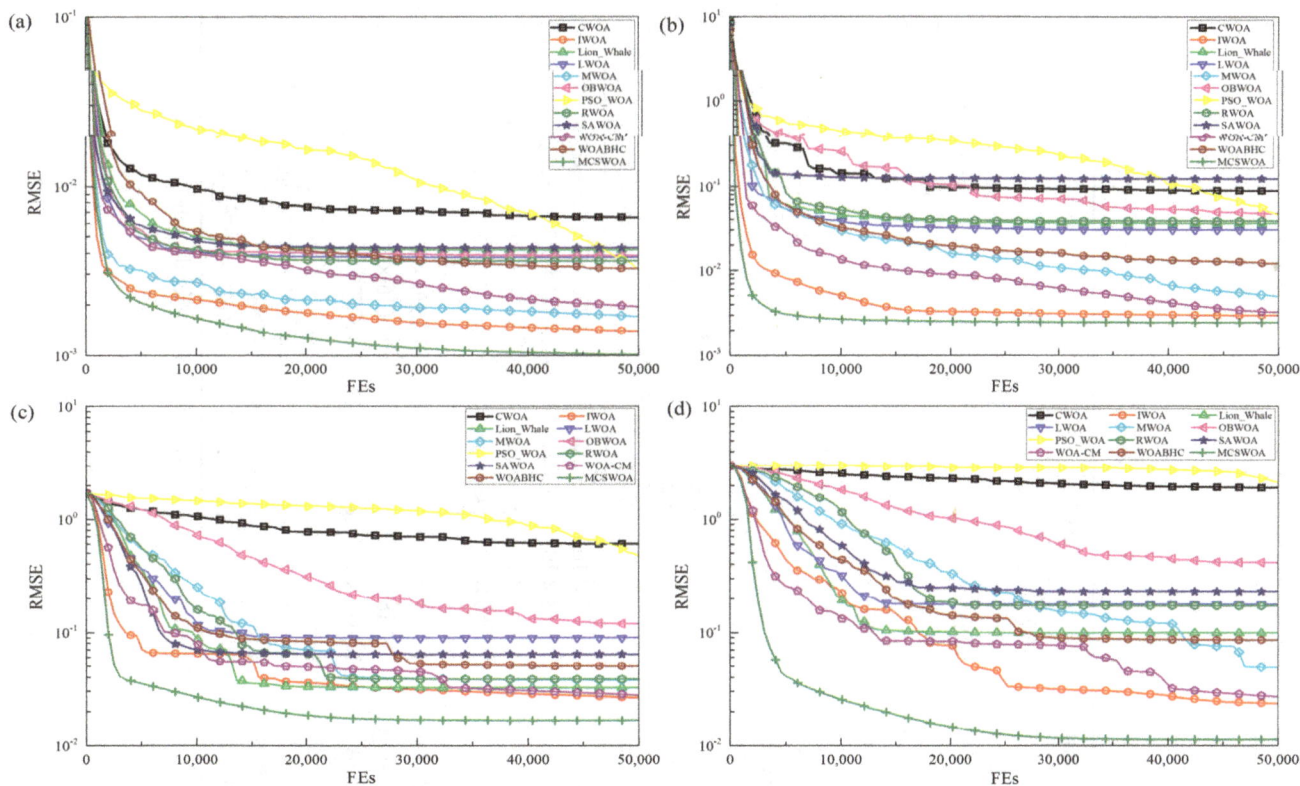

Figure 6. Convergence curves of MCSWOA with advanced WOA variants. (**a**) Case 2; (**b**) Case 4; (**c**) Case 7; (**d**) Case 9.

Table 11. Comparison with some advanced WOA variants (Mean ± Std. dev.).

Algorithm	Case 1	Case 2	Case 3	Case 4	Case 5
CWOA	6.5608×10^{-3} ±7.7906×10⁻³ †	6.5015×10^{-3} ±8.3105×10⁻³ †	4.0220×10^{-2} ±3.3878×10⁻² †	8.8557×10^{-2} ±1.2848×10⁻¹ †	7.3042×10^{-2} ±1.0974×10⁻¹ †
IWOA	1.3789×10^{-3} ±5.1312×10⁻⁴ †	1.3881×10^{-3} ±2.9395×10⁻⁴ †	2.7650×10^{-3} ±4.6827×10⁻⁴ †	2.9409×10^{-3} ±6.7649×10⁻⁴ †	2.8122×10^{-3} ±4.5910×10⁻⁴ †
Lion_Whale	3.1843×10^{-3} ±2.2032×10⁻³ †	4.1686×10^{-3} ±3.1841×10⁻³ †	8.3281×10^{-3} ±1.0314×10⁻² †	3.5424×10^{-2} ±2.8837×10⁻² †	1.2740×10^{-2} ±8.5614×10⁻³ †
LWOA	3.8223×10^{-3} ±2.5841×10⁻³ †	3.7734×10^{-3} ±3.2433×10⁻³ †	7.5580×10^{-3} ±1.0062×10⁻² †	2.9719×10^{-2} ±2.6177×10⁻² †	2.3724×10^{-2} ±5.9677×10⁻² †
MWOA	1.4352×10^{-3} ±3.8523×10⁻⁴ †	1.6923×10^{-3} ±5.4619×10⁻⁴ †	3.4700×10^{-3} ±1.4623×10⁻³ †	4.9057×10^{-3} ±2.7372×10⁻³ †	1.9707×10^{-1} ±9.6087×10⁻² †
OBWOA	3.0937×10^{-3} ±2.1925×10⁻³ †	3.8497×10^{-3} ±2.0783×10⁻³ †	1.1591×10^{-2} ±1.1474×10⁻² †	4.6378×10^{-2} ±3.6616×10⁻² †	4.1245×10^{-2} ±8.1091×10⁻² †
PSO_WOA	2.5317×10^{-3} ±1.0688×10⁻³ †	3.1643×10^{-3} ±1.0202×10⁻² †	6.1397×10^{-3} ±2.5857×10⁻³ †	4.4845×10^{-2} ±5.8766×10⁻² †	2.5510×10^{-2} ±4.2215×10⁻² †
RWOA	3.5386×10^{-3} ±2.5903×10⁻³ †	3.5906×10^{-3} ±2.5699×10⁻³ †	1.1432×10^{-2} ±1.2700×10⁻² †	3.7910×10^{-2} ±2.8936×10⁻² †	1.4016×10^{-2} ±8.1452×10⁻³ †
SAWOA	3.9103×10^{-3} ±3.2763×10⁻³ †	4.2854×10^{-3} ±2.7580×10⁻³ †	1.0367×10^{-2} ±1.1587×10⁻² †	1.2186×10^{-1} ±6.1004×10⁻¹ †	1.0681×10^{-2} ±5.7482×10⁻³ †
WOA–CM	1.8057×10^{-3} ±9.4142×10⁻⁴ †	1.9303×10^{-3} ±6.4609×10⁻⁴ †	3.0553×10^{-3} ±1.1304×10⁻³ †	3.2230×10^{-3} ±1.0350×10⁻³ †	2.7473×10^{-3} ±6.1670×10⁻⁴ †
WOABHC	2.4830×10^{-3} ±1.4878×10⁻³ †	3.2285×10^{-3} ±1.7139×10⁻³ †	5.7079×10^{-3} ±5.9201×10⁻³ †	1.2009×10^{-2} ±1.3695×10⁻² †	1.7371×10^{-2} ±7.5501×10⁻² †
MCSWOA	$\mathbf{9.8602 \times 10^{-4}}$ ±4.8373×10⁻¹⁰	$\mathbf{1.0078 \times 10^{-3}}$ ±3.7224×10⁻⁵	$\mathbf{2.4252 \times 10^{-3}}$ ±3.3927×10⁻⁷	$\mathbf{2.4377 \times 10^{-3}}$ ±1.3424×10⁻⁵	$\mathbf{1.7311 \times 10^{-3}}$ ±1.0774×10⁻⁶

Algorithm	Case 6	Case 7	Case 8	Case 9	Case 10
CWOA	4.1656×10^{-2} ±6.0841×10⁻² †	6.0298×10^{-1} ±7.0364×10⁻¹ †	2.4784×10^{-1} ±4.8707×10⁻¹ †	1.9090 ±1.0622 †	1.1956 ±1.2143 †
IWOA	2.8068×10^{-3} ±6.1560×10⁻⁴ †	2.6344×10^{-2} ±6.5869×10⁻³ †	2.4487×10^{-2} ±6.0914×10⁻³ †	2.3385×10^{-2} ±1.1011×10⁻² †	2.0908×10^{-2} ±8.4553×10⁻³ †
Lion_Whale	1.1758×10^{-2} ±7.4784×10⁻³ †	3.2231×10^{-2} ±1.1794×10⁻² †	3.2987×10^{-2} ±1.2592×10⁻² †	9.8029×10^{-2} ±3.4184×10⁻¹ †	2.9543×10^{-2} ±1.3201×10⁻² †
LWOA	1.1983×10^{-2} ±7.8099×10⁻³ †	8.9130×10^{-2} ±2.7403×10⁻¹ †	3.7172×10^{-2} ±1.5830×10⁻² †	1.7580×10^{-1} ±5.0999×10⁻¹ †	3.1649×10^{-2} ±1.4127×10⁻² †
MWOA	1.6428×10^{-1} ±7.9106×10⁻² †	3.7422×10^{-2} ±1.2357×10⁻² †	3.8979×10^{-2} ±1.0903×10⁻² †	4.8405×10^{-2} ±1.0824×10⁻¹ †	3.2497×10^{-2} ±1.2601×10⁻² †
OBWOA	1.7604×10^{-2} ±9.8420×10⁻³ †	1.1870×10^{-1} ±3.3200×10⁻¹ †	9.8499×10^{-2} ±2.7772×10⁻¹ †	4.0964×10^{-1} ±8.0655×10⁻¹ †	2.9575×10^{-2} ±1.2568×10⁻² †
PSO_WOA	2.0718×10^{-2} ±1.1155×10⁻² †	4.6616×10^{-1} ±6.7103×10⁻¹ †	1.2651×10^{-1} ±1.2418×10⁻¹ †	2.1068 ±1.1111 †	2.1078 ±1.0319 †
RWOA	1.3841×10^{-2} ±8.0800×10⁻³ †	3.8377×10^{-2} ±2.1557×10⁻² †	3.5678×10^{-2} ±1.2714×10⁻² †	1.7159×10^{-1} ±5.0952×10⁻¹ †	2.7693×10^{-2} ±1.2816×10⁻² †
SAWOA	1.3133×10^{-2} ±8.9981×10⁻³ †	6.3576×10^{-2} ±1.9534×10⁻¹ †	3.8696×10^{-2} ±3.0127×10⁻² †	2.2969×10^{-1} ±5.8559×10⁻¹ †	3.7130×10^{-2} ±1.2907×10⁻² †
WOA–CM	2.9571×10^{-3} ±6.6077×10⁻⁴ †	2.7691×10^{-2} ±1.0487×10⁻² †	2.7232×10^{-2} ±1.0768×10⁻² †	2.6774×10^{-2} ±1.6707×10⁻² †	2.9099×10^{-2} ±1.3264×10⁻² †
WOABHC	1.7284×10^{-2} ±8.0367×10⁻³ †	4.9880×10^{-2} ±7.6648×10⁻³ †	4.6525×10^{-2} ±1.1413×10⁻² †	8.4911×10^{-2} ±2.9295×10⁻¹ †	4.1759×10^{-2} ±1.0658×10⁻² †
MCSWOA	$\mathbf{1.7296 \times 10^{-3}}$ ±5.4724×10⁻⁶	$\mathbf{1.6632 \times 10^{-2}}$ ±2.6486×10⁻⁵	$\mathbf{1.6640 \times 10^{-2}}$ ±2.8956×10⁻⁵	$\mathbf{1.1187 \times 10^{-2}}$ ±9.1358×10⁻⁶	$\mathbf{1.1190 \times 10^{-2}}$ ±8.4623×10⁻⁶

† denotes MCSWOA is significantly better than the compared algorithm according to the Wilcoxon's rank sum test at 5% significance difference.

5.3.4. Comparison with Advanced Non−WOA Variants

The performance of MCSWOA was further verified by some advanced non−WOA variants. Thirteen algorithms consisting of BLPSO [43], CLPSO [44], CSO [45], DBBO [46], DE/BBO [47], GOTLBO [14], IJAYA [17], LETLBO [48], MABC [49], ODE [50], SATLBO [15], SLPSO [51], and TLABC [24] were employed for comparison in this subsection. The result of Wilcoxon's rank sum test tabulated in Table 12 shows that MCSWOA performed very competitively and outperformed all of the other 13 algorithms on 9 cases except Case 4, on which MCSWOA was surpassed by ODE and DBBO, and tied by TLABC. Considering the standard deviation values, the comparison result was similar to that of the mean values of RMSE, which validated the good robustness of MCSWOA. Similarly, the Friedman test result given in Figure 7 shows that MCSWOA won the first ranking again, followed by TLABC, IJAYA, SATLBO, LETLBO, GOTLBO, ODE, DE/BBO, DBBO, CLPSO, MABC, BLPSO, SLPSO, and CSO. In addition, the convergence curves in Figure 8 reveal again that MCSWOA obtained a competitively fast convergence speed throughout the whole evolutionary process although it was temporarily surpassed by ODE at the early stage.

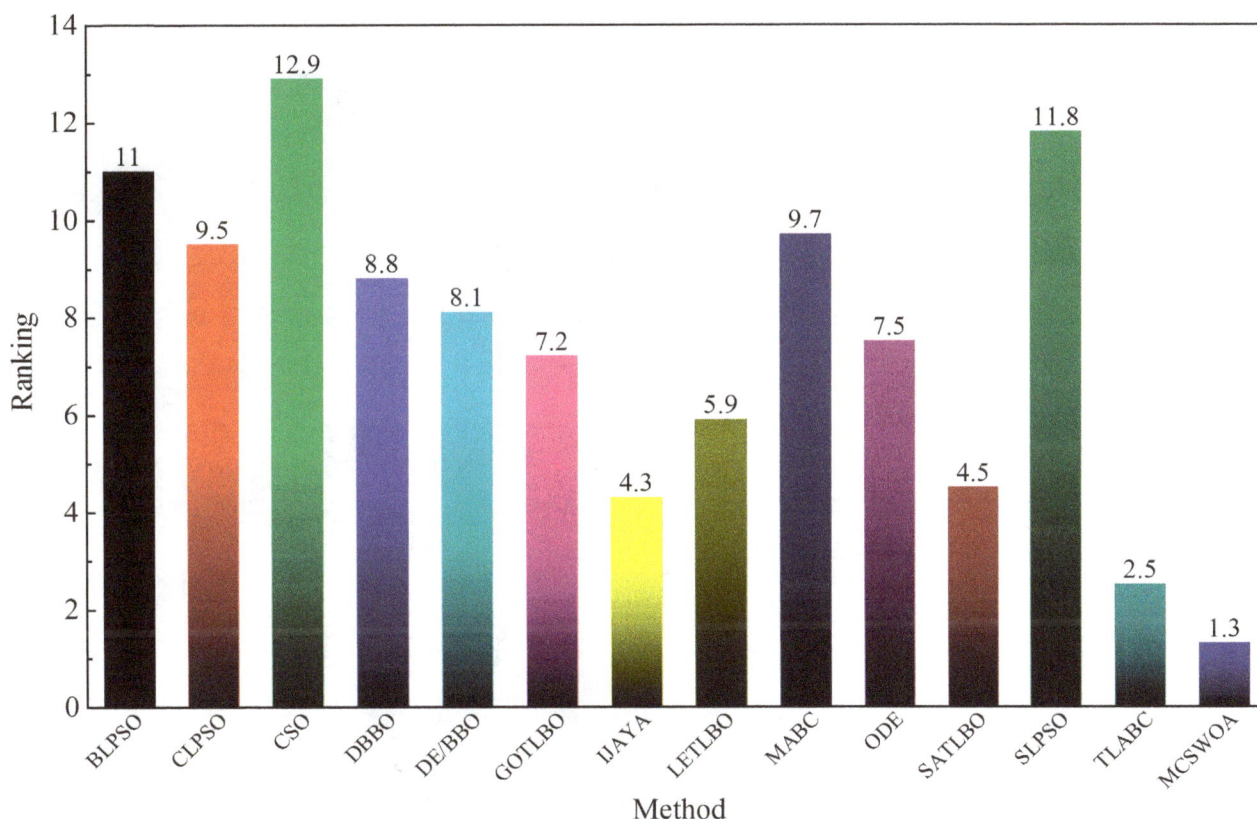

Figure 7. Friedman test result of MCSWOA with advanced non−WOA variants.

Table 12. Comparison with some advanced non−WOA variants (Mean ± Std. dev.).

Algorithm	Case 1	Case 2	Case 3	Case 4	Case 5
BLPSO	1.9021×10^{-3} ±1.8505×10^{-4} +	2.0514×10^{-3} ±2.7912×10^{-4} +	2.4898×10^{-3} ±2.7678×10^{-5} +	2.5112×10^{-3} ±5.4421×10^{-5} +	5.2325×10^{-3} ±1.1639×10^{-3} +
CLPSO	1.1194×10^{-3} ±1.0940×10^{-4} +	1.2102×10^{-3} ±1.2533×10^{-4} +	2.4833×10^{-3} ±3.3208×10^{-5} +	2.5561×10^{-3} ±6.5265×10^{-5} +	3.9131×10^{-3} ±9.9804×10^{-4} +
CSO	1.7135×10^{-3} ±3.7256×10^{-4} +	2.3968×10^{-3} ±5.0421×10^{-4} +	2.4779×10^{-3} ±6.1374×10^{-5} +	2.4703×10^{-3} ±3.3601×10^{-5} +	3.6956×10^{-2} ±5.2404×10^{-2} +
DBBO	1.2829×10^{-3} ±2.5357×10^{-4} +	1.0515×10^{-3} ±1.0529×10^{-4} +	2.4255×10^{-3} ±1.8443×10^{-6} +	2.4257×10^{-3} ±2.1496×10^{-6} ‡	1.5373×10^{-2} ±1.3834×10^{-2} +
DE/BBO	1.1196×10^{-3} ±1.1647×10^{-4} +	1.1190×10^{-3} ±1.5390×10^{-4} +	2.4332×10^{-3} ±5.3545×10^{-5} +	2.4536×10^{-3} ±5.7504×10^{-5} +	3.7298×10^{-3} ±2.9966×10^{-3} +
GOTLBO	1.0777×10^{-3} ±1.0248×10^{-4} +	1.1211×10^{-3} ±1.1785×10^{-4} +	2.4710×10^{-3} ±8.6113×10^{-5} +	2.5120×10^{-3} ±1.4228×10^{-4} +	2.7002×10^{-3} ±2.9037×10^{-4} +
IJAYA	1.0116×10^{-3} ±3.9701×10^{-5} +	1.0375×10^{-3} ±6.5079×10^{-5} +	2.4402×10^{-3} ±1.7719×10^{-5} +	2.4547×10^{-3} ±2.8211×10^{-5} +	2.2691×10^{-3} ±3.7081×10^{-4} +
LETLBO	1.0118×10^{-3} ±2.9676×10^{-5} +	1.0565×10^{-3} ±1.0299×10^{-4} +	2.4517×10^{-3} ±4.1189×10^{-5} +	2.4607×10^{-3} ±4.1340×10^{-5} +	2.3621×10^{-3} ±3.3351×10^{-3} +
MABC	1.1217×10^{-3} ±1.5006×10^{-4} +	1.1301×10^{-3} ±1.1174×10^{-4} +	2.4592×10^{-3} ±3.4902×10^{-5} +	2.4913×10^{-3} ±4.6322×10^{-5} +	1.2849×10^{-2} ±7.4066×10^{-3} +
ODE	1.1306×10^{-3} ±1.3390×10^{-4} +	1.0152×10^{-3} ±7.3670×10^{-5} +	2.4265×10^{-3} ±7.2112×10^{-6} +	2.4255×10^{-3} ±1.5214×10^{-6} +	3.2435×10^{-3} ±1.6449×10^{-3} +
SATLBO	9.9236×10^{-4} ±7.7023×10^{-6} +	1.0196×10^{-3} ±4.4399×10^{-5} +	2.4503×10^{-3} ±8.8712×10^{-5} +	2.5334×10^{-3} ±2.4232×10^{-4} +	1.9681×10^{-3} ±1.6428×10^{-4} +
SLPSO	1.6741×10^{-3} ±3.8943×10^{-4} +	2.2540×10^{-3} ±6.0816×10^{-4} +	2.5069×10^{-3} ±1.8101×10^{-4} +	2.4713×10^{-3} ±4.0124×10^{-5} +	1.2625×10^{-2} ±5.1388×10^{-3} +
TLABC	9.9237×10^{-4} ±1.5009×10^{-5} +	1.0325×10^{-3} ±6.4577×10^{-5} +	2.4255×10^{-3} ±9.5526×10^{-7} +	2.4339×10^{-3} ±9.0969×10^{-6} ≈	1.8665×10^{-3} ±1.0099×10^{-3} +
MCSWOA	$\mathbf{9.8602 \times 10^{-4}}$ ±8.4873×10^{-10}	$\mathbf{1.0078 \times 10^{-3}}$ ±3.7224×10^{-5}	$\mathbf{2.4252 \times 10^{-3}}$ ±3.2927×10^{-7}	$\mathbf{2.4377 \times 10^{-3}}$ ±1.3424×10^{-5}	$\mathbf{1.7311 \times 10^{-3}}$ ±1.0774×10^{-6}

Algorithm	Case 6	Case 7	Case 8	Case 9	Case 10
BLPSO	5.0586×10^{-3} ±1.2686×10^{-3} +	4.7472×10^{-2} ±3.2271×10^{-3} +	4.4430×10^{-2} ±5.2342×10^{-3} +	4.4674×10^{-2} ±5.5602×10^{-3} +	4.3783×10^{-2} ±4.7164×10^{-3} +
CLPSO	4.2857×10^{-3} ±1.0083×10^{-3} +	2.6297×10^{-2} ±6.0504×10^{-3} +	3.0761×10^{-2} ±8.4038×10^{-3} +	1.2006×10^{-1} ±7.4285×10^{-2} +	1.2302×10^{-1} ±8.9533×10^{-2} +
CSO	1.5507×10^{-2} ±7.6428×10^{-3} +	3.8608×10^{-1} ±4.6849×10^{-1} +	1.3560×10^{-1} ±2.4327×10^{-1} +	1.3952 ±6.9328×10^{-1} +	9.5533×10^{-1} ±8.0496×10^{-1} +
DBBO	1.3809×10^{-2} ±9.4018×10^{-3} +	1.5307×10^{-1} ±2.0137×10^{-1} +	7.4939×10^{-2} ±8.1393×10^{-2} +	3.5746×10^{-2} ±2.0978×10^{-2} +	3.4309×10^{-2} ±8.1225×10^{-2} +
DE/BBO	4.6286×10^{-3} ±3.1740×10^{-3} +	3.2601×10^{-2} ±7.9176×10^{-3} +	3.2281×10^{-2} ±7.4126×10^{-3} +	3.3622×10^{-1} ±5.2313×10^{-1} +	2.7941×10^{-1} ±4.3527×10^{-1} +
GOTLBO	3.3486×10^{-3} ±6.6655×10^{-4} +	2.1023×10^{-2} ±2.9156×10^{-3} +	2.6143×10^{-2} ±6.4333×10^{-3} +	1.9831×10^{-2} ±5.5072×10^{-3} +	2.5341×10^{-2} ±9.1729×10^{-3} +
IJAYA	2.5200×10^{-3} ±5.1689×10^{-4} +	1.7273×10^{-2} ±4.0886×10^{-3} +	1.7915×10^{-2} ±1.6640×10^{-3} +	1.2786×10^{-2} ±1.5584×10^{-3} +	1.3658×10^{-2} ±2.4658×10^{-3} +
LETLBO	2.8076×10^{-3} ±8.0176×10^{-4} +	2.2716×10^{-2} ±1.9207×10^{-2} +	1.9306×10^{-2} ±2.8808×10^{-3} +	3.1644×10^{-2} ±3.5249×10^{-3} +	2.4674×10^{-2} ±1.9033×10^{-3} +
MABC	1.1607×10^{-2} ±7.3824×10^{-3} +	4.1445×10^{-2} ±1.0439×10^{-2} +	4.0201×10^{-2} ±1.1824×10^{-2} +	3.7567×10^{-2} ±8.9141×10^{-3} +	3.4091×10^{-2} ±1.1119×10^{-2} +
ODE	3.0783×10^{-3} ±1.3525×10^{-3} +	4.5691×10^{-2} ±5.5273×10^{-2} +	3.4596×10^{-2} ±3.5109×10^{-2} +	1.2531 ±4.2568×10^{-1} +	1.2490 ±3.5744×10^{-1} +
SATLBO	2.0176×10^{-3} ±1.6428×10^{-4} +	1.7206×10^{-2} ±9.1397×10^{-4} +	1.7356×10^{-2} ±9.3366×10^{-4} +	1.6181×10^{-2} ±9.9094×10^{-3} +	1.9837×10^{-2} ±1.2493×10^{-2} +
SLPSO	9.5470×10^{-3} ±5.4545×10^{-3} +	1.3935×10^{-1} ±1.8024×10^{-1} +	6.4134×10^{-2} ±7.0877×10^{-2} +	3.6172×10^{-1} ±3.2445×10^{-1} +	3.9282×10^{-1} ±3.9592×10^{-1} +
TLABC	1.9030×10^{-3} ±1.0096×10^{-4} +	1.6806×10^{-2} ±2.3608×10^{-4} +	1.6773×10^{-2} ±9.1609×10^{-5} +	1.1691×10^{-2} ±7.1799×10^{-4} +	1.1892×10^{-2} ±1.3444×10^{-3} +
MCSWOA	$\mathbf{1.7296 \times 10^{-3}}$ ±5.4724×10^{-6}	$\mathbf{1.6632 \times 10^{-2}}$ ±2.6486×10^{-5}	$\mathbf{1.6640 \times 10^{-2}}$ ±2.8956×10^{-5}	$\mathbf{1.1187 \times 10^{-2}}$ ±9.1358×10^{-6}	$\mathbf{1.1190 \times 10^{-2}}$ ±8.4623×10^{-6}

†, ≈, and ‡ denote MCSWOA is respectively better than, equal to, and worse than the compared algorithm according to the Wilcoxon's rank sum test at 5% significance difference.

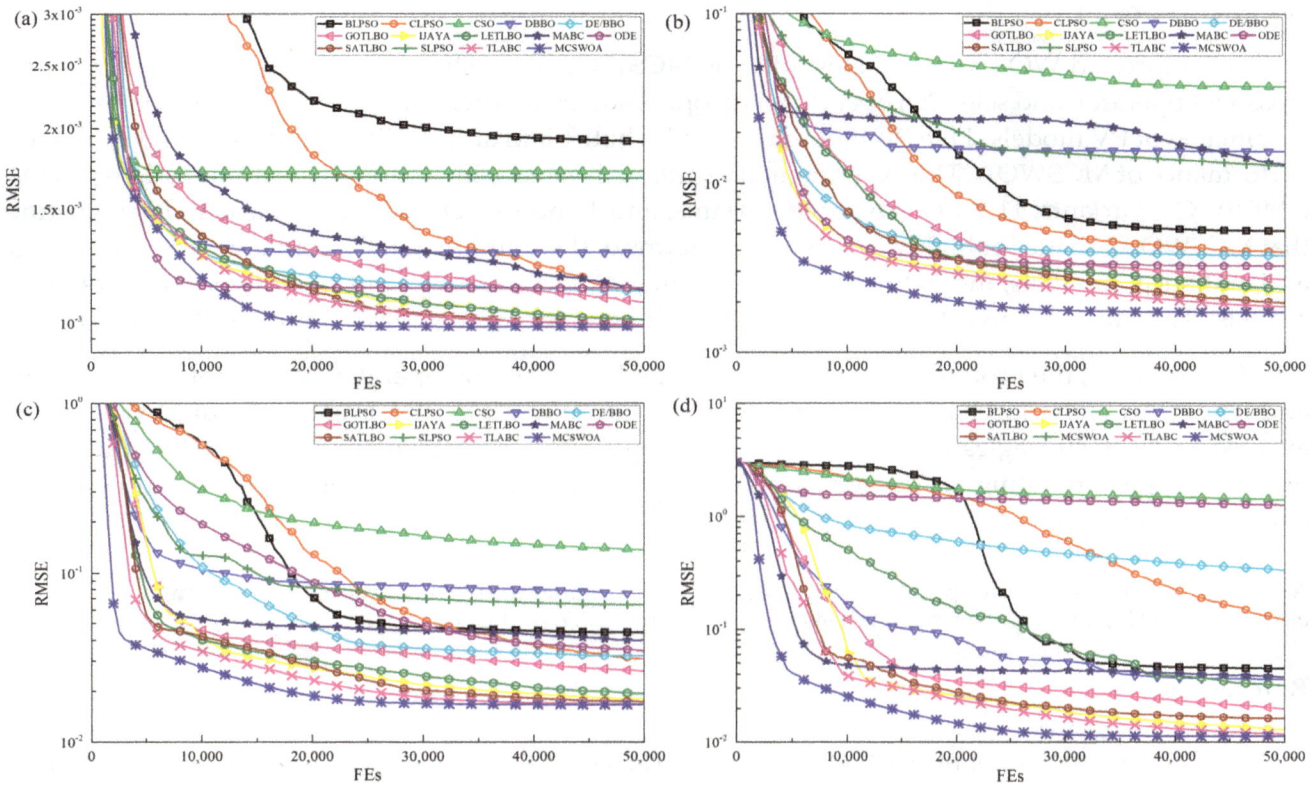

Figure 8. Convergence curves of MCSWOA with advanced non−WOA variants. (**a**) Case 1; (**b**) Case 5; (**c**) Case 8; (**d**) Case 9.

6. Discussions

In this work, we present modified search strategies, crossover operator, and selection operator to enhance the performance of MCSWOA. In the modified search strategies, WOA/rand/1 strategy focuses on the exploration, while WOA/current−to−best/1 strategy emphasizes the exploitation. They can cooperate well to achieve a good ratio between exploration and exploitation. In the crossover operator, each dimension of each donor individual has the same chance of deriving from two search strategies, which can further promote the balance between exploration and exploitation. In the selection operator, only comparative or better individuals can survive to the next iteration, which makes the population either gain quality improvement or maintain the current quality level, but never get worse. Experiments have been conducted on five PV types modeled by both SDM and DDM. From the experimental results and comparisons, we can summarize that:

(1)   MCSWOA obtains better results on most of the cases except Case 4, which can be explained by the no free lunch theorem [52]. According to the theorem, there is no "one size fits all" method that always wins all cases.

(2)   The convergence curves show that MCSWOA converges the fastest overall throughout the whole evolutionary process, which indicates that it achieves an excellent balance between exploration and exploitation.

(3)   The crossover operator contributes the most to MCSWOA, followed by the selection operator and modified search strategies. Nevertheless, each component is indispensable, and missing anyone will deteriorate the performance MCSWOA significantly.

(4)   Comparing the results of SDM and DDM, it concludes that not every equivalent circuit model is suitable for every PV type. Notwithstanding, the differences are very small. In addition, the DDM is harder to optimize under the same stopping condition (i.e., the same value of Max_FEs) because it has seven unknown parameters whereas the SDM has only five.

7. Conclusions

An improved WOA variant referred to as MCSWOA by integrating modified search strategies, crossover operator, and selection operators is proposed to extract accurate values for involved unknown parameters of PV models. Five PV types modeled by both SDM and DDM are employed to validate the performance of MCSWOA. The experimental results compared with various algorithms (original WOA, 6 MCSWOA variants, 11 WOA advanced variants, and 13 non−WOA advanced variants) demonstrate that MCSWOA is better or highly competitive in terms of the solution quality, convergence performance, and statistical analysis, indicating that it can achieve more accurate and reliable parameters of PV models. Therefore, MCSWOA is a promising candidate for parameter extraction of PV models.

In this work, the proposed MCSWOA is verified at one given operating condition for a PV type, and its performance still has room to improve. In future work, on the one hand, adaptive learning and local search strategies will be used to further enhance its performance and, on the other hand, other PV types operating at different irradiances and temperatures will be employed to verify the enhanced performance.

Author Contributions: Conceptualization, G.X.; Writing—original draft preparation, G.X.; Writing—review and editing, J.Z., D.S., and L.Z.; Formal analysis, X.Y.; Resources, X.Y. and G.Y.

References

1. Hayat, M.B.; Ali, D.; Monyake, K.C.; Alagha, L.; Ahmed, N. Solar energy—A look into power generation, challenges, and a solar-powered future. *Int. J. Energy Res.* **2019**, *43*, 1049–1067. [CrossRef]
2. Islam, M.R.; Mahfuz-Ur-Rahman, A.M.; Muttaqi, K.M.; Sutanto, D. State-of-the-Art of the Medium-Voltage Power Converter Technologies for Grid Integration of Solar Photovoltaic Power Plants. *IEEE Trans. Energy Conver.* **2019**, *34*, 372–384. [CrossRef]
3. National Energy Administration. Introduction to the Operation of Grid Connected Renewable Energy in the First Quarter of 2019. Available online: http://www.nea.gov.cn/2019-04/29/c_138021561.htm (accessed on 25 May 2019). (In Chinese)
4. Chin, V.J.; Salam, Z.; Ishaque, K. Cell modelling and model parameters estimation techniques for photovoltaic simulator application: A review. *Appl. Energy* **2015**, *154*, 500–519. [CrossRef]
5. Ishaque, K.; Salam, Z.; Taheri, H.; Shamsudin, A. A critical evaluation of EA computational methods for Photovoltaic cell parameter extraction based on two diode model. *Sol. Energy* **2011**, *85*, 1768–1779. [CrossRef]
6. Gao, X.; Cui, Y.; Hu, J.; Xu, G.; Wang, Z.; Qu, J.; Wang, H. Parameter extraction of solar cell models using improved shuffled complex evolution algorithm. *Energy Convers. Manag.* **2018**, *157*, 460–479. [CrossRef]
7. Gomes, R.C.M.; Vitorino, M.A.; Fernandes, D.A.; Wang, R. Shuffled Complex Evolution on Photovoltaic Parameter Extraction: A Comparative Analysis. *IEEE Trans. Sustain Energy* **2016**, *8*, 805–815. [CrossRef]
8. Yeh, W.C.; Huang, C.L.; Lin, P.; Chen, Z.; Jiang, Y.; Sun, B. Simplex Simplified Swarm Optimization for the Efficient Optimization of Parameter Identification for Solar Cell Models. *IET Renew. Power Gen.* **2018**, *12*, 45–51. [CrossRef]
9. Rezaee Jordehi, A. Enhanced leader particle swarm optimisation (ELPSO): An efficient algorithm for parameter estimation of photovoltaic (PV) cells and modules. *Sol. Energy* **2018**, *159*, 78–87. [CrossRef]
10. Nunes, H.G.G.; Pombo, J.A.N.; Mariano, S.J.P.S.; Calado, M.R.A.; Souza, J.A.M.F. A new high performance method for determining the parameters of PV cells and modules based on guaranteed convergence particle swarm optimization. *Appl. Energy* **2018**, *211*, 774–791. [CrossRef]
11. Ma, J.; Man, K.L.; Guan, S.U.; Ting, T.O.; Wong, P.W.H. Parameter estimation of photovoltaic model via parallel particle swarm optimization algorithm. *Int. J. Energy Res.* **2016**, *40*, 343–352. [CrossRef]
12. Muangkote, N.; Sunat, K.; Chiewchanwattana, S.; Kaiwinit, S. An advanced onlooker-ranking-based adaptive differential evolution to extract the parameters of solar cell models. *Renew. Energy* **2019**, *134*, 1129–1147. [CrossRef]
13. Chen, X.; Yu, K.; Du, W.; Zhao, W.; Liu, G. Parameters identification of solar cell models using generalized oppositional teaching learning based optimization. *Energy* **2016**, *99*, 170–180. [CrossRef]

14. Yu, K.; Chen, X.; Wang, X.; Wang, Z. Parameters identification of photovoltaic models using self-adaptive teaching-learning-based optimization. *Energy Convers. Manag.* **2017**, *145*, 233–246. [CrossRef]

15. Xiong, G.; Zhang, J.; Shi, D.; Yuan, X. Application of Supply-Demand-Based Optimization for Parameter Extraction of Solar Photovoltaic Models. *Complexity* **2019**, *2019*, 3923691. [CrossRef]

16. Xiong, G.; Zhang, J.; Yuan, X.; Shi, D.; He, Y. Application of Symbiotic Organisms Search Algorithm for Parameter Extraction of Solar Cell Models. *Appl. Sci.* **2018**, *8*, 2155. [CrossRef]

17. Yu, K.; Liang, J.J.; Qu, B.Y.; Chen, X.; Wang, H. Parameters identification of photovoltaic models using an improved JAYA optimization algorithm. *Energy Convers. Manag.* **2017**, *150*, 742–753. [CrossRef]

18. Oliva, D.; Ewees, A.A.; Aziz, M.A.E.; Hassanien, A.E.; Cisneros, M.P. A Chaotic Improved Artificial Bee Colony for Parameter Estimation of Photovoltaic Cells. *Energies* **2017**, *10*, 865. [CrossRef]

19. Fathy, A.; Rezk, H. Parameter estimation of photovoltaic system using imperialist competitive algorithm. *Renew. Energy* **2017**, *111*, 307–320. [CrossRef]

20. Benkercha, R.; Moulahoum, S.; Taghezouit, B. Extraction of the PV modules parameters with MPP estimation using the modified flower algorithm. *Renew. Energy* **2019**, *143*, 1698–1709. [CrossRef]

21. Ram, J.P.; Babu, T.S.; Dragicevic, T.; Rajasekar, N. A new hybrid bee pollinator flower pollination algorithm for solar PV parameter estimation. *Energy Convers. Manag.* **2017**, *135*, 463–476. [CrossRef]

22. Mughal, M.A.; Ma, Q.; Xiao, C. Photovoltaic Cell Parameter Estimation Using Hybrid Particle Swarm Optimization and Simulated Annealing. *Energies* **2017**, *10*, 1213. [CrossRef]

23. Muhsen, D.H.; Ghazali, A.B.; Khatib, T.; Abed, I.A. Extraction of photovoltaic module model's parameters using an improved hybrid differential evolution/electromagnetism−like algorithm. *Sol. Energy* **2015**, *119*, 286–297. [CrossRef]

24. Chen, X.; Xu, B.; Mei, C.; Ding, Y.; Li, K. Teaching–learning–based artificial bee colony for solar photovoltaic parameter estimation. *Appl. Energy* **2018**, *212*, 1578–1588. [CrossRef]

25. Mirjalili, S.; Lewis, A. The Whale Optimization Algorithm. *Adv. Eng. Softw.* **2016**, *95*, 51–67. [CrossRef]

26. Oliva, D.; Aziz, M.A.E.; Hassanien, A.E. Parameter estimation of photovoltaic cells using an improved chaotic whale optimization algorithm. *Appl. Energy* **2017**, *200*, 141–154. [CrossRef]

27. Abd Elaziz, M.; Oliva, D. Parameter estimation of solar cells diode models by an improved opposition-based whale optimization algorithm. *Energy Convers. Manag.* **2018**, *171*, 1843–1859. [CrossRef]

28. Xiong, G.; Zhang, J.; Shi, D.; He, Y. Parameter extraction of solar photovoltaic models using an improved whale optimization algorithm. *Energy Convers. Manag.* **2018**, *174*, 388–405. [CrossRef]

29. Xiong, G.; Zhang, J.; Yuan, X.; Shi, D.; He, Y.; Yao, G. Parameter extraction of solar photovoltaic models by means of a hybrid differential evolution with whale optimization algorithm. *Sol. Energy* **2018**, *176*, 742–761. [CrossRef]

30. Muhsen, D.H.; Ghazali, A.B.; Khatib, T.; Abed, I.A. A comparative study of evolutionary algorithms and adapting control parameters for estimating the parameters of a single-diode photovoltaic module's model. *Renew. Energy* **2016**, *96*, 377–389. [CrossRef]

31. Kichou, S.; Silvestre, S.; Guglielminotti, L.; Mora-López, L.; Muñoz-Cerón, E. Comparison of two PV array models for the simulation of PV systems using five different algorithms for the parameters identification. *Renew. Energy* **2016**, *99*, 270–279. [CrossRef]

32. Storn, R.; Price, K. Differential evolution–a simple and efficient heuristic for global optimization over continuous spaces. *J. Glob. Optim.* **1997**, *11*, 341–359. [CrossRef]

33. Yu, K.; Qu, B.; Yue, C.; Ge, S.; Chen, X.; Liang, J. A performance-guided JAYA algorithm for parameters identification of photovoltaic cell and module. *Appl. Energy* **2019**, *237*, 241–257. [CrossRef]

34. Prasad, D.; Mukherjee, A.; Shankar, G.; Mukherjee, V. Application of chaotic whale optimization algorithm for transient stability constrained optimal power flow. *IET Sci. Meas. Technol.* **2017**, *11*, 1002–1013. [CrossRef]

35. Venkata Krishna, J.; Apparao Naidu, G.; Upadhayaya, N. A Lion-Whale optimization-based migration of virtual machines for data centers in cloud computing. *Int. J. Commun. Syst.* **2018**, *31*, e3539. [CrossRef]

36. Ling, Y.; Zhou, Y.; Luo, Q. Lévy flight trajectory-based whale optimization algorithm for global optimization. *IEEE Access* **2017**, *5*, 6168–6186. [CrossRef]

37. Sun, Y.; Wang, X.; Chen, Y.; Liu, Z. A modified whale optimization algorithm for large-scale global optimization problems. *Expert Syst. Appl.* **2018**, *114*, 563–577. [CrossRef]

38. Trivedi, I.N.; Jangir, P.; Kumar, A.; Jangir, N.; Totlani, R. A novel hybrid PSO-WOA algorithm for global numerical functions optimization. In *Advances in Computer and Computational Sciences*; Bhatia, S., Mishra, K., Tiwari, S., Singh, V., Eds.; Springer: Singapore, 2018; Volume 554.

39. El-Amary, N.H.; Balbaa, A.; Swief, R.; Abdel-Salam, T. A reconfigured whale optimization technique (RWOT) for renewable electrical energy optimal scheduling impact on sustainable development applied to Damietta seaport, Egypt. *Energies* **2018**, *11*, 535. [CrossRef]

40. Reddy, M.P.K.; Babu, M.R. Implementing self adaptiveness in whale optimization for cluster head section in Internet of Things. *Clust. Comput.* **2017**, *22*, 1–12. [CrossRef]

41. Mafarja, M.; Mirjalili, S. Whale optimization approaches for wrapper feature selection. *Appl. Soft Comput.* **2018**, *62*, 441–453. [CrossRef]

42. Abed-alguni, B.H.; Klaib, A.F. Hybrid Whale Optimization and β-hill Climbing Algorithm for Continuous Optimization Problems. *Int. J. Comput. Sci. Math.* **2019**, in press.

43. Chen, X.; Tianfield, H.; Mei, C.; Du, W.; Liu, G. Biogeography-based learning particle swarm optimization. *Soft Comput.* **2017**, *21*, 7519–7541. [CrossRef]

44. Liang, J.J.; Qin, A.K.; Suganthan, P.N.; Baskar, S. Comprehensive learning particle swarm optimizer for global optimization of multimodal functions. *IEEE Trans. Evol. Comput.* **2006**, *10*, 281–295. [CrossRef]

45. Cheng, R.; Jin, Y. A competitive swarm optimizer for large scale optimization. *IEEE Trans. Cybern.* **2015**, *42*, 191–204. [CrossRef]

46. Boussaïd, I.; Chatterjee, A.; Siarry, P.; Ahmed-Nacer, M. Two-stage update biogeography-based optimization using differential evolution algorithm (DBBO). *Comput. Oper. Res.* **2011**, *38*, 1188–1198. [CrossRef]

47. Gong, W.; Cai, Z.; Ling, C. DE/BBO: A hybrid differential evolution with biogeography-based optimization for global numerical optimization. *Soft Comput.* **2010**, *15*, 645–665. [CrossRef]

48. Zou, F.; Wang, L.; Hei, X.; Chen, D. Teaching-learning-based optimization with learning experience of other learners and its application. *Appl. Soft Comput.* **2015**, *37*, 725–736. [CrossRef]

49. Gao, W.; Liu, S. A modified artificial bee colony algorithm. *Comput. Oper. Res.* **2012**, *39*, 687–697. [CrossRef]

50. Rahnamayan, S.; Tizhoosh, H.R.; Salama, M.M. Opposition-based differential evolution. *IEEE Trans. Evol. Comput.* **2008**, *12*, 64–79. [CrossRef]

51. Cheng, R.; Jin, Y. A social learning particle swarm optimization algorithm for scalable optimization. *Inf. Sci.* **2015**, *291*, 43–60. [CrossRef]

52. Wolpert, D.H.; Macready, W.G. No free lunch theorems for optimization. *IEEE Trans. Evol. Comput.* **1997**, *1*, 67–82. [CrossRef]

3

Coastal Wind Measurements using a Single Scanning LiDAR

Susumu Shimada [1,*], Jay Prakash Goit [1,2], Teruo Ohsawa [3], Tetsuya Kogaki [1] and Satoshi Nakamura [4]

[1] Renewable Energy Research Center, National Institute of Advanced Industrial Science and Technology, Koriyama 963-0298, Japan; jay.goit@hiro.kindai.ac.jp (J.P.G.); kogaki.t@aist.go.jp (T.K.)

[2] Department of Mechanical Engineering, Kindai University, Higashi-Hiroshima, Hiroshima 739-2116, Japan

[3] Graduate School of Maritime Sciences, Kobe University, Kobe 658-0022, Japan; ohsawa@port.kobe-u.ac.jp

[4] Coastal and Estuarin Environment Division, Port and Airport Research Institute, Yokosuka 239-0826, Japan; nakamura_s@p.mpat.go.jp

* Correspondence: susumu.shimada@aist.go.jp

Abstract: A wind measurement campaign using a single scanning light detection and ranging (LiDAR) device was conducted at the Hazaki Oceanographical Research Station (HORS) on the Hazaki coast of Japan to evaluate the performance of the device for coastal wind measurements. The scanning LiDAR was deployed on the landward end of the HORS pier. We compared the wind speed and direction data recorded by the scanning LiDAR to the observations obtained from a vertical profiling LiDAR installed at the opposite end of the pier, 400 m from the scanning LiDAR. The best practice for offshore wind measurements using a single scanning LiDAR was evaluated by comparing results from a total of nine experiments using several different scanning settings. A two-parameter velocity volume processing (VVP) method was employed to retrieve the horizontal wind speed and direction from the radial wind speed. Our experiment showed that, at the current offshore site with a negligibly small vertical wind speed component, the accuracy of the scanning LiDAR wind speeds and directions was sensitive to the azimuth angle setting, but not to the elevation angle setting. In addition to the validations for the 10-minute mean wind speeds and directions, the application of LiDARs for the measurement of the turbulence intensity (TI) was also discussed by comparing the results with observations obtained from a sonic anemometer, mounted at the seaward end of the HORS pier, 400 m from the scanning LiDAR. The standard deviation obtained from the scanning LiDAR measurement showed a greater fluctuation than that obtained from the sonic anemometer measurement. However, the difference between the scanning LiDAR and sonic measurements appeared to be within an acceptable range for the wind turbine design. We discuss the variations in data availability and accuracy based on an analysis of the carrier-to-noise ratio (CNR) distribution and the goodness of fit for curve fitting via the VVP method.

Keywords: coastal wind measurements; scanning LiDAR; plan position indicator; velocity volume processing; Hazaki Oceanographical Research Station

1. Introduction

The global offshore wind energy market has been continuously growing with 4.5 GW of new wind turbines installed in the year of 2018, bringing the total cumulative installations to 23 GW [1]. Although offshore wind energy, which currently represents a global share of four percent of the total cumulative wind power generation, is significantly smaller compared to the onshore wind market, offshore wind has huge potential and is poised to grow with a stable addition from Europe and significant contributions from the emerging markets of Asia. According to the Global Wind Energy Council

(GWEC), the annual offshore installations is expected to exceed 6 GW in the near future [1]. For instance, the interest in offshore wind installations in Japan was stimulated by the recently announced maritime renewable energy policy of the Japanese government [2]. However, offshore wind energy poses several challenges right from the early phase of development. One of the foremost challenges faced by the developer is the accurate and economic means of wind resource assessment at the prospective offshore wind farm sites. Unlike onshore sites, where a meteorological mast (met mast) can be constructed with relative ease, building offshore met masts for the same purpose can involve serious technical and financial challenges. As an alternative to met masts, wind resource assessment using light detection and ranging (LiDAR) has received significant interest in the wind energy community [3–7]. The current work aims to investigate the performance of a scanning Doppler LiDAR for wind resource assessment in the nearshore region, which is considered promising for the development of fixed-bottom offshore wind farms.

In several emerging markets, including Japan, most offshore wind farms are likely to be constructed just a few kilometers off the coast, because the water depth increases rapidly further offshore and the development becomes more costly. Therefore, accurate characterization of the coastal wind is crucial. In this regard, Shimada et al. [7] conducted measurements using two profiling LiDARs at the landward and seaward ends near the coast and found that the wind speeds could increase by up to 120% at a distance of 2 km off the coast. They also showed that the differences between the onshore and offshore winds were less pronounced at higher altitudes. In the follow-up study, they performed mesoscale simulations assimilating profiling LiDAR observations from the coast to reduce the uncertainty in nearshore wind resource predictions [8].

Contrary to the vertical profiling LiDARs, scanning Doppler LiDARs can accommodate measurement ranges from 3 to 10 km [9]. Therefore, they offer the potential for assessing offshore wind from the coast, thus obviating the need for offshore met masts or floating LiDARs [10–12]. To date, scanning LiDARs have been used for the measurements of wind fields and, most importantly, turbulence in single or multi-LiDAR modes [13–16]. Scanning LiDARs have also been used in the measurements of flow fields around a wind turbine or to evaluate wind turbine power performance [17,18]. However, these studies and other similar works in the literature mainly conducted short-term measurement campaigns, while actual wind resource assessment requires long-term wind statistics, e.g., mean wind speeds over a period of a few months to one year.

Several projects on the application of scanning LiDARs for wind resource assessment are introduced here [19–22]. Cameron et al. [19] reported validation results using a dual scanning LiDAR system in Dublin Bay. Two scanning LiDARs used in the study were at distances of 15.9 and 8.8 km from the validation target. Courtney et al. [20] and Simon and Courtney [22] conducted validation campaigns at a test site at the Danish Technical University, testing several scanning patterns from a point 1.5 km from the target met mast. The results indicate that a scanning configuration with an azimuth width of 45° and a scan rate of 3°/s showed the best performance. Coutts et al. [21] also showed validation results with a distance of approximately 1.8 km at a German test site. Some of the reports have shown good agreement with observations obtained from met masts. However, details of the validation methods employed, such as the horizontal wind retrieving algorithm used in the analysis, have not been made available. Thus, the usefulness of a scanning LiDAR for wind resource assessment over coastal waters has remained ambiguous.

In the current study, we conducted an offshore wind measurement campaign with a single scanning Doppler LiDAR installed at a coast in central Japan. The horizontal wind speeds and wind directions were retrieved from the LiDAR-measured radial wind speeds using the velocity volume processing method. One of the objectives of this study was to investigate the effect of the scan parameters, such as the azimuth range and elevation angle on the quality of measured wind speeds. Furthermore, the effects of the wind direction and data availability on the accuracy of the retrieved wind speeds is also analyzed. The ability of a scanning LiDAR to measure the turbulence intensity (TI) was also investigated. To that end, scanning LiDAR measurements were compared against the data from a profiling LiDAR and a sonic anemometer at the site.

The paper is organized as follows. In Section 2, we describe the measurement site and devices. The measurement cases and velocity retrieval techniques are also presented in this section. In Sections 3 and 4, we present our results and related discussions. The main conclusions of this work are summarized in Section 5.

2. Materials and Methods

2.1. Experimental Setup

The measurement campaign was conducted from January to September 2019 at the Hazaki Oceanographic Research Station (HORS) [23]. The HORS is located at the coastline of the Pacific Ocean in Ibaraki prefecture, Japan. This region has a rectilinear coastline running from 150° to 330° and is surrounded by flat terrain with mixed vegetation. Figure 1 shows the instruments used in the study and the positions on the HORS pier where they were placed. The pier is 427 m long and at a height of 7 m above sea level (ASL). As 1.5 MW wind turbines with a hub height of 64.5 m and a rotor diameter of 62 m align along the coast, the winds on the pier at a wind direction of approximately 207° are directly influenced by the wind turbine wake. Detailed wind conditions on the HORS research platform can be found in Shimada et al. (2018) [7].

Figure 1. Experimental setup.

Table 1 summarizes the list of the measurement devices used in this study. The scanning LiDAR (Windcube 100s, hereafter referred to as the 100s) was deployed on the roof of a 3.5-m-high observational facility located at the landward end of the HORS pier. The dimensions of the scanning LiDAR were approximately 0.8 m × 1.0 m × 1.2 m and it weighed 235 kg. This LiDAR had a measurement range from 50 to 3000 m, with a horizontal resolution, which can be set between 25 and 100 m. Further specifications of the 100s are available on the manufacturer's website [9]. For validation of the wind speeds and directions obtained from the 100s, the offshore measurements from a sonic anemometer (hereafter referred to as SA) and a vertical profiling LiDAR (Windcube V1, hereafter referred to as V1) installed at the seaside end of the pier were used. The measurements from the V1 were mainly used for the validation of the 10-minute mean wind speeds, while those from the sonic anemometer (SA) were used to assess the accuracy of the turbulence intensity (TI) measurements.

Table 1. List of the instruments used in the study.

Instrument	Location on the Pier	Height Above Sea Level	Measurement Parameters (Sampling Interval)
Scanning light detection and ranging (LiDAR) (Windcube 100s (100s))	Landward end	10.5 m	Radial wind speeds (1 Hz)
Vertical profiling LiDAR (Windcube V1 (V1))	Seaward end	7 m	Horizontal and vertical wind speeds and directions at heights of 40–200 m (1 Hz)
Sonic anemometer (Young Model 81000)	Seaward end	10 m	Horizontal and vertical wind speeds, virtual temperature (4 Hz)

2.2. Leveling Calibration of the Scannig LiDAR

It is best to have the scanning LiDAR installed in a perfectly level position, as a tilted device results in differences between the target and the actual measurement points. In particular, a tilt angle in the vertical direction can significantly affect the accuracy of wind speed measurements taken with a scanning LiDAR due to the strong vertical wind shear that exists within the surface layer. The impact of a tilt angle increases linearly with the distance from the device. For instance, a device deployed with a tilt angle of 1° leads to vertical height errors of 17.5 and 35 m, respectively, at the ranges of 1000 and 2000 m. This will consequently introduce extra uncertainty into the wind speed measurements.

Field measurement environments differ from laboratory environments, and it appears difficult to deploy a scanning LiDAR with a tilt angle error of less than 0.1° (the impact of which can be assumed to be negligibly small), even using extremely precise digital levels. Therefore, a calibration method referred to as hard target calibration (HTC) [19] was used for the leveling calibration. In HTC, the tilt angle is generally estimated by analyzing the carrier-to-noise ratio (CNR) values, which are used for checking the quality of measurements. Generally, the CNR values peak near the device, then decrease as the distance increases. Moreover, it is also known that scanning LiDARs record dramatically higher CNR values when the laser is reflected by obstacles such as a meteorological tower or light house.

By using characteristics of the CNR recorded by scanning LiDARs when striking an obstacle, the pitch and roll angles of the device, which are the tilt angles in the x and y directions, respectively, can be estimated from the difference between the actual location of the obstacle and the position detected by the CNR distribution of the scanning LiDAR. After calculating the pitch and roll angles, they are added to the original LiDAR-registered azimuth and elevation angles as an offset. This approach, using the reflection from obstacles, is applicable with fewer constraints over land, but is difficult to apply to offshore wind measurements due to the difficulty of finding suitable targets offshore.

Therefore, in the current study, another characteristic of the CNR distribution of the LiDAR was used to estimate the pitch and roll angles misalignment. At the time of installation, we noticed that the CNR appeared to rapidly decline when the laser beam hit the sea surface. Figure 2 shows a snapshot of the CNR distribution obtained from the 100s with an azimuth width of 45° and for an elevation angle of 0°. The Japanese Local Standard Time (LST), which is +9 hours from UTC, is used in this study. The high CNR values at the x and y positions of 0 and 400 m were due to the reflection from the observational hut on the pier. Much lower CNR values are observed in the upper right region of the figure. As an elevation angle of 0° was set in this case and no obstacle was present, this must be the result of the laser beam hitting the water surface in this region. From this CNR distribution, we were able to estimate the pitch and roll error using planar fitting, as the horizontal and vertical positions of the LiDAR device are the given information. The equation of the planar fitting is given by

$$z = x \tan \theta_x + y \tan \theta_y + z_0 \tag{1}$$

where z_0 is the deployment height of the 100s (= 11.5 m ASL), and θ_x and θ_y are the pitch and roll angles. For the LiDAR installation in the current study, a pitch angle of $-0.092°$ and a roll angle of

−0.217° were obtained by applying this method. In Figure 2, the contour of the measurement heights, taking into account the pitch and roll angles derived from the planar equation, is overlaid. The pitch and roll angles estimated here were used when calculating the height at each measurement point from the azimuth and elevation angles in the wind speed retrieval process. In contrast to the HTC approach, we call this approach soft target calibration (STC).

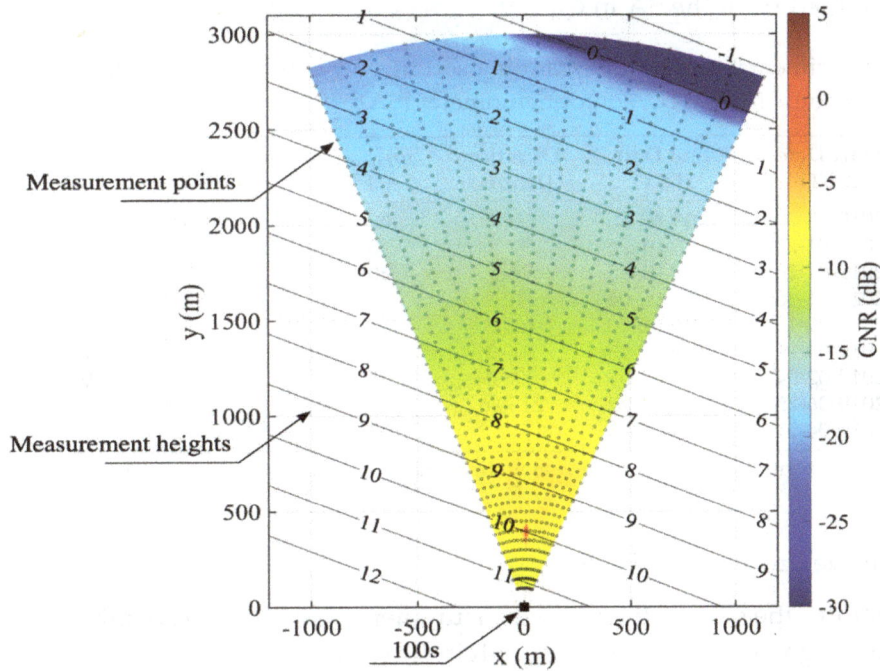

Figure 2. Snapshot of the carrier-to-noise ratio (CNR) distribution obtained from the 100s with an azimuth width of 45° and an elevation angle of 0° for the period 2019-05-11 09:05:33 to 09:05:47 LST.

2.3. Scanning Mode and the Velocity Volume Processing (VVP) Method

2.3.1. Scanning Mode

Four scanning modes were available for the 100s: the plan position indicator (PPI) mode, range height indicator (RHI) mode; Doppler beam swinging (DBS) mode, and fixed mode. The choice of scanning mode depends on the purpose of the measurements. The RHI mode and the fixed mode are frequently used for wind turbine wake measurements [15,16] and virtual met tower measurements [24], respectively. The current study uses the PPI mode, as it is compatible with the VVP method [25–27] for retrieving horizontal wind speeds and directions from the LiDAR-measured radial wind speeds.

In the PPI mode, the scanning LiDAR measures the radial wind speeds by sweeping the scanning head over a range of azimuth angles while the elevation angle is fixed. In this study, a scanning range of up to 3000 m, with a radial resolution of 50 m and a data accumulation time of 1 s, were employed as the scan settings. The data accumulation time is the integration interval used to derive each radial wind speed. As the azimuth and elevation angle settings, such as the scanning width and scan rate, would be the primary factors affecting the accuracy of the wind speed measurements, the PPI measurements were carried out using different azimuth and elevation angles in this study.

Table 2 lists the experimental duration and scan settings for each experimental case. The reference sensor (V1 or SA) used for each validation is shown in the U_{ref} column. The evaluation height for each case is also described in the table. The impacts of the azimuth angle settings were investigated in the experiments for Cases 1 through 5. In Cases 1 to 4, the accuracies of the wind speeds at a height of 47 m were compared for four different azimuth widths. In Case 5, the settings were identical to those in Case 4 except for the scan rate, which was reduced to 1° (from 3°) so that the number of data at the given radial position increased. This allowed us to investigate the effect of the number of input data on the quality of wind speed computed using the VVP method. In Case 6 and Case 7, the impact of the elevation angle on the measurement accuracy was investigated by comparing the results for different elevation angles. Case 8 had identical settings to those of Case 2, but ran for a longer period, to assess the impact of the wind direction. In addition to the validation of the 10-minute mean wind speeds and wind directions, the accuracy of the TI values was also examined by comparing them against the measurements obtained from the SA in Case 9.

Table 2. The experimental period and scanning settings for each case. The Japanese Local Standard Time (LST) is +9 hours from UTC, sonic anemometer (SA).

Case	Start Date (LST)	End Date (LST)	Duration (Hours)	φ_{range} (°)	ω (°/s)	θ (°)	U_{ref}	Height (m)
1	2019-02-20	2019-02-25	119.6	60	3	5.07	V1	47
2	2019-01-22	2019-01-26	97.5	45	3	5.07	V1	47
3	2019-01-26	2019-01-30	96.1	30	3	5.07	V1	47
4	2019-01-30	2019-02-04	112.5	15	3	5.07	V1	47
5	2019-02-13	2019-02-20	169.3	15	1	5.07	V1	47
6	2019-04-19	2019-04-23	100.2	45	3	14.11	V1	107
7	2019-02-04	2019-02-08	102.5	45	3	26.05	V1	207
8	2019-06-25	2019-09-22	2150.6	45	3	5.07	V1	47
9	2019-04-23	2019-06-25	1512.2	45	3	0.00	SA	7

φ_{range}: azimuth angle range; ω: scan rate; θ: elevation angle; U_{ref}: reference sensor for the validation; Height: validation height.

2.3.2. Velocity Volume Processing (VVP)

Figure 3 illustrates the schematic of the measurement positions and heights for Case 1. The small red square and the diagonal line passing through indicate the 100s and the coastline, respectively. The heading angle of the 100s and the along- and across-wind direction sectors are shown in the pie chart in the upper left corner. The radial wind speeds were measured along the line of the colored dots, with the measurement line moving to the next position at an interval of 1 s. For this case (i.e., Case 1), it takes roughly 20 s to complete a cycle; the device then repeats the operation. A set of radial wind speeds for an arc at the same radial distance was used to compute the horizontal wind speeds and directions in the process of wind vector retrieval using the VVP method.

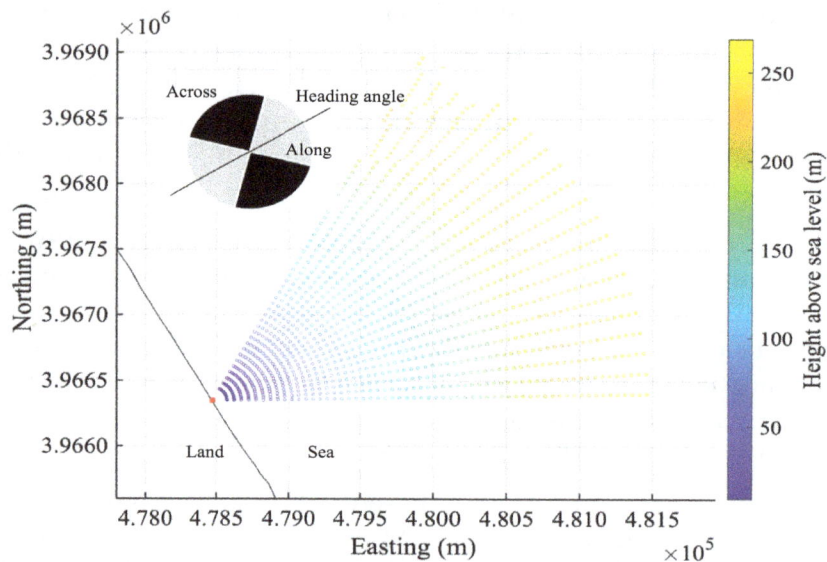

Figure 3. Measurement of horizontal and vertical positions of the radial wind speed for Case 1. The red square indicates the location of the 100s. The pie chart shows the definition of the along and across wind direction sectors.

As noted earlier, the radial wind speeds obtained from the 100s were converted to horizontal wind speeds using the VVP method, which is discussed next. The radial wind speed can be expressed as the projection of the actual velocity vector along the radial direction, i.e.,

$$Vr = u \cos \theta \sin \varphi + v \cos \theta \cos \varphi + w \cos \theta, \tag{2}$$

where Vr is the radial wind speed, u, v, and w are the horizontal and vertical wind speed components, θ is the elevation angle, and φ is the azimuth angle. In the original VVP, the wind speed is assumed to vary linearly about the analysis point, thus resulting in derivative terms of each wind speed component along all three spatial directions [26,27]. Therefore, in total, there will be 12 parameters on the right-hand side of Equation (2). As it is unnecessary to have the full set of these parameters for wind resource assessment considered in this study, we used an assumption to simplify the equation, following previous studies [25]. If we assume that the wind field is homogeneous in time and space for one scan cycle, and that the vertical component w is negligibly small compared to the horizontal wind speed components, Equation (2) can be written as

$$Vr = u \cos \theta \sin \varphi + v \cos \theta \cos \varphi. \tag{3}$$

As Vr, θ, and φ are obtained from the 100s, u and v can be simply computed using the linear-least-squares method. The equation used in the least square method can be expressed as the following cost functional, which can be minimized for u and v:

$$J = \frac{1}{2} \sum_{i=1}^{N} (Vr' - (u \sin \varphi + v \cos \varphi))^2, \tag{4}$$

where N is the number of scans during a sweep and $Vr' = Vr/\cos\theta$. The condition for the minimization the cost functional is that its gradient with respect to the velocity is 0, i.e.,

$$\frac{\partial J}{\partial u} = 0, \ \frac{\partial J}{\partial v} = 0. \tag{5}$$

This leads to the following equation that is solved in this study:

$$\begin{bmatrix} \sum \sin^2 \varphi & \sum \cos \varphi \cdot \sin \varphi \\ \sum \cos \varphi \cdot \sin \varphi & \sum \cos^2 \varphi \end{bmatrix} \begin{bmatrix} u \\ v \end{bmatrix} = \begin{bmatrix} \sum \sin \phi \cdot Vr' \\ \sum \cos \phi \cdot Vr' \end{bmatrix}. \tag{6}$$

We tested other more sophisticated numerical approaches for solving Equation (3), such as a non-linear fitting method; however, the results are not significantly different for the above discussed approach. This can be attributed to the simplicity of the equation.

Before retrieving the horizontal wind speeds and directions using the VVP method, some pre-processing of the raw radial wind speed data was necessary. The raw measurements recorded by the V1 included not only instantaneous radial wind speeds but also a confidence index indicating the reliability of the various measurements diagnosed by the hardware. To screen out unreliable data, observations without a confidence flag were excluded from the analysis. As a pre-processing step for the curve fitting, a linear interpolation in the vertical direction was applied to the radial wind speed to adjust for the measurement height difference between the reference sensor (V1 or SA) and the 100s. The pitch and roll angle offsets described in Section 2.2 were also taken into account during this vertical interpolation procedure.

Following this vertical interpolation, we applied noise reduction to the measured radial wind speed. The measurement uncertainty of the scanning LiDARs was higher than that of the vertical profiling LiDARs. According to the device specifications, the 100s has a measurement uncertainty of 0.5 m/s for radial wind speed, while most vertical profiling LiDARs have an uncertainty of less than 0.2 m/s for horizontal wind speed. To reduce the noise in the raw data, a first-order low-pass filter with a recursive expression was applied. This is given by

$$\overline{Vr}_k = \alpha\overline{Vr}_{k-1} + (1-\alpha)Vr_k, \tag{7}$$

where $\overline{Vr}_{k-1}$ indicates the filtered radial wind speed at the previous time step and α (= 0.3) is a smoothing factor. This process is useful for eliminating spikes in the data time-series.

In the analysis code, the values of u and v were calculated by solving Equation (6) using the radial wind speeds for a sweep, i.e., one cycle in the PPI scan. Thus, the temporal resolution of the wind speed obtained from the 100s depended on the azimuth width and scan rate. For example, the wind speeds and directions for Case 1 (φ_{range} = 60°, ω = 3°/s) had a temporal resolution of approximately 20 s. From the instantaneous wind speeds and directions with an interval of 20 s, we calculated the 10-minute means and standard deviations. These were then compared to the reference observations from V1 or SA. As the data availability of the LiDARs depends on the atmospheric conditions, if the availability during the 10-minute interval was less than 80%, the measurements were excluded from the validation.

3. Results

3.1. Wind Condition during the Experimental Campaign

Before presenting the VVP-processed results, we briefly describe the wind characteristics of the measurement site. Figure 4 shows the time series of wind speeds and wind directions at a height of 107 m ASL obtained from V1 during the current measurement period (22 January to 23 September 2019). The data availability for V1 during this period was 91%. Figure 5 presents the distribution of these wind speeds and wind directions in the form of occurrence frequencies and a wind rose. Figure 5a also presents the Weibull probability density function (PDF) obtained by fitting to the following Weibull function:

$$f(U) = \frac{k}{A}\left(\frac{U}{A}\right)^{k-1}\exp\left[-\left(\frac{U}{A}\right)^k\right], \tag{8}$$

where U indicates the horizontal wind speed, k is the shape parameter and A is the scale parameter. For the current site, k = 2.09 and A = 7.9. Furthermore, a mean wind speed of 7.0 m/s was observed. A north-easterly wind was dominant during the experimental period, though a south-westerly wind was also significant. A typhoon approached HORS on 8 to 9 September 2019, when strong winds with more than a 35 m/s 10-minute mean were observed.

As described in Shimada et al. (2018) [7], at the HORS site, winds coming from 335° to 145° were categorized as sea winds (from sea to land sectors), while those from 155° to 325° were categorized as land winds (from land to sea sectors). Notably, land winds coming from between 175° and 230° were influenced by the neighboring wind turbines situated on the coastline. As a homogeneous wind field was assumed in the VVP method, the winds blowing from the wake sector may reduce the measurement accuracies, relative to the other sectors. Consequently, observations with a wind direction that was directly disturbed by the wake of the wind turbines were excluded from the validation.

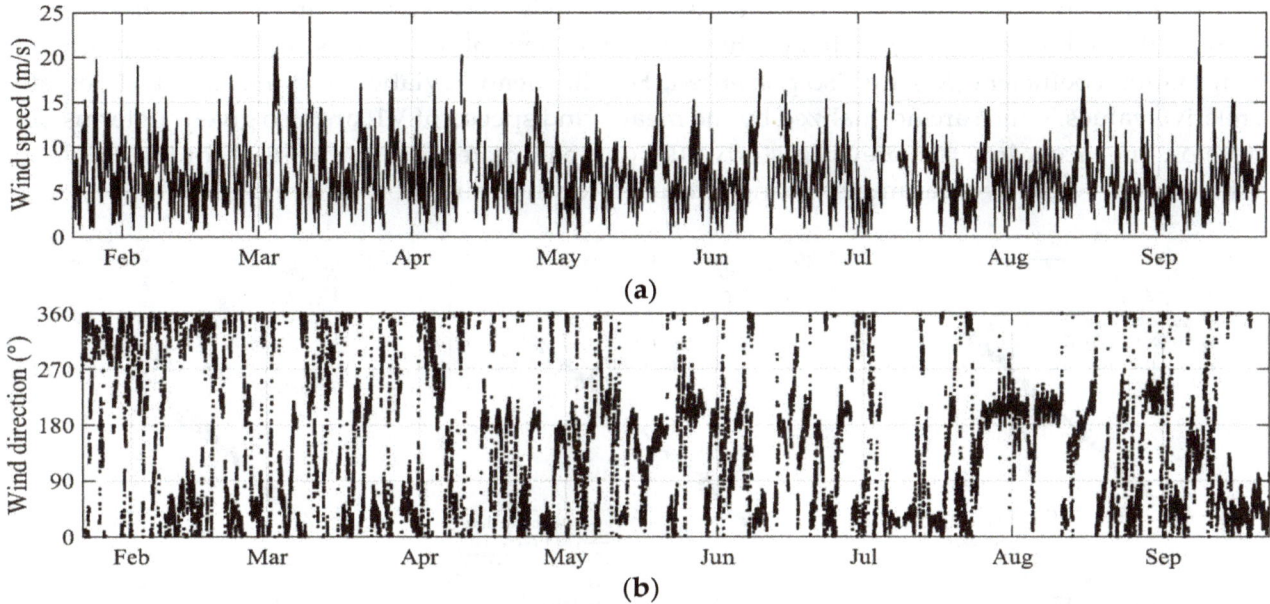

Figure 4. Time series of (**a**) wind speeds and (**b**) wind directions obtained from V1 at a height of 107 m above sea level (ASL) for the period from January to September 2019 at Hazaki Oceanographical Research Station (HORS).

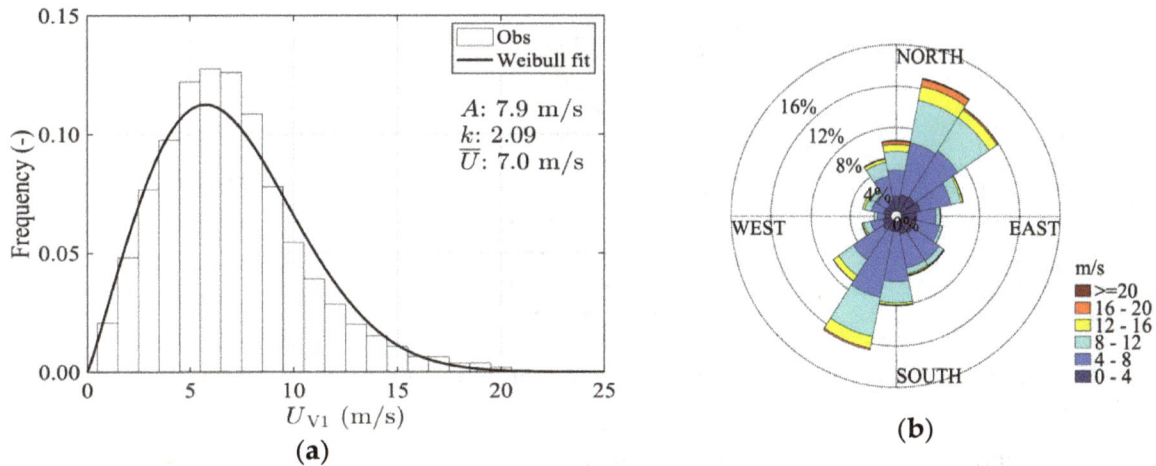

Figure 5. Distributions of (**a**) the wind speeds and (**b**) wind directions at a height of 107 m ASL for the period from January to September 2019 at HORS.

3.2. Validation of 10-Minute Mean Wind Speeds and Wind Directions

3.2.1. Sensitivity to the Azimuth Angle Settings

Following the validation approaches recommended in previous studies [10,28,29], the accuracy of the offshore wind measurements using a single scanning LiDAR and the VVP method was quantitatively evaluated. The observations recorded during low wind conditions, when the V1 wind speeds were less than 2 m/s, were excluded from the analysis. Finally, as the measurements of V1 for wind directions around 207° were directly influenced by the wind turbines situated on the coast, the observations with a wind direction range of 207° ± 10° were also excluded from the analysis.

Figure 6 shows the scatter plots comparing wind speeds from the V1 (U_{V1}) and 100s (U_{100s}) for Cases 1–5. In order to quantitatively compare the accuracies between experiment cases, the deviations of the V1 and 100s wind speeds ($\varepsilon_U = U_{100s} - U_{V1}$) were first calculated. The mean deviation and the standard deviation of the deviation were then computed. Table 3 presents the number of samples (N), the mean wind speeds of the V1 ($\overline{U}_{V1}$) and 100s ($\overline{U}_{100s}$), the mean deviation ($\overline{\varepsilon}_U$), and the standard deviation of the deviation ($\sigma(\varepsilon_U)$). In addition, the coefficient of the regression line ($y = mx$) and the determination coefficient (R^2) are also presented. For the mean deviation and the standard deviation, the relative values, which are normalized by the mean wind speeds of V1, are also given. In terms of the regression line, according to a previous study [10], one-parameter fitting was used for the validations of wind speed, while two-parameter fitting was used for the validations of wind direction.

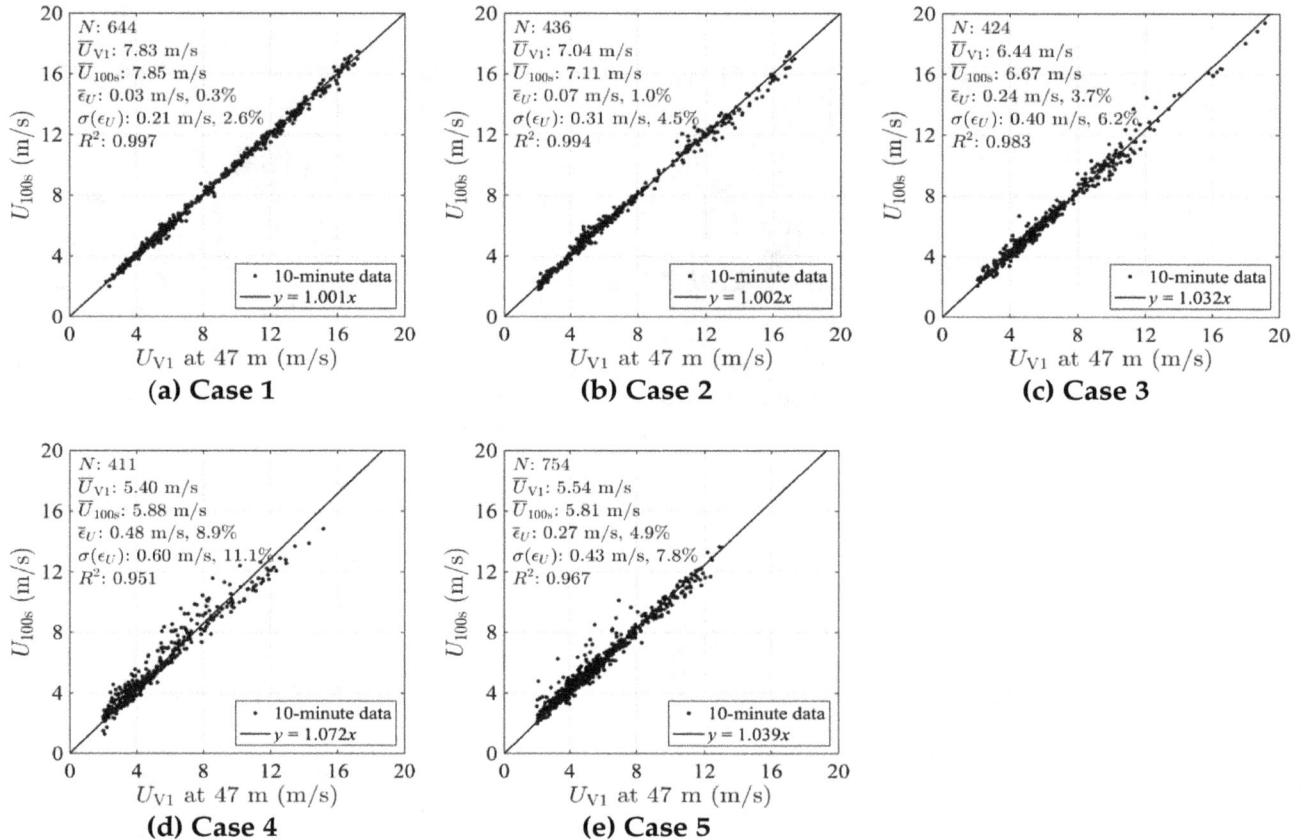

Figure 6. Scatter plots of the V1 and 100s wind speeds for **(a)** Case 1, **(b)** Case 2, **(c)** Case 3, **(d)** Case 4, and **(e)** Case 5.

Table 3. Statistics of the 10-minute wind speed from 100s for Cases 1–9.

Case	U_{ref}	N (-)	$\overline{U}_{\text{ref}}$ (m/s)	$\overline{U}_{100s}$ (m/s)	$\overline{\varepsilon}_U$ (m/s, %)	$\sigma(\varepsilon_U)$ (m/s, %)	$y = mx$ m	R^2 (-)
1	V1	644	7.83	7.85	0.03, 0.3%	0.21, 2.6%	1.001	0.997
2	V1	436	7.04	7.11	0.07, 1.0%	0.31, 4.5%	1.002	0.994
3	V1	424	6.44	6.67	0.24, 3.7%	0.40, 6.2%	1.032	0.983
4	V1	411	5.40	5.88	0.48, 8.9%	0.60, 11.1%	1.072	0.951
5	V1	754	5.54	5.81	0.27, 4.9%	0.43, 7.8%	1.039	0.967
6	V1	602	6.20	6.23	0.03, 0.5%	0.17, 2.7%	1.004	0.998
7	V1	531	8.19	8.20	0.01, 0.1%	0.23, 2.8%	1.001	0.995
8	V1	8039	5.92	5.97	0.04, 0.7%	0.24, 4.0%	1.006	0.992
9	SA	3557	5.08	4.90	−0.17, −3.4%	0.55, 10.8%	0.959	0.963

The comparisons between the experiments with different azimuth ranges (Cases 1 to 4) show that the accuracy of the VVP-based wind speed retrieval was significantly affected by the azimuth angle range. The results for Case 1 and Case 2, which had azimuth angle ranges of 60° and 45°, were found to have good agreement with the V1 wind speeds. On the other hand, the measurement accuracy decreased as the azimuth range narrowed, as shown in the figures for Case 3 and Case 4. In Case 5, the settings are identical to those in Case 4 except for the scan rate. The slower scan rate in Case 5 was used to increase the number of radial wind speeds data in the curve fitting. It was found that the results for Case 5 were relatively improved compared to Case 4, but the points for this case were still more scattered than those for Case 1 and Case 2.

Figure 7 shows the wind direction plots. Here, $\overline{D}_{V1}$ and $\overline{D}_{100s}$ indicate the mean wind directions for V1 and 100s, respectively. The statistics for the mean wind direction for all of the experiments are described in Table 4. In calculating the error metrics for wind direction $\overline{\varepsilon}_D$, $\sigma(\varepsilon_D)$, $y = mx + b$, and R^2), the direction difference ε_D was limited to values between $-180°$ and $+180°$ to cancel the effect of periodicity between $0°$ and $360°$. Similar to the results obtained for wind speeds, the wind direction accuracy of the 100s depended strongly on the azimuth range. When the azimuth ranges were less than $45°$, the wind direction accuracy of the 100s decreased. A comparison between Case 4 ($\omega = 3°/s$) and Case 5 ($\omega = 1°/s$) showed that decreasing the scan rate so that more radial wind speed data can be used for curve fitting did not improve the wind direction accuracy of the 100s.

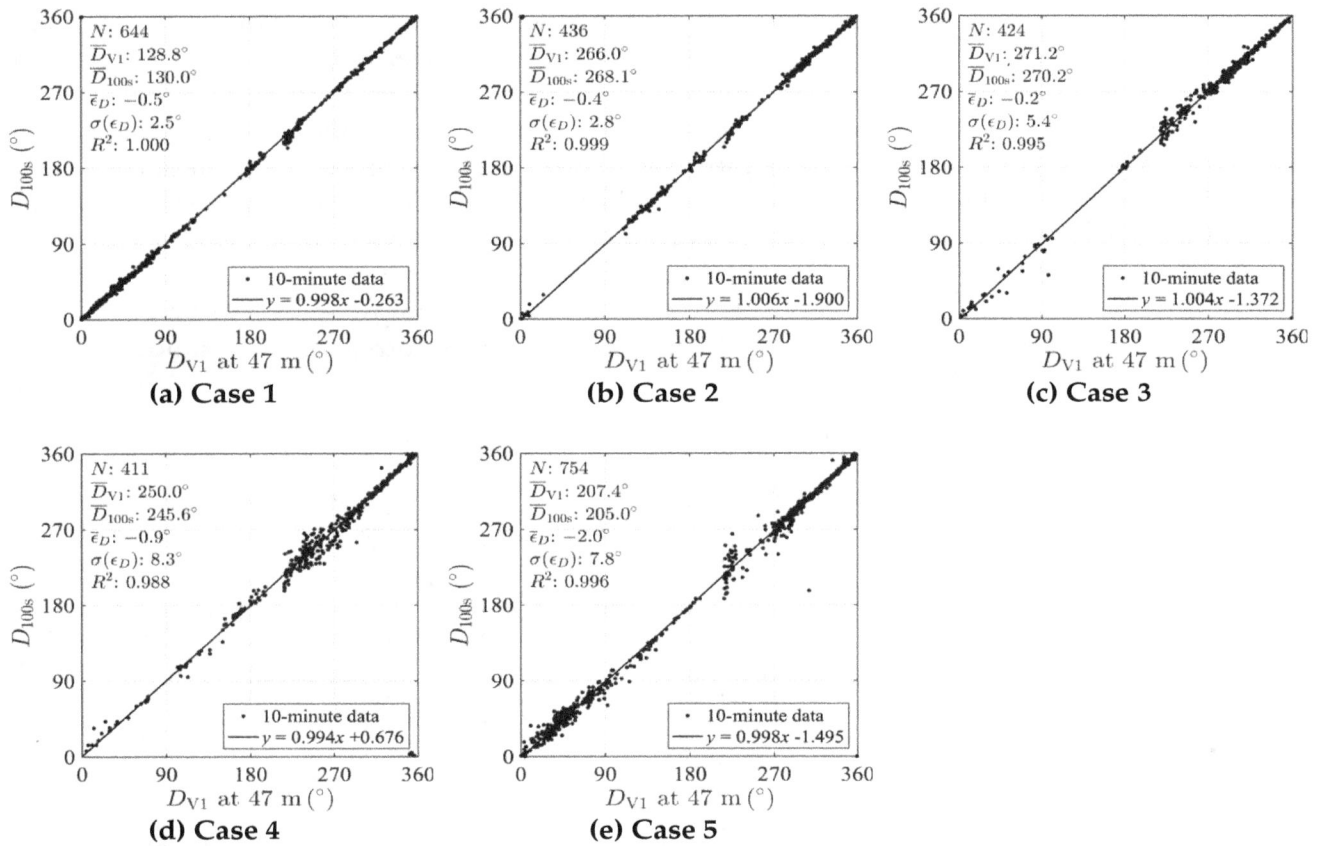

Figure 7. Scatter plots of the V1 and 100s wind directions for **(a)** Case 1, **(b)** Case 2, **(c)** Case 3, **(d)** Case 4, and **(e)** Case 5.

Table 4. Statistics of the 10-minute wind direction from 100s for Cases 1–9.

Case	U_{ref}	N (-)	$\overline{D}_{V1}$ (°)	$\overline{D}_{100s}$ (°)	$\overline{\varepsilon}_D$ (°)	$\sigma(\varepsilon_D)$ (°)	$y = mx + b$ m	b	R^2 (-)
1	V1	644	128.8	130.0	−0.5	2.5	0.998	−0.263	1.000
2	V1	436	266.0	268.1	−0.4	2.8	1.006	−1.900	0.999
3	V1	424	271.2	270.2	−0.2	5.4	1.004	−1.372	0.995
4	V1	411	250.0	245.6	−0.9	8.3	0.994	0.676	0.988
5	V1	754	207.4	205.0	−2.0	7.8	0.998	−1.495	0.996
6	V1	540	84.6	83.4	−0.5	2.0	0.999	−0.354	0.999
7	V1	505	185.7	186.8	−0.3	2.5	1.000	−0.178	1.000
8	V1	8039	114.9	114.0	−0.7	3.7	0.997	−0.329	0.998
9	SA	3557	158.7	157.8	1.4	12.1	0.995	2.254	0.984

Figures 6 and 7 indicate that a wider azimuth angle produced better results for scanning LiDARs with the VVP method. The question arises as to how large of an azimuth angle is reasonable for wind resource assessment. In terms of LiDAR accuracy, previous studies [19,29] established acceptance criteria based on the International Electrotechnical Commission (IEC) standard 61400-12-1 [28], which included a technical note for remote sensing devices used for the power performance testing of wind turbines. Strictly speaking, the details of the criteria presented in the literature differ slightly. However, it is commonly accepted that for both wind speed and wind direction accuracy, the slope of the regression line should be between 0.98 and 1.02 and the determination coefficient should at least be greater than 0.98. Case 1 and Case 2 satisfy these acceptance criteria. The indication is that an azimuth width of more than 45° would be preferable for offshore wind measurements using the 100s and the VVP method.

3.2.2. Sensitivity to Elevation Angle Setting

Figures 8 and 9 show the scatter plots of the V1 and 100s wind speeds and directions for Case 6 ($\theta = 14.11°$) and Case 7 ($\theta = 26.05°$). The evaluation heights for Case 6 and Case 7 are 107 and 207 m ASL, respectively. The azimuth angle settings in these cases were identical to Case 2, i.e., $\varphi_{range} = 45°$. The two-parameter fitting VVP, which ignores the vertical wind speed component (*cf.* Equation (6)) included in the original equation, was used for the analysis. Thus, if the vertical component of wind speeds was not negligibly small compared to the horizontal components, the accuracy of the wind speeds and directions should decrease as the elevation angle increases. However, these figures show that the accuracy of the 100s wind speeds and directions aloft did not deteriorate as the elevation angle increased. In fact, the accuracies for Case 6 and Case 7 increased in comparison to Case 2. This may be associated with the tendency of the wind to become more homogeneous at higher altitudes.

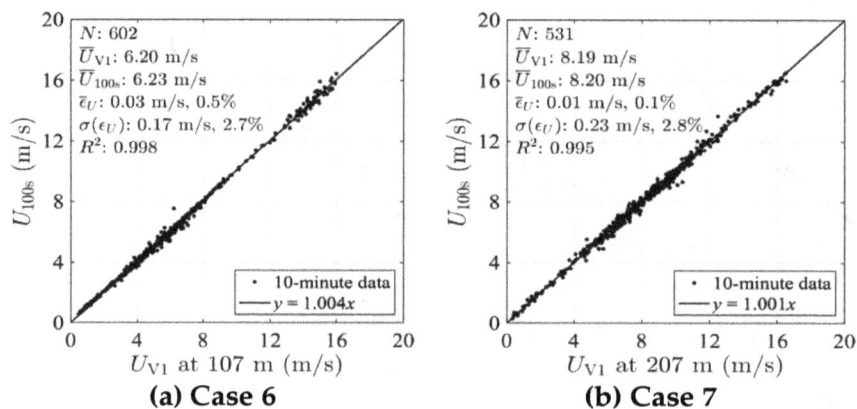

Figure 8. Scatter plots between the V1 and the 100s wind speeds for (**a**) Case 6 and (**b**) Case 7.

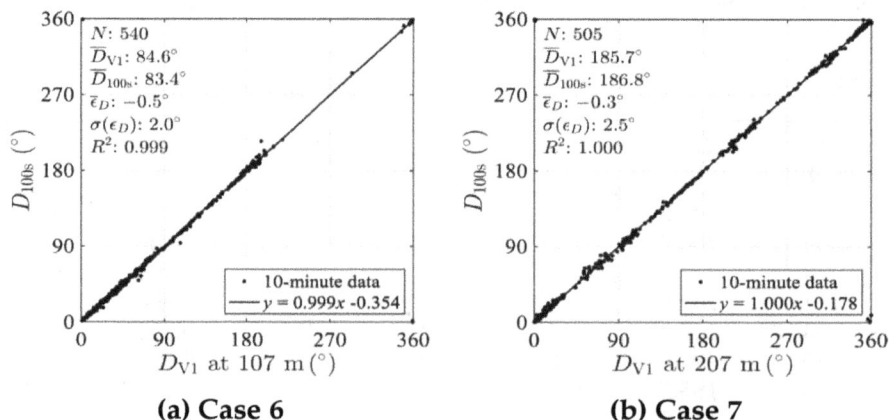

Figure 9. Scatter plots of the V1 and 100s wind directions for (**a**) Case 6 and (**b**) Case 7.

Figure 10 shows the time series of the horizontal and vertical wind speed components observed by V1 at a height of 107 m ASL for the period from 5 to 8 February 2019. The w component of velocity was not directly observed by V1, as it is calculated from a transformation conversion of the line-of-sight wind speed using the DBS method. From Figure 10, it appears that the magnitude of w is mostly less than 10% of the horizontal wind speed. Thus, these results indicate that neglecting the vertical wind speed component in the two-parameter fitting VVP is quite acceptable for measuring offshore winds, which have a small vertical wind speed component. Therefore, we concluded that the elevation angle setting between 5° and 26° did not produce any significant difference in the measurement accuracy for the same distance from the position of the scanning LiDAR.

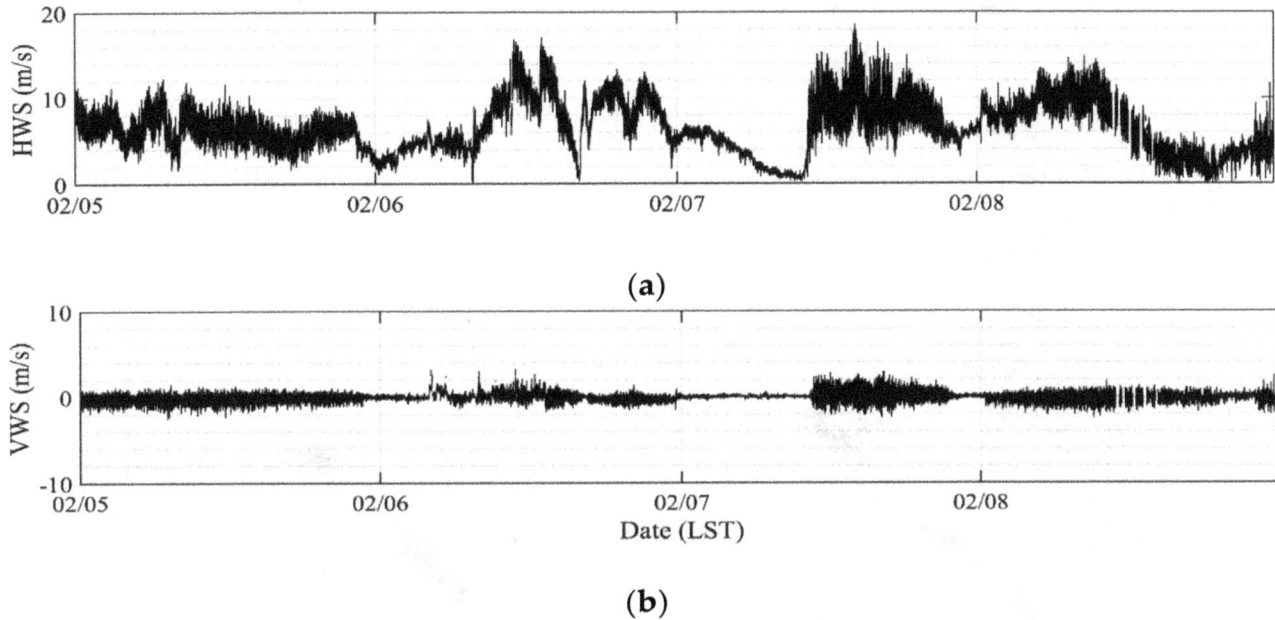

(a)

(b)

Figure 10. Time series of (**a**) horizontal wind speeds and (**b**) vertical wind speeds, obtained from V1 at a height of 107 m ASL for the period from 5 to 8 February 2019.

3.2.3. Impact of Wind Direction on Wind Speed Accuracy

Previous studies [19] indicated that the accuracy of wind speeds obtained from a single scanning LiDAR were significantly affected by the wind direction. The accuracy of the wind speeds with an across-wind direction, defined as perpendicular to the LiDAR heading angle ±45°, might be lower than those with an along-wind direction, defined as parallel to the LiDAR heading angle ±45°(cf. Figure 3 for the definitions). In order to investigate the impact of wind direction on the accuracy of the retrieved horizontal wind speeds, an experiment with the same setting as Case 2 but covering a longer period of time (62 days) was conducted and is defined as Case 8 in Table 2.

Scatter plots of the V1 and 100s wind speeds and wind directions for Case 8 are shown in Figure 11. The 100s wind speeds have a mean deviation of 0.04 m/s (0.7%), a standard deviation of deviation of 0.24 m/s (4.0%), and a determination coefficient of 0.992; the 100s wind directions have a mean deviation of −0.7°, a standard deviation of deviation of 3.8°, and a determination coefficient of 0.998. To identify the impact of wind direction, Figure 12 shows two scatter plots for wind speed—one for along direction cases (as defined above) and one for across direction cases, based on the wind direction obtained from V1. As shown here, the results for the along-wind direction exhibit slightly better results in the standard deviation of the deviation, but no significant differences were observed between the two cases.

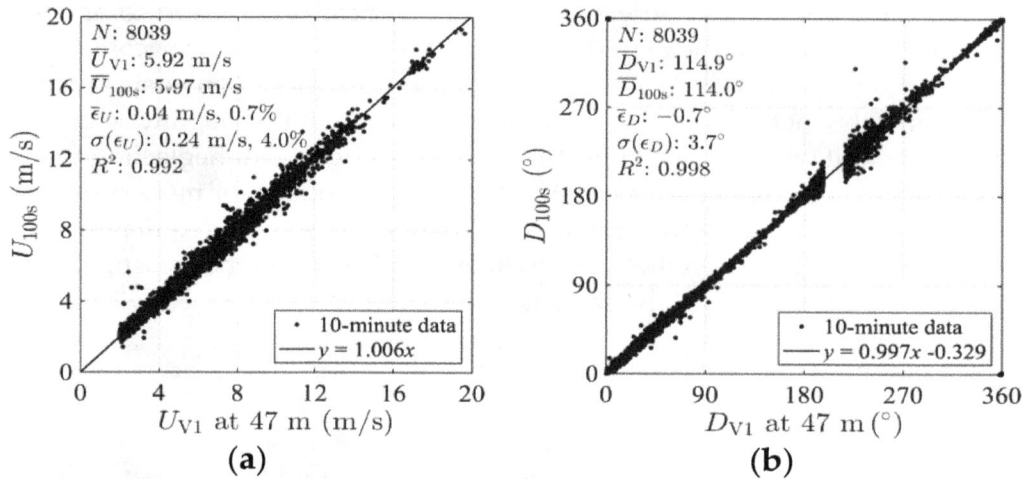

Figure 11. Scatter plots of (**a**) wind speeds and (**b**) wind directions for V1 and 100s for Case 8 (all directions).

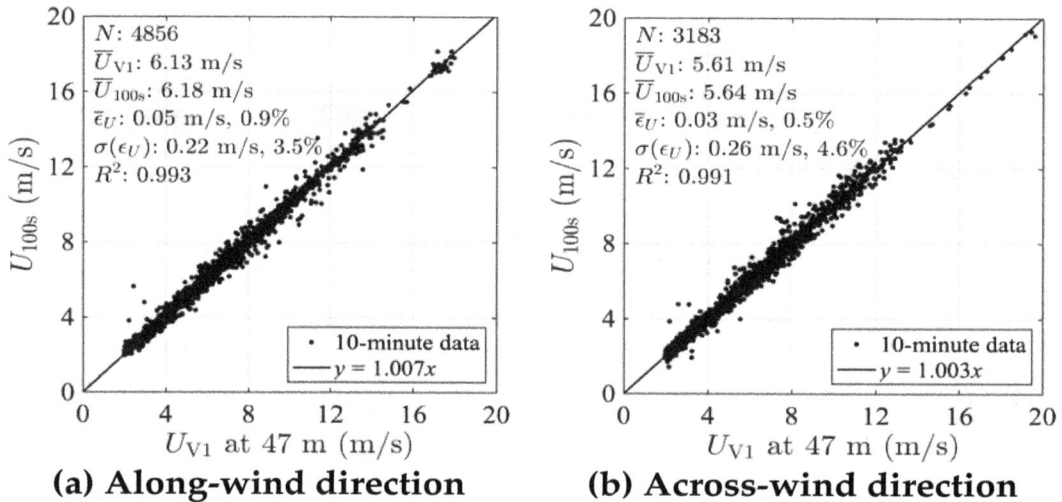

(a) Along-wind direction　　　　**(b) Across-wind direction**

Figure 12. Scatter plots of (**a**) wind speeds with an along-wind direction (parallel to the LiDAR heading angle ±45°) and (**b**) an across-wind direction (perpendicular to the LiDAR heading angle ±45°).

To investigate the influence of the wind direction in greater detail, the ratios of the V1 wind speeds and the 100s wind speeds were plotted as a function of the V1 wind direction at a height of 47 m (Figure 13). The bin-averaged values with a 5° bin width and the associated 1 standard deviation bounds were also plotted. The along and across direction classifications were indicated by the dark solid and dashed horizontal lines. The blue and red arrows, respectively, indicate wind blowing from the sea to the land and wind blowing from the land to the sea. As can be seen here, most of the bin-averaged values were plotted for directions between 335° and 145°, with the winds coming from the sea sectors, had a ratio close to 1.0. On the other hand, the bin-averaged ratios for directions between 155° and 325°, with the winds coming from the land sectors, were more widely scattered. The ratios between 175° and 230° were greatly scattered due to the influence from the nearby wind turbines. Thus, the accuracy of the wind speed measurements appears to be less sensitive to whether the wind is an along or across wind compared to whether the wind is an onshore or offshore wind. This result indicates how important a homogeneous wind field is to the accurate observation of horizontal wind speeds and directions using a scanning LiDAR with the VVP method.

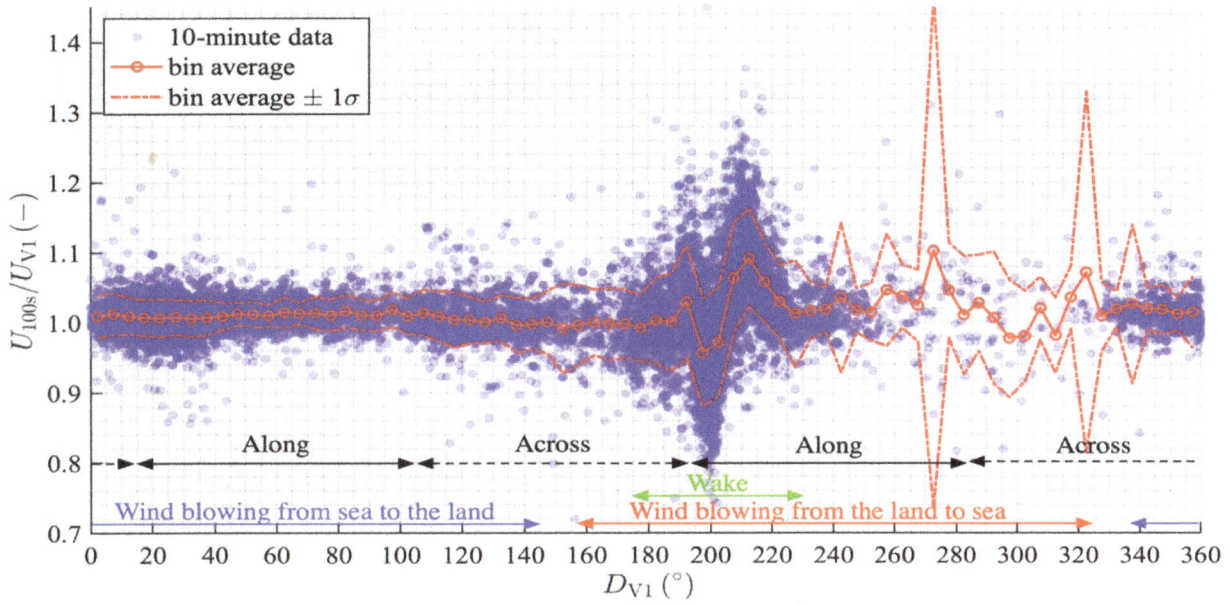

Figure 13. Ratio of 100s to V1 wind speeds for Case 8, as a function of the wind direction obtained from V1 at a height of 47 m ASL.

3.3. Data Availability and Accuracy Indicators

Overall, the wind measurements using the 100s and the VVP method agreed with the V1 values in the current study. Our validation results suggest that using offshore wind measurements via a single scanning LiDAR would be a promising way to assess the nearshore wind conditions. However, the comparisons above were made with references located only 400 m from the scanning LiDAR position. In reality, the wind turbines would be situated a few kilometers off the coast even for nearshore wind farms. Accordingly, the data availability further offshore and the indicators for assessing the data quality were investigated by analyzing the CNR value and the goodness of fit obtained from the curve fitting in the VVP method.

3.3.1. CNR Variation on the Measurement Range

As shown in previous studies [30], the CNR can serve as a good indicator for evaluating the reliability of LiDAR-generated measurement data. Generally, the fraction of the unreliable contaminated data increased as the CNR value decreased. Before estimating the variation in data availability and accuracy associated with the measurement range, the relationship between the CNR value and the accuracy of the 100s wind speed were investigated. Here, the relative absolute errors (RAEs) for the 10-minute means were employed as an index of accuracy. The definition of RAE is given by

$$\mathrm{RAE} = \frac{|u_{100s} - u_{v1}|}{u_{v1}} \times 100. \tag{9}$$

Figure 14 shows a plot of the RAEs for the 100s as a function of the CNR values for Case 8. The red circles and error bars indicate the bin-averaged and the standard deviation values, respectively. As expected, the CNR values were obviously correlated with the RAEs. The accuracy of the 100s tended to gradually decline, as the CNR took on values below −23 dB. The accuracy worsened significantly when the CNR value fell below −30 dB.

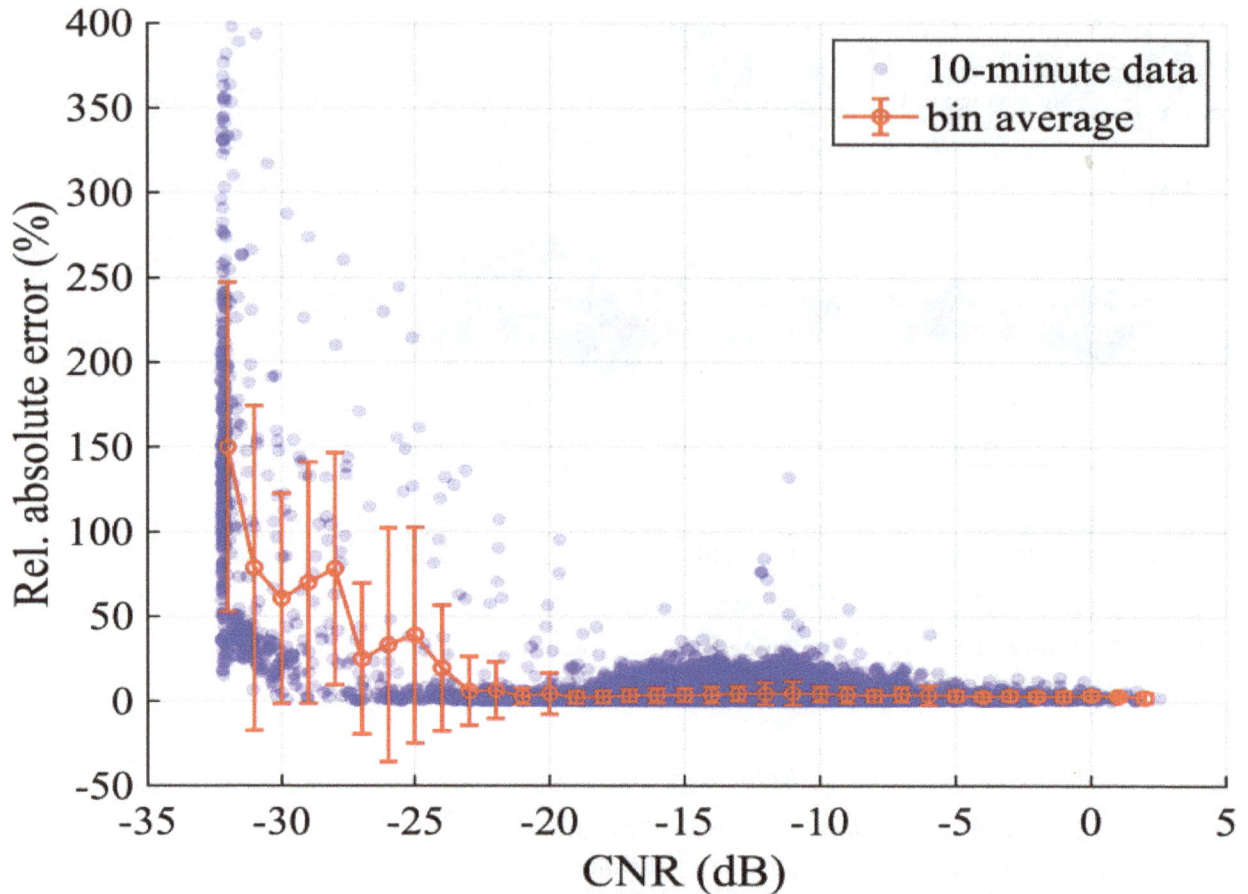

Figure 14. Scatter plot of relative absolute errors (RAE) as a function of the sweep averaged CNR for Case 8. The red circles and error bars indicate the bin-averaged values and the standard deviations, respectively, at a bin width of 1 dB for each CNR range.

Figure 15 shows the frequency distribution of the CNR values for measurement ranges from 500 to 3000 m as determined from the Case 8 data collected over a three-month period. As shown, the peak frequency in the CNR distribution at a 500 m range occurred around −15 dB. As the distance increased, the peak frequency shifted toward lower CNR values. The CNR distribution for the 3000 m distance had one peak around −25 dB and the other smaller peak around −33 dB. While it is preferable to set a higher CNR threshold to increase the quality of data used for actual assessment, a higher CNR threshold will reduce the data availability, especially further away from the LiDAR position. For example, if a CNR value of −30 dB is used as the data reliability threshold, 33% of the raw measurements at a distance of 3000 m should be excluded from the analysis. The greatest number of usable observations are available at the 500 m distance.

Figure 16 shows the data availability for various CNR thresholds as a function of the distance from the 100s for Case 8. As illustrated in the figure, the CNR threshold of −30 dB was set, and we can expect data availability of 85%, 80%, and 70% at distances of 2000, 2500, and 3000 m, respectively. According to the report from the Carbon Trust on the acceptance criteria for floating LiDAR technologies [10], a data availability of 85% is required in order to qualify as an acceptable method to tradition measurement with met masts. Given this requirement, a distance of 2000 m would appear to be the maximum effective range for the application of offshore wind measurements using the current scanning LiDAR. The CNR variation along the measurement range is likely influenced by factors including the specific device, height, and atmospheric conditions.

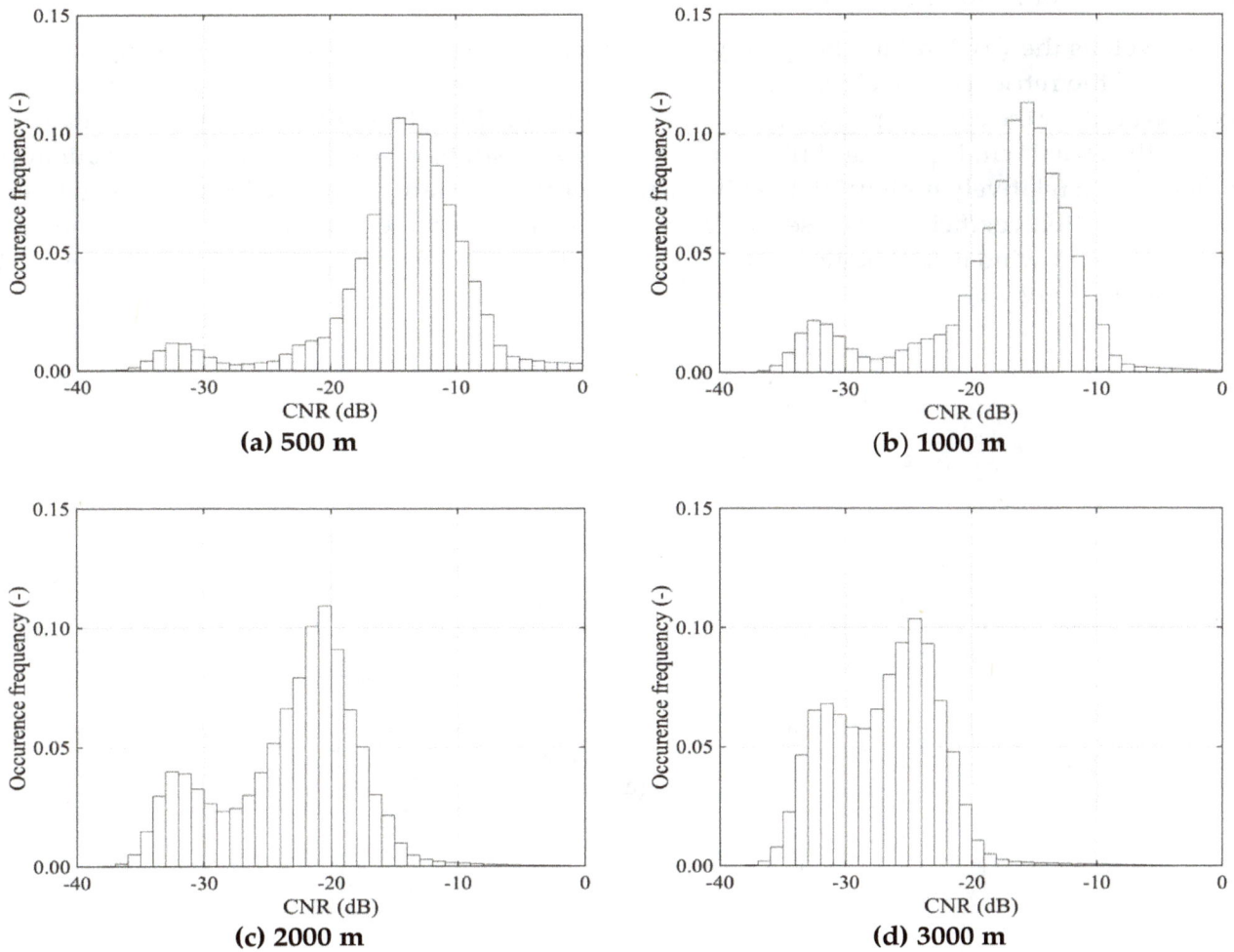

Figure 15. Occurrence frequency of the CNR values at distances of (**a**) 500 m, (**b**) 1000 m, (**c**) 2000 m and (**d**) 3000 m from the LiDAR position. Analyzed for Case 8.

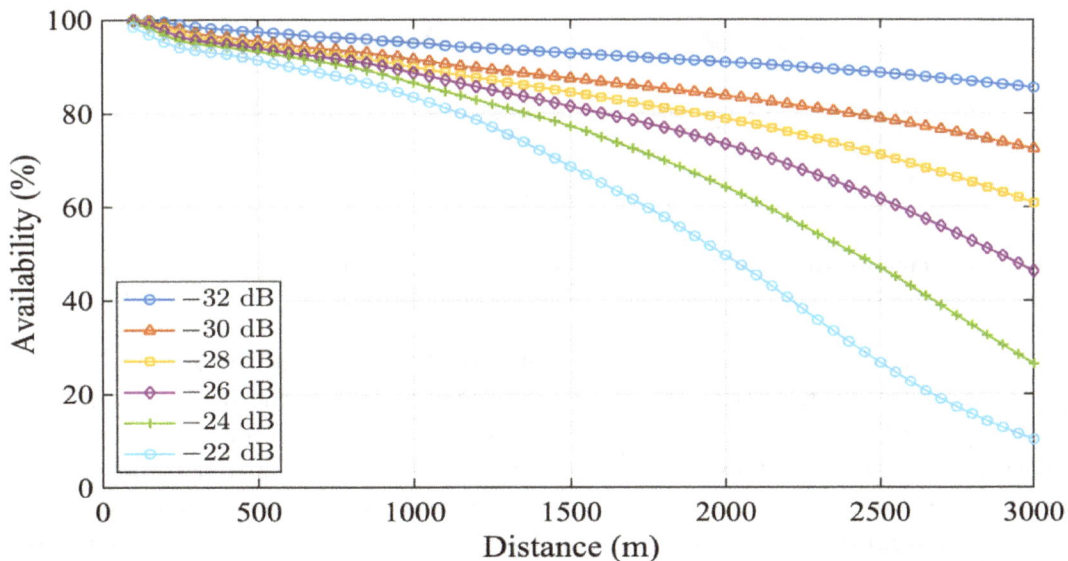

Figure 16. Data availability for various CNR thresholds (−32, −30, −28, −26, −24, and −22 dB) as a function of the distance from 100s for Case 8.

3.3.2. Goodness of Fit for the Curve Fitting

As well as the CNR value, the quality of the curve fitting is an indicator for investigating the accuracy of the retrieved velocities with the VVP method. For this examination, we analyzed the goodness of fit for the curve fitting performed for the VVP method. Figure 17 shows the instantaneous value of the radial wind speed and fitted curve for two illustrative cases. In one, the determination coefficient was relatively high (0.992), while in another it was relatively low (0.285). Cases in which the determination coefficient is close to 1.0 are compatible with the assumption used in the VVP method—that the wind field is homogeneous in time and space—whereas cases in which the coefficient value is low are not.

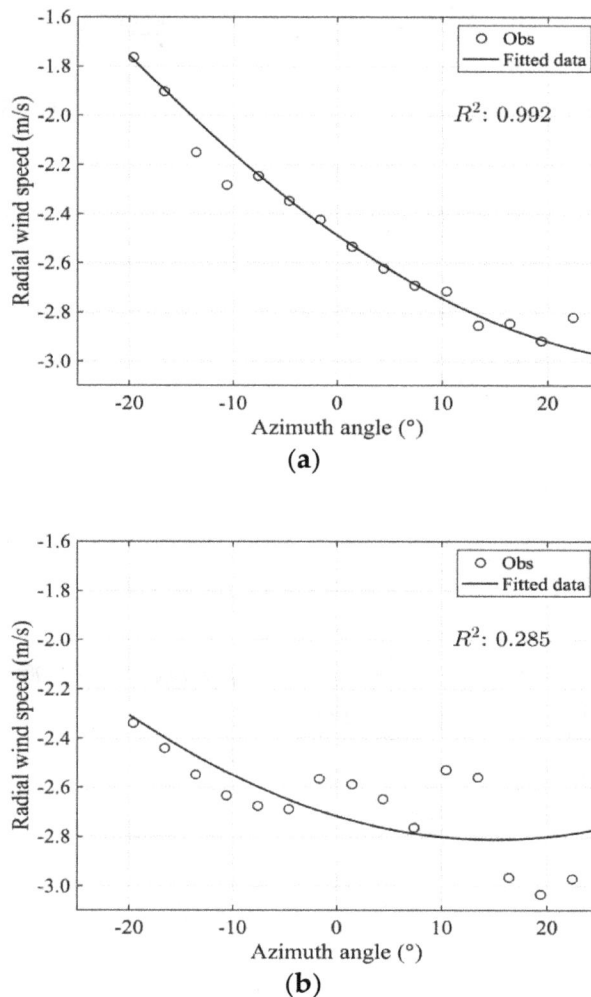

Figure 17. Instantaneous radial wind speeds as a function of azimuth angles and the corresponding fits for the cases of (**a**) high and (**b**) low determination coefficients.

Similar to the analysis of the relationship between the CNR values and the accuracy of the wind speed measurement (cf. Figure 14), the relationship between the RAE and the coefficient of determination (R^2) for Case 8 was investigated and the results are presented in Figure 18. The red circles with error bars in Figure 18 indicate the bin-averaged values and their standard deviation. As can be seen here, the RAE values increase rapidly when the value of R^2 falls below 0.2. This result indicates that if the quality of the curve fitting decreases, the accuracy of the measurements also decreases. In other words, the determination coefficient derived from the curve-fitting process serves as an indicator for estimating the quality of the horizontal wind speed measurements using a single scanning LiDAR with the VVP method.

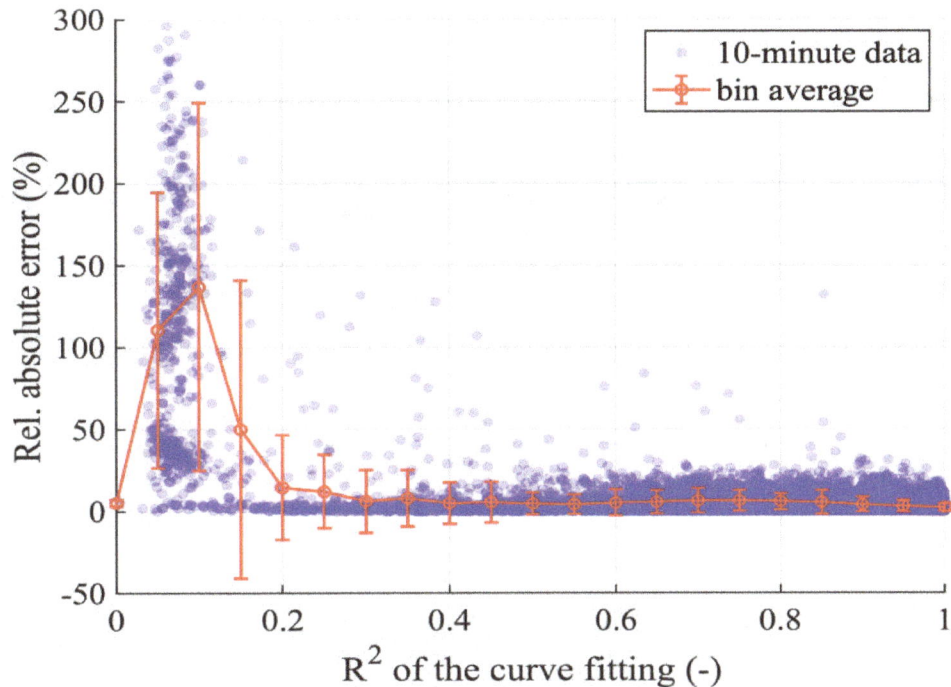

Figure 18. Distribution of the RAE as a function of the 10-minute mean R^2 from the VVP method. Results are presented for Case 8.

4. Discussion

For assessing offshore wind conditions, in addition to the mean wind speeds and wind directions, the accuracy of the turbulence intensity (TI) measurements using a scanning LiDAR is also crucial. This parameter is tightly associated with wind turbine design. We discussed the applicability of a single scanning LiDAR for TI measurements by comparing the observations from the 100s with those from the SA at a height of 10 m ASL. Figure 19 shows the time series of the 10-minute mean and standard deviation of wind speed values from the SA and 100s for the period of one week from 1 to 7 May 2019. This corresponds to Case 9 in Table 2. As the SA measurements were strongly influenced by the presence of the neighboring observational hut, observations with wind directions between 49° and 89° were excluded from the analysis.

The match between the time series of the mean wind speeds recorded by the 100s and the time series of the mean wind speeds recorded by the SA was not quite as good as it was for the 100s–V1 comparisons. In comparing the mean 100s wind speeds to the mean SA wind speeds, we found that the 100s wind speeds compared to the SA wind speeds had a mean deviation of −0.17 m/s (−3.4%) and a standard deviation of 0.55 m/s (10.8%), and the linear regression equation was y = 0.959x, with a determination coefficient of 0.963. The comparatively poor match may be due to the non-homogeneous nature of the wind field near the surface.

Figure 19 indicates that measuring the standard deviations of wind speeds with scanning LiDARs was more challenging than measuring the 10-minute mean wind speeds. We found that, for a time scale of 2 h, the standard deviation of the 100s values followed a trend similar to that of the standard deviations obtained from the SA. However, large differences were observed for the shorter time scales. The sonic anemometer used in this study had a 0.25-second temporal interval (= 4 Hz), while the 100s required approximately 15 s (= 0.067 Hz) to measure horizontal wind speeds and directions with the VVP method. As a result, the 100s collected only 40 samples in a 10-minute duration, while the SA collected 2400 samples during the same time. This difference may explain why the 100s time series of standard deviation values shows greater fluctuations than that the SA time series, as well as the difference in the measurement volumes between the 100s and SA.

(a)

(b)

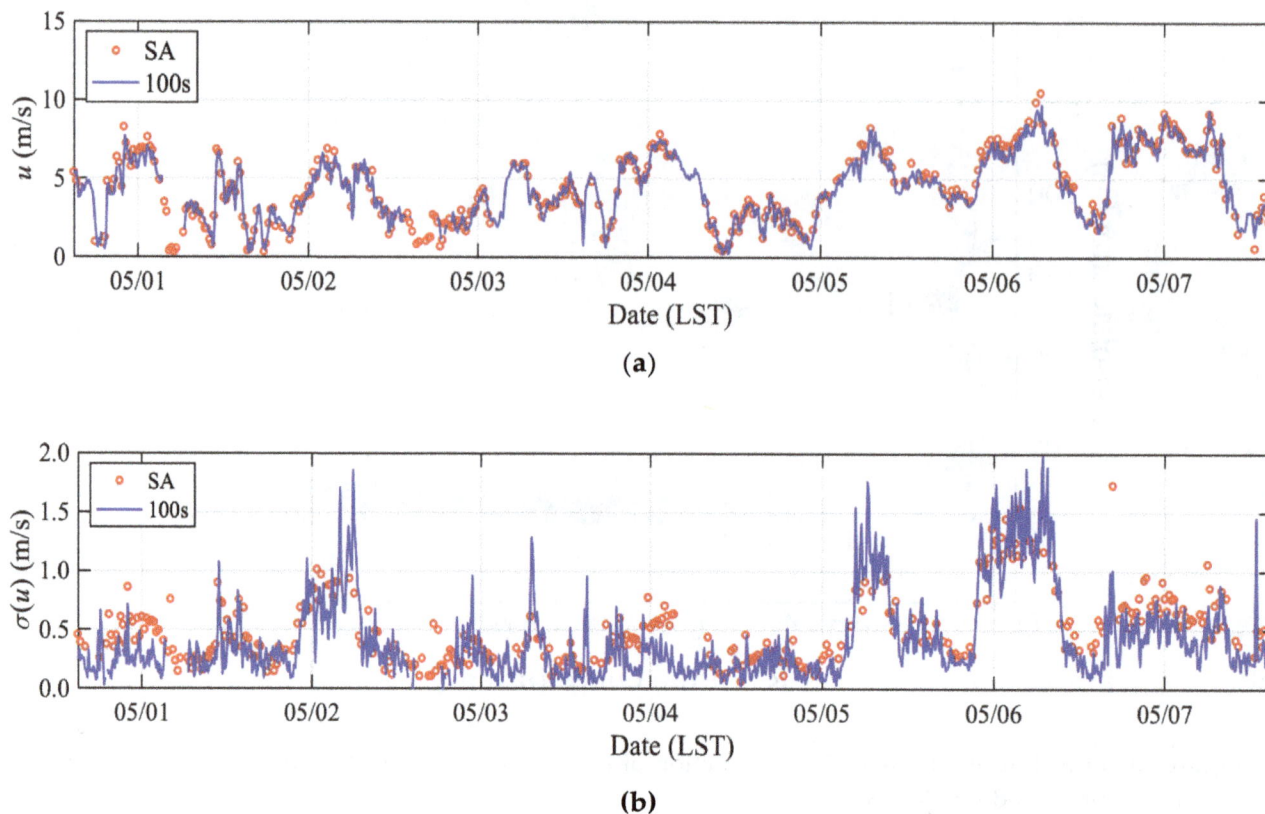

Figure 19. Time series of the wind speeds of (**a**) 10-minute mean and (**b**) standard deviation, obtained from the SA and 100s for the period from 1 to 7 May 2019 (Case 9).

The accuracy of the standard deviations obtained from the 100s was worse than that of the mean wind speeds. However, the instantaneous values of the standard deviation of wind speeds are not directly used for wind turbine design. In practice, TI, which is the 10-minute standard deviation of wind speed, normalized by the mean wind speed, is first calculated. The 90% quantile values of TI with a bin width of 1 or 2 m/s are then plotted as a function of the mean wind speed to evaluate the turbulence condition. Figure 20a shows the relationships between the 10-minute mean wind speed and TI of the SA and 100s for Case 9. The dark lines indicate the IEC standards [30] for classifying wind turbines based on site-specific turbulence conditions. The number of samples for each bin is shown in Figure 20b.

The TI values produced by the 100s were slightly larger than those by the SA for mean wind speeds in the range of 5–8 m/s; for a mean wind speed of 9–11 m/s, the 100s values were lower. We found that the difference between the SA and 100s values was comparable to or less than the difference between the IEC categories for most ranges. Given that the comparisons here were with an SA installed near the surface, the accuracy of the standard deviation measurements for the 100s at the height of a wind turbine hub might well improve, as might the accuracy of the mean wind speeds and directions. Consequently, the turbulence intensity obtained from the 100s with the VVP method would appear to have potential as an alternative to in situ wind measurements using meteorological masts. However, this is just a speculation based on the data from two months. The accuracy of the TI measurements from a single scanning LiDAR would be impacted by the scanning configurations. Thus, further investigations are necessary to reach any conclusions regarding the accuracy of scanning LiDAR-measured turbulence intensity at the hub height level of utility scale wind turbines.

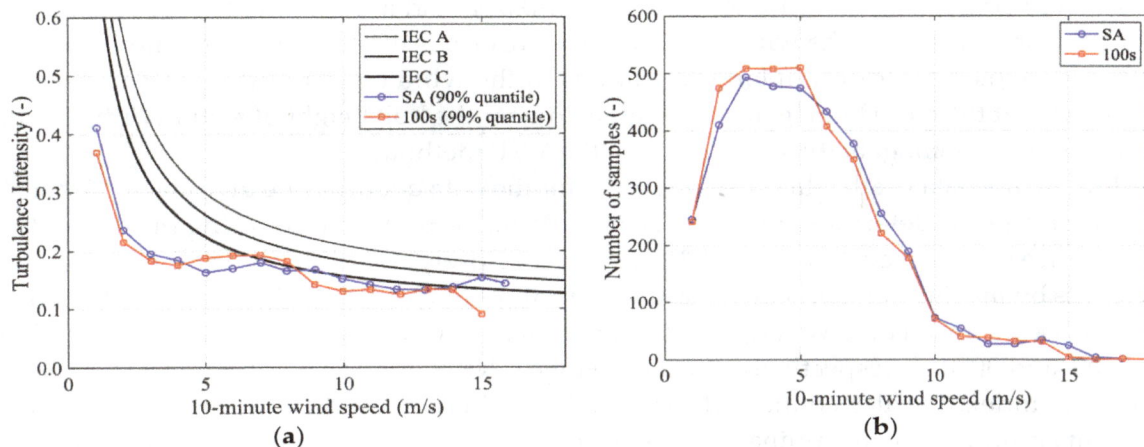

Figure 20. (a) 90% quantiles of turbulence intensity (TI) from SA and 100s, as a function of the 10-minute mean wind speed, and (b) the number of samples for each wind speed range for Case 9.

5. Conclusions

This study investigated the accuracy of offshore wind speed measurements using a single scanning LiDAR and a two-parameter curve-fitting VVP method. The accuracies of the 10-minute mean wind speeds and directions at heights of 47, 107, and 207 m ASL, that were obtained from a scanning LiDAR positioned at the water's edge, were quantitatively examined by comparing the scanning LiDAR values to observations from a vertical profiling LiDAR. The accuracy of the LiDAR-generated turbulence intensity value, which is a parameter used in wind turbine design, was also investigated by comparing the scanning LiDAR values with the observations from an SA mounted on the pier. A methodology for estimating the pitch and roll angles of the instrument after deployment was also described. The results obtained from the experiments can be summarized as follows:

In Case 1, the 100s mean wind speeds were found to have a mean deviation of 0.03 m/s (0.3% of the mean value), a standard deviation of deviation of 0.21 m/s (2.6%), and a determination coefficient of 0.997. The 100s wind direction values were found to have a mean deviation of −0.5°, a standard deviation of deviation of 2.5°, and a determination coefficient of 1.000. These values satisfied the criteria used in vertical and floating LiDAR verification by the IEC. We also investigated the influences of the azimuth width on the wind speed and direction accuracy. As a result, the accuracy of the 100s was found to decline with decreases in the azimuth width (Cases 2–4). In Case 5, an attempt was made to improve the accuracy by decreasing the scan rate for a smaller azimuth range. However, we found that this did not increase the accuracy. An azimuth width of more than 45° was preferred for use with the VVP method.

During a three-month long validation (Case 8), the impact of the wind direction on the accuracy of the wind speed retrievals was investigated. Previous studies indicated that the accuracy for winds in the across-wind sectors (winds perpendicular to the LiDAR heading angle) was worse than that for winds in the along-wind sectors (winds parallel to the LiDAR heading angle). However, we found no clear differences in wind speed accuracy between the sectors. The measurement accuracy differences between the land- and sea-sector winds were also examined. We found that the wind speed measurements for the sea sector showed distinctly better results. The poorer results for the land sector may be due to the non-homogeneous wind field induced by the land surface.

We also investigated the impact of the elevation angle. The results from the relevant cases (Case 6 and Case 7) showed that the wind speed obtained from the 100s was insensitive to the elevation angle setting. This may be due to the vertical wind speed being negligibly small compared to the horizontal wind speed in coastal areas. We also found that the accuracy of the wind speed and direction increased as the height increased. The wind speed at the height of 107 m ASL had a mean bias of 0.03 m/s (0.5%) and a determination coefficient of 0.998, while the (less accurate) wind speeds measured near the surface showed a mean deviation of −0.18 m/s (−3.5%) and a determination coefficient of 0.963 against the SA observations. The wind field near the surface is likely to be much more disturbed by surface factors, which can result in large variations in the wind conditions in time and space. Our results suggest that the more homogeneous wind field near the hub height of a wind turbine would be compatible with the homogeneity assumption in the VVP method.

In addition, the data availability and indicators for the data quality were investigated by analyzing the CNR value and the determination coefficients calculated in the curve-fitting process of the VVP method. We first confirmed that the CNR value was closely associated with the accuracy of the measurements by making comparisons with the vertical profiling LiDAR observations. The additional analysis for Case 8 showed that we can expect a data availability level of 85%, 80%, and 70% at distances of 2000, 2500, and 3000 m, respectively, if we employ a CNR value of −30 dB as the threshold for the data reliability index. We also confirmed that the determination coefficient was connected with the measurement accuracy. The determination coefficient obtained via the curve fitting process in the VVP method was an additional indicator for assessing the data quality.

Finally, we also examined the measurement accuracy of TI by comparing the scanning LiDAR-based measurements with the SA measurements near the surface in Case 9. We found that the differences between the standard deviations of the SA and 100s wind speeds were much larger than the differences between the 10-minute mean wind speeds and directions. Due to the coarse temporal resolution and large measurement volume of the 100s, the 10-minute standard deviations from the 100s were expected to be more scattered than those from the SA. In addition, the 90% quantiles of the standard deviations for the SA and 100s measurements were compared. As a result, the difference in the TI was less than the difference between the IEC standard categories. Thus, we found that using a single scanning LiDAR for offshore wind measurements had potential not only for wind resource assessment but also for assessing the site-specific conditions considered in wind turbine design.

Overall, the offshore wind measurements using a single scanning LiDAR with the VVP method showed good performance when compared to the reference observations obtained at a distance of 400 m from the scanning LiDAR. An attempt to evaluate the data availability further offshore showed that this measurement method would be equally as accurate within a range from 2000 to 2500 m. Accordingly, we argue that using a single scanning LiDAR with the VVP offers a promising and cheaper alternative to in situ observations using offshore meteorological masts and floating LiDAR technologies. For future research, we plan to conduct measurement campaigns with an offshore met mast and a vertical profiling LiDAR located 2 to 3 km away from the two scanning LiDARs installed on the coast to compare the performance with dual LiDAR measurements.

Author Contributions: Conceptualization, S.S. and J.P.G.; Data curation, S.S., J.P.G., T.O., T.K. and S.N.; Funding acquisition, S.S. and T.K.; Methodology, S.S. and J.P.G.; Project administration, S.S.; Writing—original draft, S.S. and J.P.G.; Writing—review and editing, T.O., T.K. and S.N. All authors have read and agreed to the published version of the manuscript.

Abbreviations

100s	Windcube 100s
ASL	Above Sea Level
CNR	Carrier-to-Noise Ratio
DBS	Doppler Beam Swinging
HORS	Hazaki Oceanographic Research Station
HTC	Hard Target Calibration
IEC	International Electrotechnical Commission
LiDAR	Light Detection and Ranging
METI	Ministry of Economy, Trade and Industry
PPI	Plan Position Indicator
RAE	Relative Absolute Error
RHI	Range Height Indicator
SA	Sonic Anemometer

STC	Soft Target Calibration
TI	Turbulence Intensity
V1	Windcube V1
VVP	Velocity Volute Processing

References

1. Global Wind Energy Council (GWEC). *Global Wind Report 2019*; GWEC: Maharashtra, India, 2019.

2. METI. Promising Sea Areas and Sites Selected for Targeted Promotion. Available online: https://www.meti. go.jp/english/press/2019/0730_001.html (accessed on 12 August 2019).

3. Barthelmie, R.J.; Wang, H.; Doubrawa, P.; Pryor, S. *Best Practice for Measuring Wind Speeds and Turbulence Offshore through In-Situ and Remote Sensing Technologies*; Report to the Department of Energy as Partial Fulfilment of Grant EE0005379; Cornell University: Ithaca, NY, USA, 2016.

4. Goit, J.P.; Shimada, S.; Kogaki, T. Can LiDARs Replace Meteorological Masts in Wind Energy? *Energies* **2019**, *12*, 3680. [CrossRef]

5. Hasager, C.B.; Sjoholm, M. Editorial for the Special Issue "Remote Sensing of Atmospheric Conditions for Wind Energy Applications". *Remote Sens.* **2019**, *11*, 781. [CrossRef]

6. Sathe, A.; Banta, R.; Pauscher, L.; Vogstad, K.; Schlipf, D.; Wylie, S. *Estimating Turbulence Statistics and Parameters from Ground-and nAcelle-Based Lidar Measurements: IEA Wind Expert Report*; DTU Wind Energy: Roskilde, Denmark, 2015.

7. Shimada, S.; Takeyama, Y.; Kogaki, T.; Ohsawa, T.; Nakamura, S. Investigation of the Fetch Effect Using Onshore and Offshore Vertical LiDAR Devices. *Remote Sens.* **2018**, *10*, 1408. [CrossRef]

8. Shimada, S.; Takeyama, Y.; Kogaki, T.; Ohsawa, T.; Nakamura, S.; Kawaguchi, K. Enhanced offshore wind simulations with WRF using LiDAR observation nudging. *J. JWEA* **2018**, *42*, 17–24. (In Japanese)

9. Leosphere. Scanning Windcube. Available online: https://www.leosphere.com (accessed on 11 March 2020).

10. Carbon Trust. *Roadmap for the Commercial Acceptance of Floating LiDAR Technology*; Carbon Trust: London, UK, 2018.

11. Gottschall, J.; Wolken-Mohlmann, G.; Viergutz, T.; Lange, B. Results and conclusions of a floating-lidar offshore test. *Enrgy Procedia* **2014**, *53*, 156–161. [CrossRef]

12. Hsuan, C.-Y.; Tasi, Y.-S.; Ke, J.-H.; Prahmana, R.A.; Chen, K.-J.; Lin, T.-H. Validation and Measurements of Floating LiDAR for Nearshore Wind Resource Assessment Application. *Energy Procedia* **2014**, *61*, 1699–1702. [CrossRef]

13. Fuertes, F.C.; Iungo, G.V.; Porte-Agel, F. 3D Turbulence Measurements Using Three Synchronous Wind Lidars: Validation against Sonic Anemometry. *J. Atmos. Ocean Technol.* **2014**, *31*, 1549–1556. [CrossRef]

14. Mann, J.; Cariou, J.P.; Courtney, M.S.; Parmentier, R.; Mikkelsen, T.; Wagner, R.; Lindelow, P.; Sjoholm, M.; Enevoldsen, K. Comparison of 3D turbulence measurements using three staring wind lidars and a sonic anemometer. *Meteorol. Z.* **2009**, *18*, 135–140. [CrossRef]

15. Newsom, R.K.; Berg, L.K.; Shaw, W.J.; Fischer, M.L. Turbine-scale wind field measurements using dual-Doppler lidar. *Wind Energy* **2015**, *18*, 219–235. [CrossRef]

16. Pena, A.; Mann, J. Turbulence Measurements with Dual-Doppler Scanning Lidars. *Remote Sens.* **2019**, *11*, 2444. [CrossRef]

17. Newman, J.F.; Bonin, T.A.; Klein, P.M.; Wharton, S.; Newsom, R.K. Testing and validation of multi-lidar scanning strategies for wind energy applications. *Wind Energy* **2016**, *19*, 2239–2254. [CrossRef]

18. Wagner, R.; Vignaroli, A.; Courtney, M.; McKeown, S.; Cussons, R.; Murthy, R.K.; Boquet, M. Real world offshore power curve using nacelle mounted and scanning Doppler lidars. In Proceedings of the EWEA Offshore 2015, Copenhagen, Denmark, 10–12 March 2015.

19. Cameron, L.; Clerc, A.; Feeney, S.; Stuart, P. *Remote Wind Measurements Offshore Using Scanning LiDAR Systems*; OWA Report; Carbon Trust: London, UK, 2014.

20. Courtney, M.; Wagner, R.; Murthy, R.K.; Boquet, M. Optimized Lidar Scanning Patterns for Reduced Project Uncertainty. In Proceedings of the EWEA 2014, Barcelona, Spain, 10–13 March 2014.

21. Coutts, E.; Oldroyd, A.; Stein, D.; Boque, M.; Krishna, R.; Akhoun, M.; Espin, F.; Miguel Gonzalez Garcia, L. Cost Effective Offshore Wind Measurement. In Proceedings of the EWEA Resource Assessment 2015, Helsinki, Finland, 2–3 June 2015.

22.   Simon, E.; Courtney, M. *A Comparison of Sector-Scan and Dual Doppler Wind Measurements at Høvsøre Test Station—One Lidar or Two?* Technical Report DTU Wind Energy E-0112 (EN); DTU Wind Energy: Roskilde, Denmark, 2016.

23.   PARI. Hazaki Oceanographical Research Station (HORS). Available online: https://www.pari.go.jp/unit/edosy/en/main-facility/2.html (accessed on 10 September 2019).

24.   Calhoun, R.; Heap, R.; Princevac, M.; Newsom, R.; Fernando, H.; Ligon, D. Virtual towers using coherent Doppler lidar during the Joint Urban 2003 dispersion experiment. *J. Appl. Meteorol. Clim.* **2006**, *45*, 1116–1126. [CrossRef]

25.   Zhou, S.H.; Wei, M.; Wang, L.J.; Zhao, C.; Zhang, M.X. Sensitivity Analysis of the VVP Wind Retrieval Method for Single-Doppler Weather Radars. *J. Atmos. Ocean Technol.* **2014**, *31*, 1289–1300.

26.   Waldteufel, P.; Corbin, H. On the Analysis of Single-Doppler Radar Data. *J. Appl. Meteorol.* **1979**, *18*, 532–542. [CrossRef]

27.   Koscielny, A.J.; Doviak, R.J.; Rabin, R. Statistical Considerations in the Estimation of Divergence from Single-Doppler Radar and Application to Pre-Storm Boundary-Layer Observations. *J. Appl. Meteorol.* **1982**, *21*, 197–210. [CrossRef]

28.   International Electrotechnical Commision (IEC). *IEC 61400-12-1 Wind Energy Generation Systems—Part 12-1: Power Performance Measurements of Electricity Producing Wind Turbines*, 2nd ed.; IEC Central Office: Geneva, Switzerland, 2017.

29.   Hasager, C.B.; Stein, D.; Courtney, M.; Pena, A.; Mikkelsen, T.; Stickland, M.; Oldroyd, A. Hub Height Ocean Winds over the North Sea Observed by the NORSEWInD Lidar Array: Measuring Techniques, Quality Control and Data Management. *Remote Sens.* **2013**, *5*, 4280–4303. [CrossRef]

30.   Gryning, S.E.; Floors, R. Carrier-to-Noise-Threshold Filtering on Off-Shore Wind Lidar Measurements. *Sensors* **2019**, *19*, 592. [CrossRef] [PubMed]

4

Enhancement of Cloudless Skies Frequency over a Large Tropical Reservoir in Brazil

André R. Gonçalves [1,*], Arcilan T. Assireu [2], Fernando R. Martins [3], Madeleine S. G. Casagrande [3], Enrique V. Mattos [2], Rodrigo S. Costa [1], Robson B. Passos [2], Silvia V. Pereira [1], Marcelo P. Pes [1], Francisco J. L. Lima [1] and Enio B. Pereira [1]

[1] National Institute for Space Research, Av. dos Astronautas, 1758, São José dos Campos SP 12227-010, Brazil; rodrigo.costa@inpe.br (R.S.C.); silvia.pereira@inpe.br (S.V.P.); marcelo.pes@inpe.br (M.P.P.); francisco.lopes@inpe.br (F.J.L.L.); enio.pereira@inpe.br (E.B.P.)

[2] Federal University of Itajuba, Av. BPS, 1303, Pinheirinho, Itajubá MG 37500-903, Brazil; arcilan@unifei.edu.br (A.T.A.); enrique@unifei.edu.br (E.V.M.); robsonbarretodospassos@gmail.com (R.B.P.)

[3] Federal University of São Paulo, Rua Carvalho de Mendonça, 144, Santos SP 11070-102, Brazil; fernando.martins@unifesp.br (F.R.M.); madeleine.gacita@unifesp.br (M.S.G.C.)

* Correspondence: andre.goncalves@inpe.br

Abstract: Several studies show the effects of lake breezes on cloudiness over natural lakes and large rivers, but only few contain information regarding large flooded areas of hydroelectric dams. Most Brazilian hydropower plants have large water reservoirs that may induce significant changes in the local environment. In this work, we describe the prevailing breeze mechanism in a Brazilian tropical hydropower reservoir to assess its impacts on local cloudiness and incoming surface solar irradiation. GOES-16 visible imagery, ISCCP database products, and ground measurement sites operated by INMET and LABREN/INPE provided data for the statistical analysis. We evaluate the cloudiness frequency assuming two distinct perspectives: spatial distribution by comparing cloudiness over the water surface and areas nearby its shores, and time analysis by comparing cloudiness prior and after reservoir completion. We also evaluated the solar irradiance enhancement over the water surface compared to the border and land areas surrounding the hydropower reservoir. The results pointed out daily average cloudiness increases moving away from the reservoir in any of the four cardinal directions. When looking at the afternoon-only cloudiness (14h to 16h local time), 4% fewer clouds were observed over the flooded area during summer (DJF). This difference reaches 8% during autumn (MAM) and spring (SON). Consequently, the irradiance enhancement at the water surface compared to external areas was around 1.75% for daily average and 4.59% for the afternoon-only average. Our results suggest that floating solar PV power plants in hydropower reservoirs can be an excellent option to integrate both renewable energy resources into a hybrid power generation due to the high solar irradiance in Brazilian territory combined with the prevailing breeze mechanism in large tropical water reservoirs.

Keywords: lake breeze influence; hydropower reservoir; solar irradiance enhancement; solar energy resource

1. Introduction

Lake breeze circulation is one of the most well-known thermally-induced phenomena in mesoscale meteorology. The onset of a lake breeze depends on the ratio between the thermal to inertial driving forces, as discussed by Biggs and Graves [1]. Despite the well-described physical mechanism, the evaluation of its occurrence and strength is not a simple task due to surface heterogeneity, terrain effects, and synoptic patterns superimposed on the flow.

Rabin et al. [2] compared the surface heterogeneity (based on the Normalized Difference Vegetation Index (NDVI) data) with the Geostationary Operational Environmental Satellite (GOES) visible imagery and reported cloud-free bands downwind during the warm season over large lakes in the USA. Several other studies investigated the characteristics of lake breezes on the temperate and subtropical region. Segal et al. [3] evaluated the lake breeze phenomena in Florida (USA) using modeling and observational approaches. They reported that the atmospheric forcing induced by subsidence and suppression of the cumulus cloud mutually contribute to the increase in the cloudless frequency over the lakes. Asefi-Najafabady et al. [4] used the dual-Doppler radar to analyze 3-D flows induced by an elongated 2 km wide reservoir in Alabama (USA) during summer. The horizontal scale of the breeze circulation was approximately 10 km in both shores but was extremely sensitive to wind speed changes and direction. The authors mentioned a simultaneously cloud-free zone over the lake observed from GOES satellite. Iakunin et al. [5] have shown that the lake breeze for the Alqueva reservoir (southeast of Portugal) could be detected at a distance of more than 6 km away from the shores and at altitudes up to 300 m above the water surface based on observation and model. Crosman and Horel [6] published an extensive review of numerical studies on lake breeze dynamics providing a significant contribution to the understanding of this phenomenon. It is a consensus among researchers that factors like the sensible heat flux, synoptic wind, atmospheric stability, watershed dimensions, terrain slope, and roughness affect lake breeze occurrence.

Enhanced potential for breeze production is expected in tropical regions due to higher evapotranspiration and heat availability. For example, in the Amazon rainforest, the evapotranspiration is intense and exerts a notable influence on regional and global climate patterns, playing a significant role in cloudiness observed over large rivers and the water balance [7]. Silva Dias et al. [8] reported such an effect by studying the atmospheric circulation induced by the Tapajós River. They observed that this circulation causes the formation of shallow cumulus during the morning hours over the eastern riverside and suppresses cloud formation in the western riverside during the afternoon. Yin et al. [9] analyzed the daily cycle of cloudiness over Lake Victoria (East Africa) and its influence on lake evaporation. The authors concluded that the cloudiness varies up to 22% between day and night in the northeastern quadrant of the lake, but it varied seasonally throughout the year.

This phenomenon is not constrained to large natural lakes and rivers, as it affects large flooded areas such as hydroelectric dams. The formation of large water reservoirs to feed large hydropower plants in tropical regions may present a similar pattern of lake breeze [5,10–14]. Most Brazilian hydroelectric plants have large water reservoirs that eventually lead to lake breezes and induce extensive changes in the local environment. The replacement of the land cover by the water reservoirs causes intense thermal gradients between the flooded area and the surrounding territory that may trigger lake breeze circulations and impacts the hydrological cycle, energy balance, local cloudiness, and economic activities such as agriculture and tourism. Stivari et al. [10,15] showed that lake breeze circulation is a dominant feature in the local climate on the Lake Itaipu, the water reservoir of the Itaipu Brazil-Paraguay hydropower plant. According to these studies, the water surface is consistently colder than the land in lake borders with a thermal contrast up to −3 °C during the daytime. The lake is systematically warmer at night, presenting a thermal contrast of up to +8 °C. This thermal contrast could trigger the lake breeze circulation and, thus, inhibit the formation of shallow clouds during the day. However, to the best of our knowledge, the occurrence of this phenomenon in artificial tropical hydroelectric reservoirs and its characterization have not yet been addressed. In this way, the objective of this study is to investigate the formation of the lake breeze and the magnitude and spatial pattern of enhanced cloudless skies over a tropical reservoir in Brazil.

In this work, we describe the prevailing breeze mechanism in the artificial lake of a Brazilian hydroelectric plant in the tropical region. We investigate its impact on cloudless skies frequency and incoming surface solar irradiation based on observational data and statistical metrics. We evaluate the cloudiness frequency using two different approaches: spatial distribution by comparing cloudiness over the reservoir and in the areas nearby its shores, and time analysis by comparing cloudiness prior and after reservoir completion. Our results can provide information to foster the use of solar energy resources in hydroelectric dams of tropical regions through the technology of floating photovoltaics. Solar-hydro hybrid plants can become a great alternative to integrate both resources due to the high solar irradiance in the Brazilian territory combined with the breeze mechanism produced by many dams.

2. Materials and Methods

2.1. Water Reservoir Description

The target area for this study is the Serra da Mesa hydropower plant (14°00′ S, 48°21′ W), located in a stretched region between the states of Goiás and Tocantins (central region of Brazil) as illustrated in Figure 1.

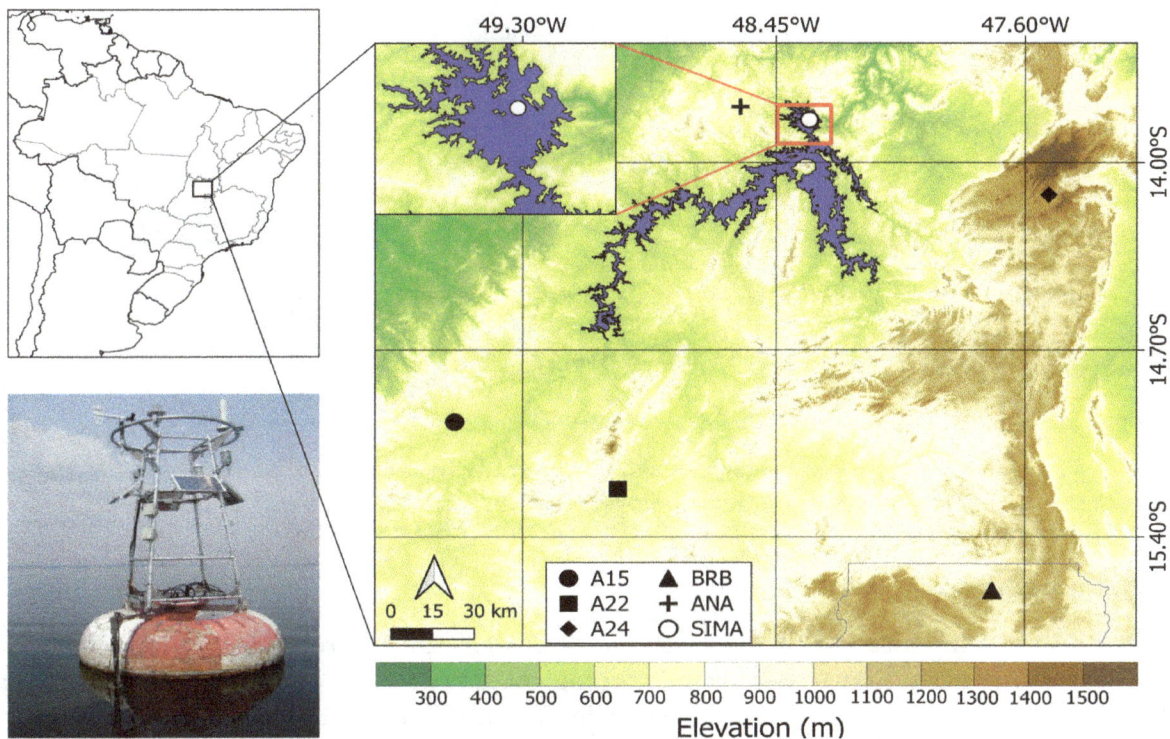

Figure 1. The geographical location of the Serra da Mesa water reservoir in the Central region of Brazil the lake geography, the locations of the ground stations and a picture of SIMA autonomous data collection buoy system in operation in the northern area of the lake.

The power plant has an installed capacity of 1275 MW, and its reservoir was established in 1998 in a region that was originally covered with tropical grassland savanna. Due to the basin's geomorphology, which is based on rugged terrain in the transition between Brazilian central elevated plains and Amazon basin lowlands, the lake presents a dendritic pattern covering a surface area of approximately 1784 km^2, with the maximum depth ~150 m. The reservoir's width is up to 8 km in the most extended portions, and the total water volume is ~54.4 billion m^3. The elevation along the nearby Tocantins River ranges from a minimum of 340 m above MSL (mean sea level) to 1100 m in the highest ridges. The regional climate is categorized as Aw–Tropical Savannah by Köppen-Geiger climate classification [16], with dry winter and rainy season occurring from November to March summing up to 1600 mm of precipitation per year [17]. The rainfall is modulated by the South American Monsoon System [18] with significant influence of other atmospheric systems such as cold fronts and squall lines. Winds are usually calm (2–3 m/s), and the monthly average temperature remains between 20 °C and 28 °C throughout the year [17].

2.2. Datasets Description

This study uses three sources of data to assess lake breeze impact on cloudiness: in situ measurements (5 years), a high-resolution short-term satellite dataset (one year), and a low-resolution long-term satellite dataset (29 years). Each of the datasets delivers complementary information on the breeze characteristics and impact from a climatological perspective. A flowchart describing the whole analysis is presented in Figure 2.

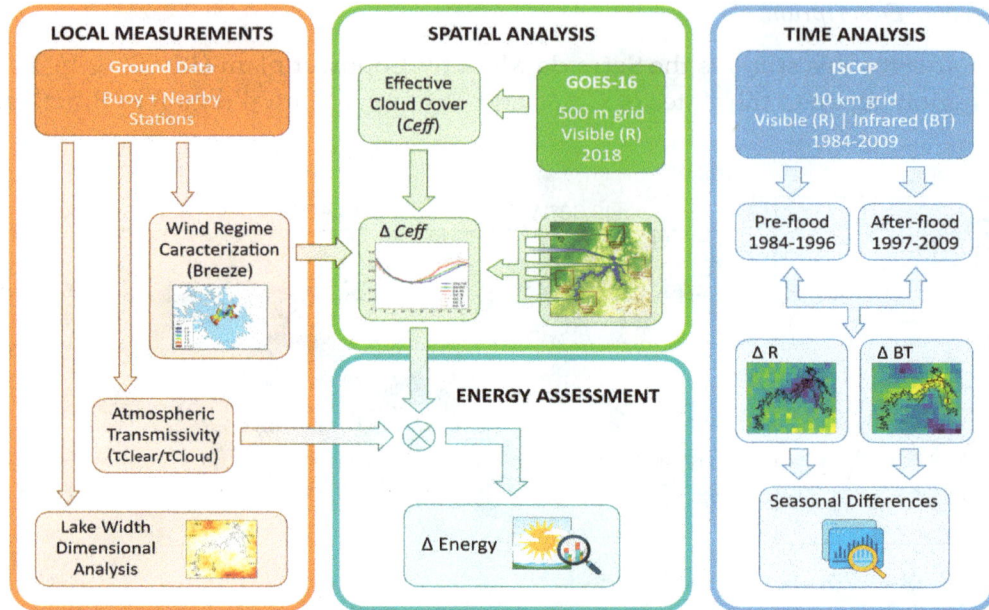

Figure 2. Flowchart of the three strategies used in this study to characterize the lake breeze: in situ measurements (5 years), a high-resolution short-term satellite dataset (one year), and a low-resolution long-term satellite dataset (29 years).

2.2.1. Ground-Based Dataset

Figure 1 shows the location of the buoy-based system for environmental monitoring (SIMA, a Brazilian Portuguese acronym) used for meteorological data acquisition, including air temperature, humidity, air pressure, wind intensity, and water temperature at 2, 5, 20, and 40 m depth. The buoy is anchored by cables attached to two train wheels, which guarantees a fixed geographic position. The observational data comprises hourly records from 2005 to 2010. The 10-min average wind data was recorded at hourly intervals. Table 1 summarizes the sensors and their technical specifications.

Table 1. Technical specifications of the sensors used in the SIMA buoy operating in the Serra da Mesa Reservoir. Adapted from Stech et al. [19].

Sensor	Manufacture	Range	Accuracy	Depth/Height
Air Temperature	Rotronic	−25 to 60 °C	±0.3 °C	3 m
Water Temperature	Yellow Spring	−5 to 60 °C	±0.15 °C	−2, −5, −20, −40 m
Wind Speed *	R.M. Young	0 to 100 ms^{-1}	±0.3 ms^{-1}	3 m
Wind Direction	R.M. Young	0 to 360°	±3°	3 m
Relative Humidity	Rotronic	0 to 100%	±1.5%	3 m
Barometric Pressure	Vaisala	500 to 1100 hPa	±0.3 hPa	3 m

* Wind direction is measured by combining the apparent wind direction from the anemometer vane and the orientation of the buoy from a compass.

Four other ground stations were used for regional estimates of climate-relevant variables for this study, such as ambient wind (10-m) and clear-sky transmissivity (through global horizontal irradiance,

GHI). These stations are operated by the SONDA project at from the National Institute for Space Research (INPE) and by the Brazilian National Meteorological Institute (INMET). Table 2 describes these data in more detail, including the SIMA buoy presented in Table 1.

Table 2. Details of SIMA buoy system and ground measurement sites near the Serra da Mesa Reservoir used to estimate the climate-relevant variables for this study.

Met. Station	Operated by	Distance to Shore (km)	Lat (°)	Lon (°)	Alt (m)	Period and Resolution	Variables Used
SIMA	INPE	offshore	−13.84	−48.33	476	2005–2010 Hourly	Table 1
ANA	ANA	15	−13.79	−48.57	694	1971–2018 Daily	Precipitation
BRB	INPE	140	−15.60	−47.71	1023	2005–2018 Minute	GHI
A15	INMET	64	−14.97	−49.53	522	2007–2018 Hourly	Wind
A22	INMET	76	−15.22	−48.98	667	2007–2018 Hourly	GHI, Wind
A24	INMET	63	−14.12	−47.52	1260	2007–2018 Hourly	GHI, Wind

2.2.2. Satellite-Based Datasets

In addition to the ground data, two satellite-derived datasets were used to evaluate the spatial and time distributions of cloudiness: the visible imagery of South America acquired by GOES−16 satellite and the Gridded Satellite images (GridSat-B1).

The National Oceanic and Atmospheric Administration (NOAA) operates the geostationary satellite GOES-16 over the Equator at 75.2° West and provides access to its visible imagery databases [20]. We used the 'RED' band images (Channel 2—central wavelength at 0.64 µm) to identify small-scale features such as river fogs and fog/clear air boundary images due to the 0.5 km spatial resolution (at the sub-satellite point). Images from the year of 2018 at one-hour intervals and native spherical grid resolution provided a highly detailed cloud pattern diagnosis over the reservoir and its shores. Despite sub-hourly imagery availability for GOES-16 in 2018, the hourly images create a satisfactory sample for cloudiness characterization.

Cloud detection from GOES-16 visible channel (0.64 µm) is impaired by surface brightness. A strategy to overcome this obstacle is to employ the Effective Cloud Cover (C_{eff}) index as a proxy for cloud coverage, as proposed in Equation (1) by Moser and Raschke [21]

$$C_{eff} = \frac{R - R_{min}}{R_{máx} - R_{min}} \tag{1}$$

where R is the visible reflectance acquired by GOES-16 at a particular pixel, and R_{min} and $R_{máx}$ are, respectively, the reflectance for cloudless and overcast sky condition. R_{min} is estimated based on a statistical analysis of satellite data observed over the same pixel in one month and is hourly dependent due to surface anisotropy effects. On the other hand, $R_{máx}$ is the maximum visible reflectance normalized by the solar zenith angle.

C_{eff} is a dimensionless coefficient, and it can assume values from zero (cloudless condition) to one (optically very thick cloudiness condition). It accounts for surface reflectivity, allowing detection of shallow cumulus clouds, typically formed in breeze fronts at the shorelines. In addition, C_{eff} has an almost linear relationship with the transmittance of solar radiation through clouds, providing a way to estimate the solar radiation incident on the surface [22]. Hourly data from the GOES-16 satellite were processed to get the C_{eff} values required to evaluate the lake breeze influence on cloudiness, and therefore the incoming solar irradiance at the surface.

The World Climate Research Program developed and delivered the second gridded satellite database (GridSat-B1) [23]. It is based on the ISCCP-B1 product of the International Satellite Cloud Climatology Project (ISCCP) [24], focusing on the global distribution of cloudiness, cloud properties, and seasonal variability. The GridSat-B1 database is gridded on a 0.07-degree surface resolution and comprises merged data by selecting the nadir-most satellite observations for each grid point. The GridSat-B1 encompasses data from three spectral band channels: infrared (around 11.0 µm), water vapor (around 6.7 μm), and visible (around 0.65 μm). The database comprises cloudiness data from 1984 to 2009 with 3-hr time resolution, allowing an assessment before and after reservoir flooding.

The infrared brightness temperature (BT) and visible reflectance (R) were proxies for cloudiness occurrence. BT is related to the energy emitted from cloud tops. In contrast, R represents the reflected solar energy from cloud tops. High clouds have low BT [25], while high R values are associated with clouds of high optical thickness. Only image pixels presenting BT lower than 0 °C (~4 km above the Mean Sea Level) were considered to minimize the effects of surface radiation contamination according to studies on cloud classification by Rossow and Garder [26] and Bottino and Ceballos [27].

The long-term database (29 years) of satellite data allows a comparison analysis of cloudiness occurrence before (1984–1996) and after (1997–2009) the Serra da Mesa reservoir construction. The difference between the observed BT "after" and "before" was assessed for the Serra da Mesa hydropower area, including the flooded and surrounding areas. The positive difference between the BT values indicates shallower clouds or less cloud occurrence. A similar procedure was adopted using reflectance data. The signal of the difference between the reflectance values "after" and "before" indicates an increase (+) or decrease (−) in the cloud optical thickness or clouds occurrence frequency.

The Kolmogorov–Smirnov two-sample homogeneity test (K–S test) was applied to prevent any misleading conclusion on the time series of cloudiness before and after filling the Serra da Mesa reservoir. This nonparametric test allows inferring, at a certain significance level (α), if two datasets have the same cumulative frequency distribution (CDF). The statistic parameter D_n is the maximum absolute differences between two CDFs, as shown in Equation (2):

$$D_n = \max\left(\left|F(x) - R(x)\right|\right) \tag{2}$$

where $R(x)$ is the reference cumulative distribution function, and $F(x)$ is the tested distribution. The K–S test rejects the null hypothesis if D_n is not within critical bounds V_c given in Equation (3) for $\alpha = 0.01$.

$$V_c = 1.63 \cdot \sqrt{\frac{n1 + n2}{n1 \cdot n2}} \tag{3}$$

where $n1$ and $n2$ are the sizes of the samples. A similar approach was used and explained in the details in Espinar et al. [28].

2.3. Lake Breeze Characterization

Unlike natural lakes, human-made reservoirs cause changes in land cover (LCC) that affect the mass, energy, and momentum exchanges between surface and atmosphere, and consequently the regional climate. The contrast between water–land surfaces, regarding the heat capacity, surface albedo, and roughness, lead to a mesoscale circulation non-existing in the region before water impoundment [29]. Segal and Arritt [30] classified a perturbed area (PA) as a contiguous region clearly distinguishable from their surroundings in terms of sensible heat flux (H), which, therefore, can induce a thermal flow similar to the sea breezes. Doran et al. [31] estimated the smallest size (L) of a PA based on the ambient wind speed (u_a) and the average potential temperature (θ) in the atmospheric boundary layer (Equation (4)).

$$L = \frac{\theta}{g\Delta(w'\theta')_s} \frac{u_a^3}{4ln(2)} \tag{4}$$

where g is the gravity acceleration and $\Delta(w'\theta')_s$ is the difference in heat flux between the PA (water reservoir) and the surrounding area.

2.4. Evaluation of Lake Breeze Influence on Incoming Shortwave Radiation

The clearness index was used to evaluate the lake breeze's effect on the incoming solar irradiance at the water surface and the land areas surrounding the Serra da Mesa reservoir. The clearness index is defined by Equation (5),

$$k_t = \frac{GHI}{GHI_0} \tag{5}$$

where GHI and GHI_0 are the incoming global horizontal solar irradiance at the surface and the top of the atmosphere, respectively. The GHI_0 is easily obtained based on the geographical location data and the solar zenith angle.

The relation between clearness index (k_t) and the effective cloud cover (C_{eff}) obtained from satellite imagery was first introduced by Moser and Raschke [21] and constitutes a critical parameter in many radiative transfer numerical algorithms [22,32,33]. The surface incoming solar irradiance depends only on the atmospheric transmittance in cloudless sky condition (τ_{clear}) when $C_{eff} = 0$. On the other hand, in the condition $C_{eff} = 1$, the atmospheric transmittance (τ_{cloud}) corresponds to an overcast sky, with maximum cloud optical thickness and absence of beam solar radiation reaching the surface. The range between these two transmittances defines the slope of the function that converts C_{eff} into global solar horizontal irradiance (GHI) at the surface (Equation (6)).

$$k_t = \frac{GHI}{GHI_0} = \{(1 - C_{eff})(\tau_{clear} - \tau_{cloud}) + \tau_{cloud}\} \tag{6}$$

The τ_{clear} values depend only on atmospheric gases, aerosols, water vapor, and air mass, while the τ_{cloud} values are a function of the maximum cloud optical depth since the cloud transmittance is the major modulating factor of the incoming solar irradiance in the overcast condition. It is reasonable to assume relatively stable values for both τ_{clear} and τ_{cloud} within similar climatic zones. From Equation (6), $k_{t\_clear} = \tau_{clear}$ for cloudless sky condition ($C_{eff} = 0$) and $k_{t\_cloud} = \tau_{cloud}$ in a totally cloudy sky condition ($C_{eff} = 1$).

We selected solar irradiance data for clear skies from three representative measurement sites to estimate typical k_{t_clear} values observed in the Serra da Mesa Reservoir region. The parameters for a simple cloudless sky model, described in Equation (7), were fitted to account for solar zenithal angle (air mass) influence in k_{t_clear} values [34],

$$k_{t\_clear} = k_{0t\_clear}.(cos(\theta_z))^{1.15} \tag{7}$$

where θ_z is the solar zenithal angle and k_{0t_clear} the clear sky transmittance for $\theta_z = 0$.

The inference of the typical k_{t_cloud} value for the Serra da Mesa reservoir region from ground measurements was challenging because the thickest clouds are rare, reducing the confidence of extreme values from k_t distribution. So, we adopted a constant value of $k_{t\_cloud} = 0.05$ based on previous irradiance modeling studies [22,35].

3. Results

3.1. Characterization of Local Climate from In Situ Data

Figure 3a shows the monthly averages of wind speed measured at the buoy measurement site and precipitation measured from 2005 to 2010 at hydrological measurement site managed by ANA (Brazilian Agency for Water Resources).

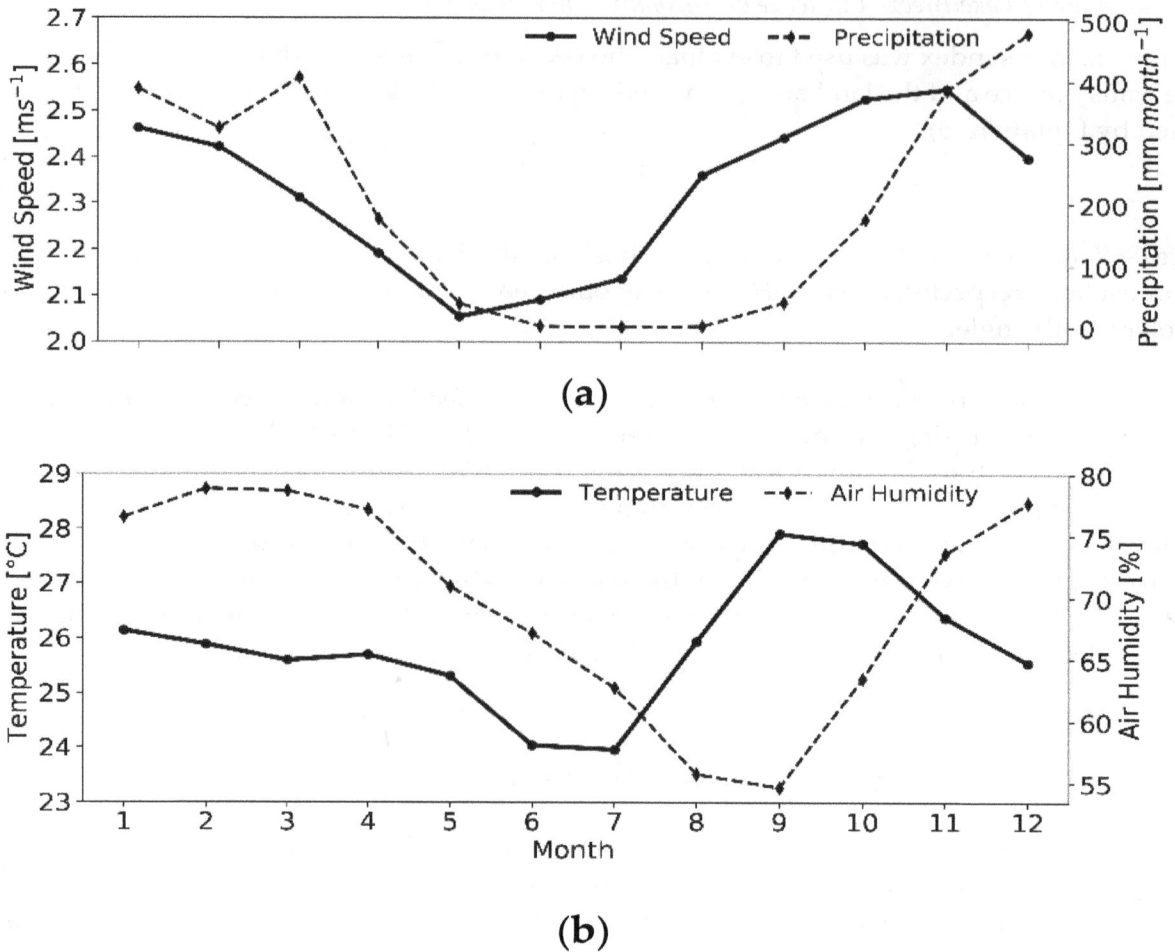

Figure 3. Climate seasonal patterns observed at the Serra da Mesa Reservoir over the period of 2005–2010: monthly mean of (**a**) precipitation (mm month^{-1}) and wind speed (ms^{-1}); (**b**) air temperature (°C) and relative humidity (%).

The precipitation shows a strong seasonal pattern defining a rainy season with precipitation larger than 300 mm per month from November to March. The precipitation is meager from May to September. The average wind speed ranged from 2.4 ms^{-1} to 2.6 ms^{-1} in the rainy season, but it decreased to 2.1 ms^{-1} at the beginning of the dry season (May to July).

The average air temperature in the rainy season ranges from 26 °C to 28 °C and breaks down to 24.5 °C in June–July (Figure 3b). The reduced daily (Figure 4a) and annual (Figure 4b) range of near-surface air temperature, known as an effect of tropical lakes on the regional climate, was previously observed in the Elqui Valley Reservoir for an arid region of Chile [36], for the great African lakes [37], and Lake Sobradinho, a large reservoir in Northeastern Brazil [14].

The relative humidity (RH) has seasonal pattern related to the air temperature, but with a small shift in the minimum RH value (57%) towards August and September. Moreover, during the rainy season, the humidity can reach up to 80% (Figure 3b).

Figure 4 shows the daily and seasonal profiles of air temperature and water temperature over the reservoir. The seasonal pattern of thermal stratification in the Serra da Mesa Reservoir has developed in spring and persisted until summer, with a mean upper amplitude of 5 °C between the surface and bottom in both seasons. During the mixing period (autumn and winter) a nearly homogeneity had been reached (Figure 4c).

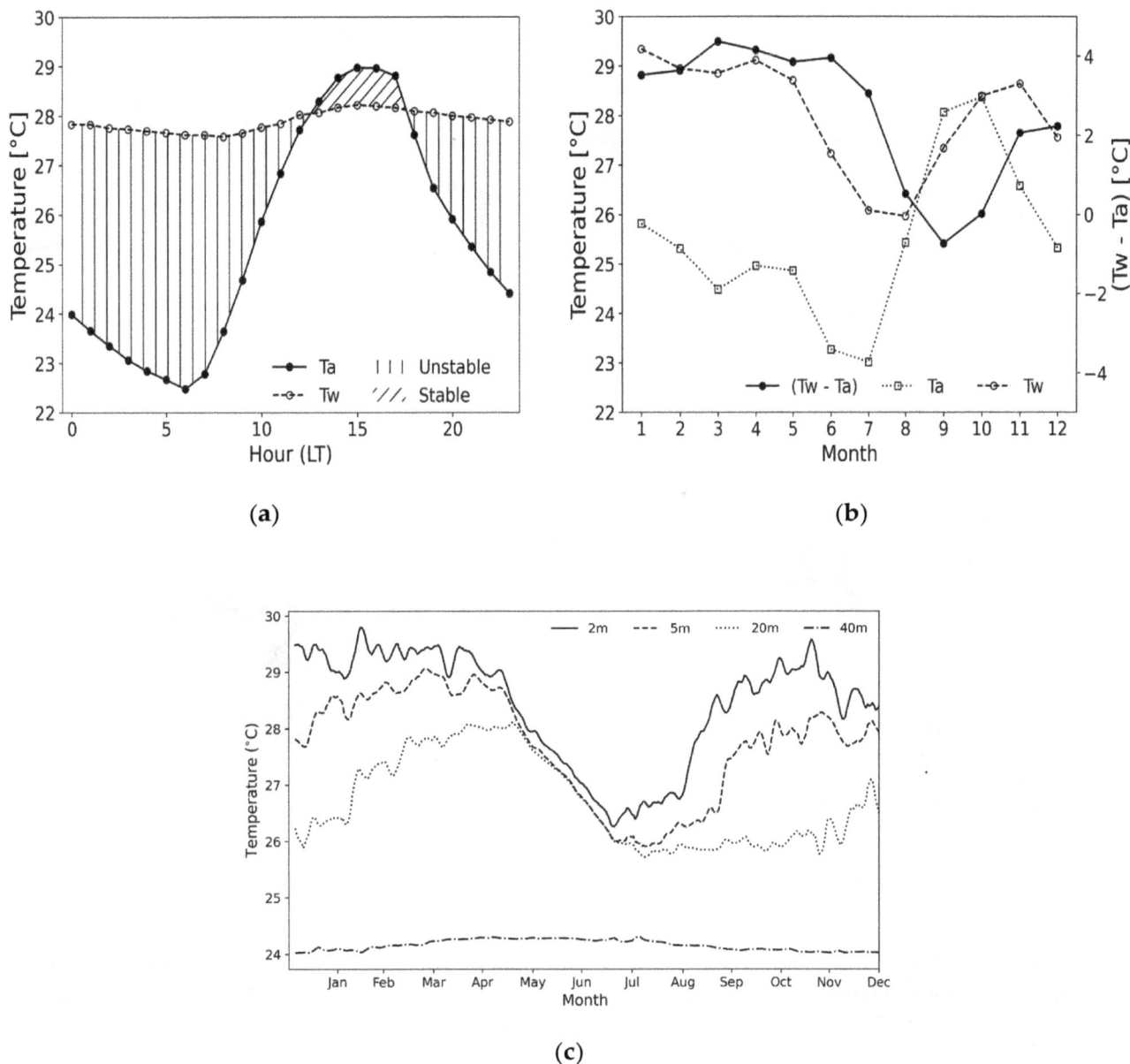

Figure 4. (a) Diurnal and (b) seasonal cycle for the air temperature (Ta) and water surface temperature (Tw) in Serra da Mesa reservoir. (c) Seasonal cycle of water temperature for various depths of the reservoir. Unstable and stable periods shown in the diurnal cycle: (a) refers to thermal stratification estimate of the bottom layer (2-m thick) of air over the water surface.

3.2. Identification of Lake Breeze from In Situ Data

The prevailing wind direction (Figure 5a) is from the Northeasterly (≈80°) in the early morning, and changes suddenly to southwesterly (≈250°) after 10 h local time (LT).

Figure 4c,d also shows the lake breeze starting around 10 h LT and wind direction remaining relatively constant from the Southwesterly (from the lake to surroundings) throughout the afternoon but veered to the Northeasterly (from surroundings to the lake) in the midafternoon (17 h LT). In general, the wind speed is low enough to allow a thermally driven secondary circulation to develop over the lake surface (Figure 5b). Such circulation is the lake breeze, as previously documented in the literature [15,29,38].

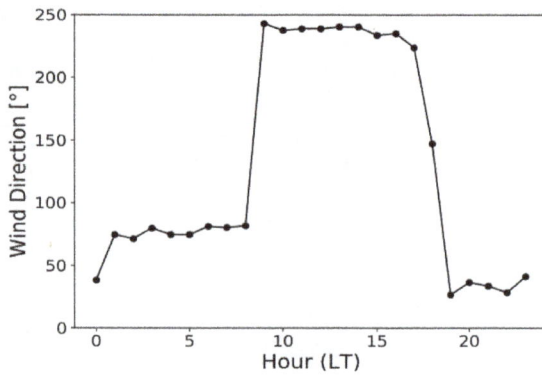

(a)

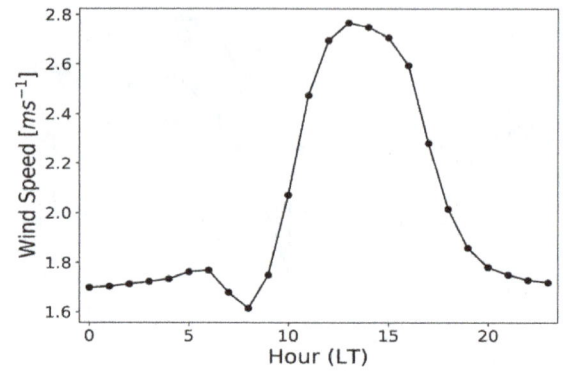

(b)

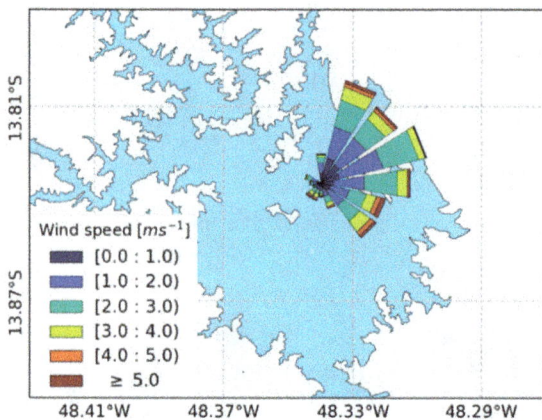

(c)

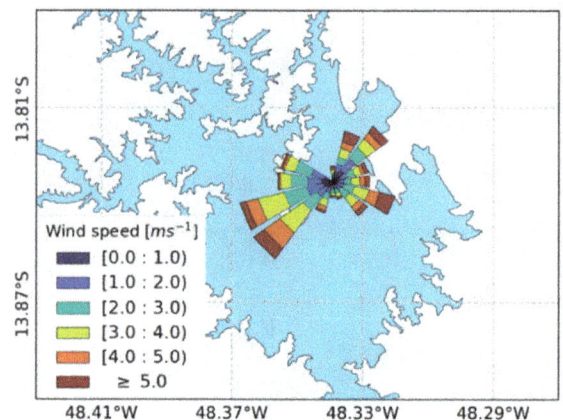

(d)

Figure 5. Diurnal cycle of mean hourly wind: (**a**) direction (mode) and (**b**) speed. Wind rose for data acquired (**c**) from 20 h till 09 h LT and (**d**) from 10 h till 17 h LT. Obtained from data acquired in the period from 2005 to 2010.

During the morning hours, the hourly mean wind speeds were 1–2 ms^{-1} and from the north–northeast direction. In the midafternoon, the wind direction shifted to a westerly simultaneously with the increase of wind speed to ~3 ms^{-1}, consistent with the development of a lake breeze circulation. As this feature is commonly observed around the entire perimeter of the inland aquatic system and for weak synoptic wind, it would be associated with a typical low-deformation lake-breeze circulation. This result describes an elementary difference between the dynamics of lake breezes and sea breezes [39]. As found by numerical simulation for gulfs and lakes elsewhere [40], circulations on each shoreline do not occur independently but interact to form a mesoscale high pressure on the surface with associated subsidence over the water. This result has been confirmed by previous studies [5,41,42] and probably defines the most relevant difference between the dynamics of lake breezes and the sea breezes.

According to Equation (4), the required reservoir width (*L*) to initiate a lake breeze depends on the third power of the wind speed, and it is inversely proportional to the difference in sensible heat flux between the lake surface and its surrounding area. Doran et al. [31] imposed a simple linear decrease of wind speed with distance to derive Equation (4). Figure 6a shows the typical diurnal cycle of the fluxes H observed during the dry season for the Brazilian Cerrado (wooded grassland, savanna) [43] and for a typical Brazilian tropical reservoir [44].

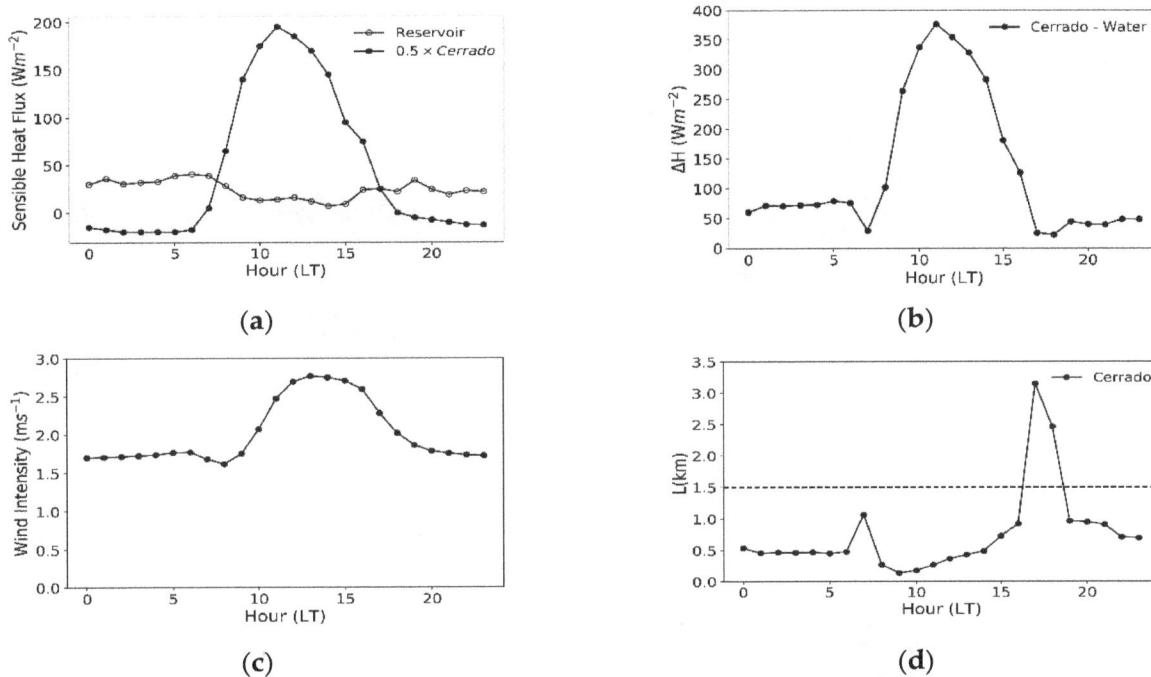

Figure 6. (a) Daily variability of sensible fluxes (H) as observed at a typical central Brazilian Cerrado area (solid line, from Miranda et al. [43]) and central Brazilian reservoir (dash-dotted, from Lorenzzetti et al. [44]) during the dry season; (b) difference between H in the reservoir and its surroundings (solid line—cerrado); (c) diurnal variability of ambient wind speed, and (d) minimum reservoir width, L, necessary to create a closed circulation. The horizontal line in panel (d) indicates the typical width of the Serra da Mesa reservoir.

The sensible heat flux averaged to 330 W/m^{-2} in the Cerrado area and only 25 W/m^{-2} over the reservoir (Figure 6a). As can be noticed, the sensible heat flux over the water remained low throughout the day (from 10 to 40 W/m^{-2}), while the maximum flux over the Cerrado exceeded 300 W/m^{-2}.

The difference between the heat fluxes over land relative to the water surface is presented in Figure 6b, while Figure 6c shows the daily cycle of the ambient wind speed (u_a) at 10 m height. For the assessment of u_a (Equation (4)), in order to differentiate the reservoir-modified wind, we considered the ambient wind speed as measured by the ridge-top site (A24 in Figure 1 and Table 2) (for details see Appendix A). The hourly values of both variables were used in Equation (4) to estimate the minimum lake width (L) required to initiate a closed atmospheric circulation. The horizontal line in Figure 6d represents the typical width of the Serra da Mesa Reservoir. There are conditions for lake breeze occurrence whenever the estimated values for L are below the horizontal line. The required lake width is larger than the typical width of the Serra da Mesa Reservoir only between 17 h and 19 h LT (Local Time).

The strength of the synoptic wind, together with its direction relative to the shoreline, also has a prevailing influence on the inland penetration of the lake breeze [42,45–48], though other factors such as orography, lake–land temperature gradient, and even soil moisture can be distinguished [49]. These factors, added to the fact of wind speed that was low enough (Figure 6c), allowed a thermally driven secondary circulation to begin to develop over the Serra da Mesa Reservoir. A weak to moderate onshore synoptic wind favors the formation of surface divergence areas over the water, while increasing wind will displace it downwind and finally inhibit breeze formation [39].

3.3. Spatial Analysis of Cloudiness on the Reservoir Area

Maps of C_{eff} were obtained from GOES-16 visible imagery based on Equation (1) with a 1-h time step for a region covering up to 100 km far from reservoir shores (hereinafter referred to as the domain region). Figure 7 shows the seasonally averaged cloud patterns over the water surface and the surrounding areas for morning and afternoon periods.

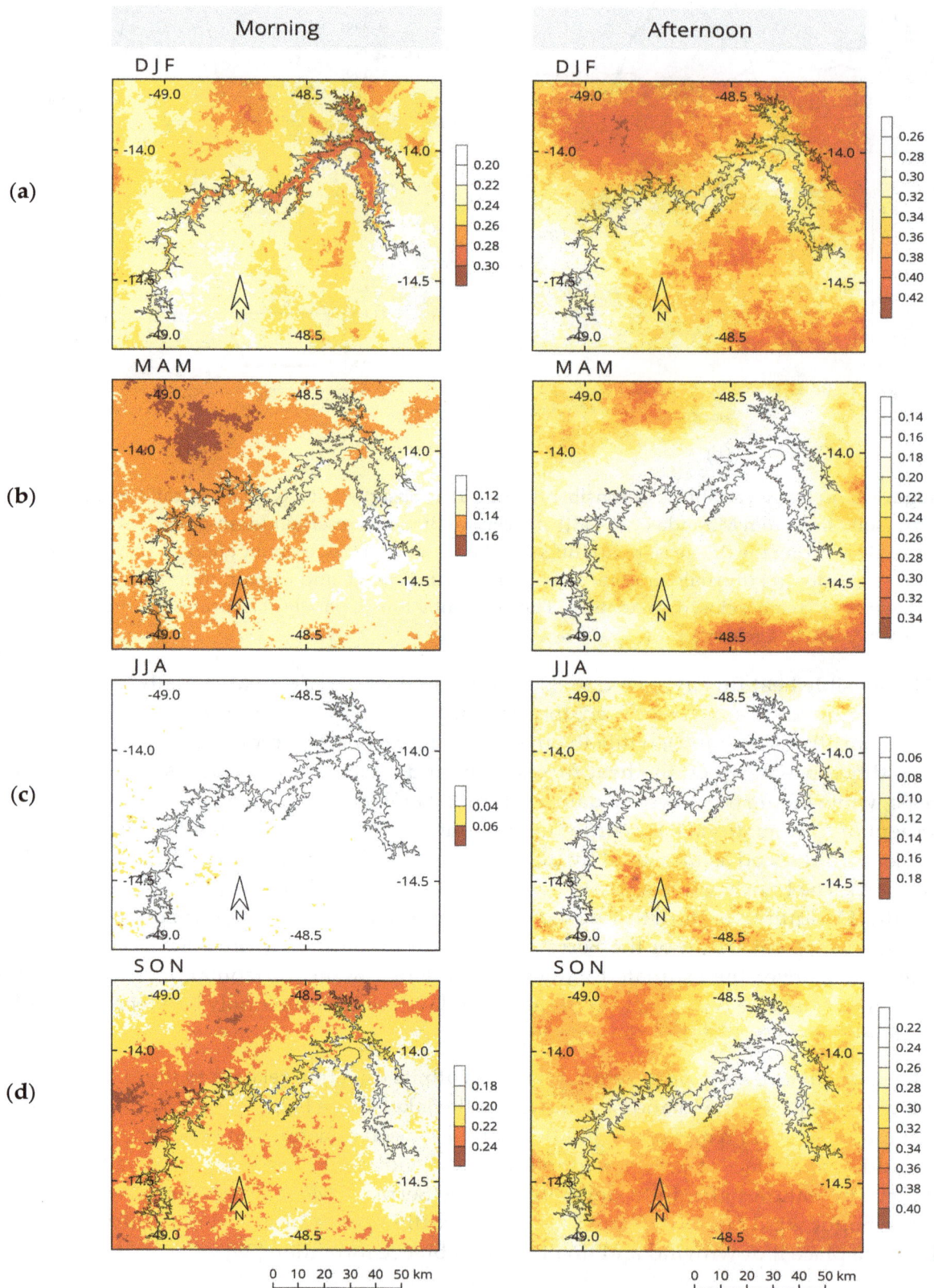

Figure 7. Seasonal maps of average C_{eff} estimates (%) over the Serra da Mesa Reservoir for the morning (9–11 h LT) and afternoon (14–16 h LT) periods for (**a**) December, January, February (DJF); (**b**) March, April, May (MAM); (**c**) June, July, August (JJA); and (**d**) September, October, November (SON).

Morning maps comprise images acquired from 09 h to 11 h LT, while afternoon maps comprise images from 14 h to 16 h LT. All maps retain the same 500 m horizontal resolution of the original satellite imagery. There is a negligible difference between cloudiness over the lake and surroundings (external) areas far from the lake borders during the morning period, apart from increased cloudiness over the lake during austral summer (DJF). This anomaly was caused by sun glint phenomenon, observed in visible satellite images acquired from January 1st to 26th and from December 5th to 31st of 2018, steadily at 10 h LT, therefore it is not related to cloudiness (for details on sun glint detection see Appendix B). No evidence of sun glint was detected for other periods, and geometrically it would be unlikely, although this possibility cannot be ruled out completely.

Compared to the surrounding dry land, the lower cloudiness over the lake is noticeable during the afternoons in the maps. The C_{eff} over the lake surface is ~0.04 (~12%) lower than over dry land during summer (DJF), and up to 0.08 (~30%) along with the fall season (MAM) and spring (SON). Nevertheless, some level of uncertainty in these values remains due to interannual variability. Remarkably, the maps show a slight southwestward drift in the cloudless signature due to prevailing Northeasterly winds in the region.

The visual observation indicated the need for statistical analysis to evaluate the lake breeze's influence on cloud cover. Thus, we took six target areas at different locations inside the domain region: the water surface, defined as the contiguous flooded area presenting at least 2 km width; the borders, outlined as the 2 km buffer from the shores; and four external (surrounding dry land) areas, distributed at cardinal regions (N, E, S, W) more than 2 km far from the reservoir shores as shown in Figure 8. These four external areas were chosen in locations presenting similar topography of the original (pre-flooded) reservoir area.

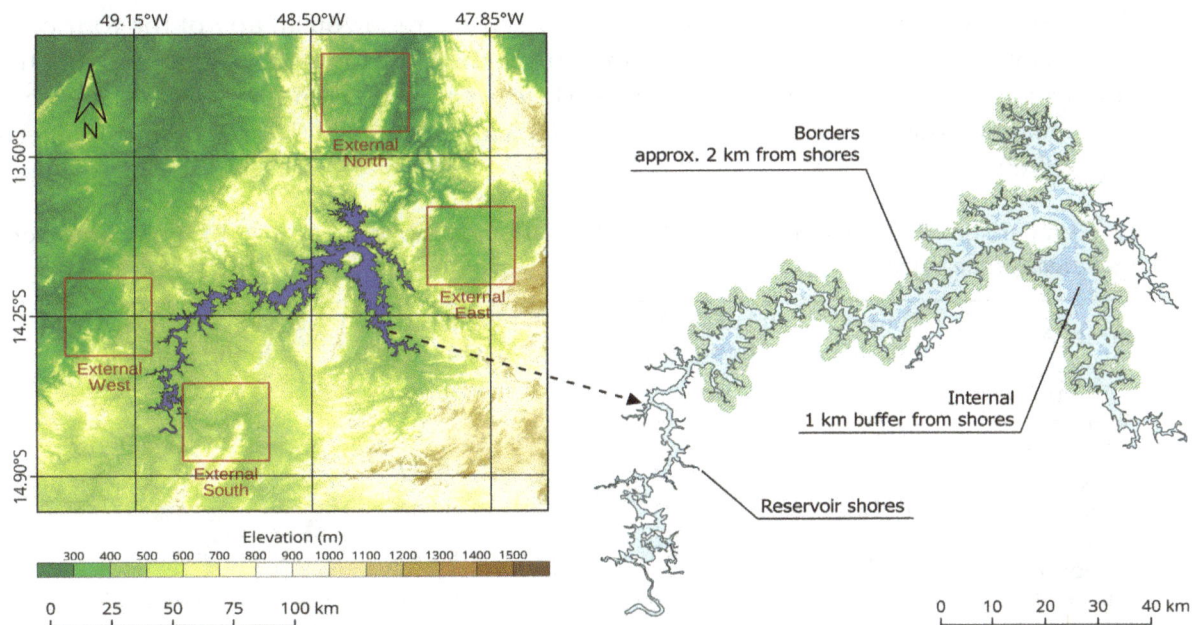

Figure 8. The topography of the region around Serra da Mesa reservoir and location of the six target areas used for cloudiness comparisons. Four external areas (**left**), the borders, and internal areas (**right**).

To acquire the representative hourly cloudiness observed by satellite in each target area, we calculated the average of hourly C_{eff} values for 50 pixels (0.25 km^2) randomly sampled in each of the six target areas (internal, borders, and external—N, E, S, W). Finally, the cloudiness in external samples was evaluated in two ways: (a) the four external areas were analyzed independently, and (b) the four external data were appended together to form one general external area, retaining the cloudiness variance observed in each target area.

Figure 9 shows the daily cycle of hourly average cloudiness for 2018.

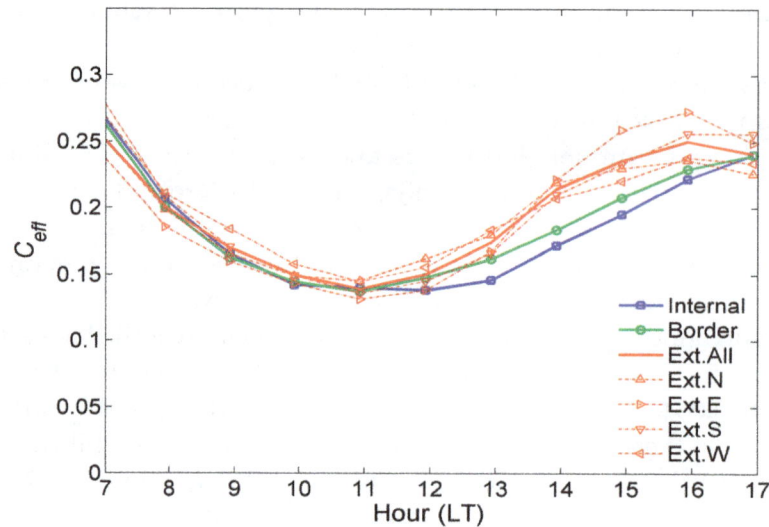

Figure 9. Hourly averaged cloudiness C_{eff} for the target areas at Serra da Mesa reservoir during the year of 2018.

The line graph endorses the lake breeze's recurrent characteristics, starting around 12 h LT and producing differences in C_{eff} values of up to 0.05 (25%) between the internal and external areas of the reservoir. The borders show an intermediate pattern, but closer to internal (water surface) than external area. Table 3 lists the detailed information, including the relative gain in cloudiness relative to the water surface. The results validate the findings from the visual observation C_{eff} from Figure 7. The cloudiness increases as we move away from the reservoir in any of the four cardinal directions. The same pattern is observed when looking at the afternoon-only cloudiness (14 h to 16 h LT). Comparing the cloudiness over the water surface area to the cloudiness over the combined external areas, one can notice a decrease of 5.7% and 19.0% in the average C_{eff} for daily and afternoon periods, respectively.

Table 3. Comparison of C_{eff} averages for daily (7–17 h LT) and afternoon (14–16 h LT) timeframes for all target areas of the Serra da Mesa Reservoir.

Target Areas	Daily (7–17 h)	Relative Difference to the Internal	Afternoon (14–16 h)	Relative Difference to the Internal
Internal	0.187	-	0.196	-
Border	0.188	0.5%	0.207	5.4%
External North	0.196	4.7%	0.229	16.4%
External East	0.197	5.0%	0.251	27.9%
External South	0.200	6.6%	0.233	18.7%
External West	0.199	6.4%	0.222	13.0%
External Combined	0.198	5.7%	0.234	19.0%

Figure 10 depicts the empirical cumulative distribution function (CDF) of C_{eff} values in both morning and afternoon periods. Figure 10a shows similar cumulative cloudiness frequencies in all target areas during the morning (09 h to 11 h LT). Figure 10b shows a detached pattern for the water area during the afternoon (14 h to 16 h LT), as expected from previous results. The conclusions are twofold: first, the reduced daily average cloudiness over the water surface is a consequence of the cloudless afternoons engendered by the lake breeze, and second, the lake breeze induced a local cloudiness anomaly, as all of the four external areas presented increased cloudiness, which would be very unlikely to occur by chance.

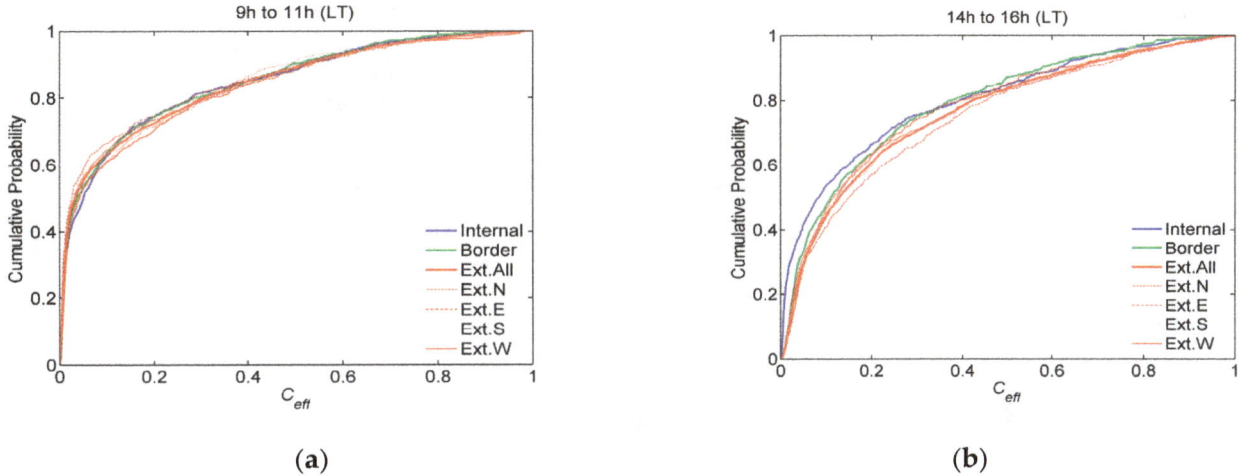

Figure 10. The cumulative distribution functions (CDF) charts for effective cloud cover C_{eff} levels for each target area during the (**a**) morning and (**b**) afternoon periods.

Table 4 summarizes the K–S test results comparing the cloudiness CDFs over the water surface area and dry land areas. The alternative hypothesis ($H1$) implies that the cloudiness CDF over external areas is lower than the CDF over the water surface, indicating that the cloudiness in the external area tends to be larger than the internal area. The p-values presented are the Type 1 error, understood as the probability of rejecting the null hypothesis considering it is true. Daily and afternoon p-values are extremely low, indicating high confidence that the distributions are different (alternative hypothesis $H1$ is true). On the other hand, there is a higher p-value of 1.81×10^{-2} for the morning, indicating increased uncertainty in assuming the null hypothesis is false.

Table 4. p-values for Kolmogorov–Smirnov (K–S) one-sided test comparing the empirical cumulative distribution functions of effective cloud cover (C_{eff}) in the wet (internal) and dry (external) areas of the reservoir. Null hypothesis, $H0 : \mu_{Internal} = \mu_{External}$, where μ denotes the mean value of the distribution from which samples were drawn.

Alternative $H1$	p-Value Daily	p-Value Morning	p-Value Afternoon
$\mu_{Internal} < \mu_{External}$	1.30×10^{-16}	1.81×10^{-2}	6.52×10^{-20}

Figure 11 shows absolute differences of cumulative cloudiness probability for internal area ($CDF_{internal}$) and external area ($CDF_{external}$) along C_{eff} values for daily, morning only, and afternoon only periods. The corresponding critical bounds, V_c, for a 99% confidence level are plotted over.

The plot shows CDF differences exceeding the critical bound for C_{eff} up to 0.20 for daily averages, and up to 0.40 for afternoon averages. It indicates a higher frequency of lower Ceff values in the internal area when compared to the external area. This result supports the hypothesis that the reservoir affects mostly shallow convection since deep convective clouds are both brighter ($C_{eff} >> 0.4$) and less susceptible to surface influence at the local scale. The K–S test results bring more evidence that cloudless skies frequency over the water surface is higher than outside the Serra da Mesa Reservoir.

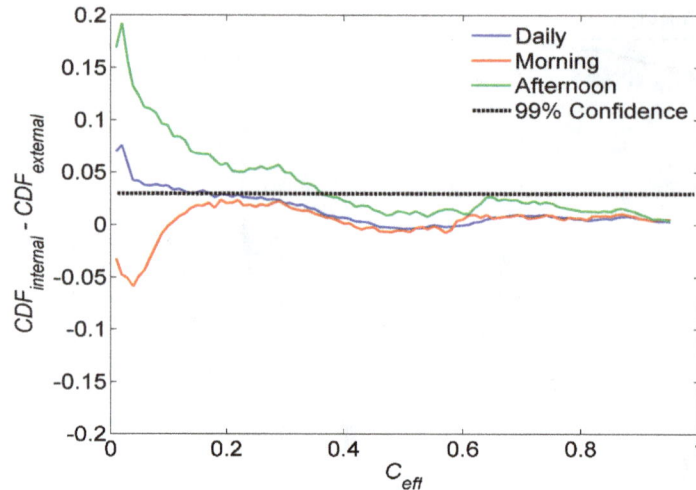

Figure 11. Absolute differences of cumulative cloudiness probability for internal area ($CDF_{internal}$) and external area ($CDF_{external}$) along C_{eff} values for Serra da Mesa hydropower plant. The horizontal line represents the one-sided test bound V_c at a 99% confidence level.

3.4. Time Analysis of Cloudiness over the Reservoir

So far, our results have shown that a spatial cloudiness pattern exists and that the reservoir probably induces it. Nevertheless, one could still argue that the internal and external areas of the reservoir are not similar enough and that other factors than the reservoir itself could impact the local cloudiness. Figure 12 presents a proxy to overcome this issue.

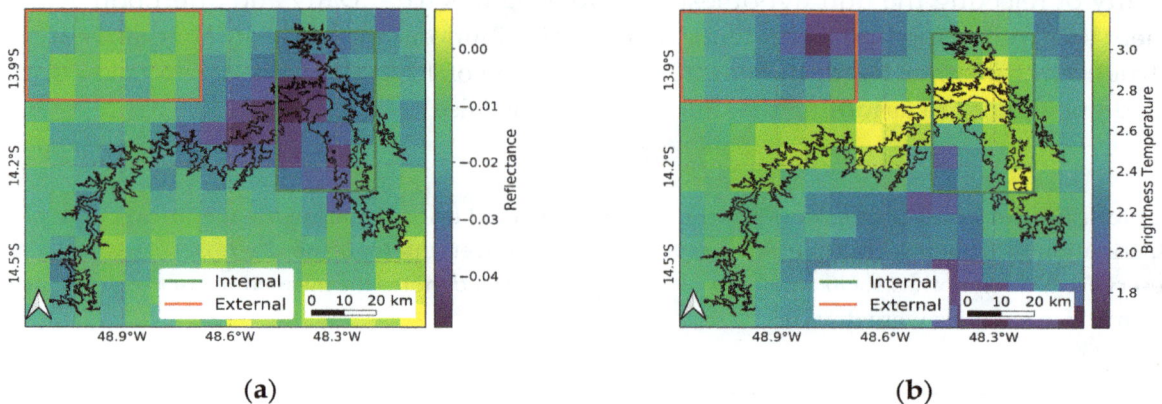

(a)

(b)

Figure 12. Mean differences (after–before) between periods after (1997–2009) and before (1984–1996) of the reservoir construction over Serra da Mesa Basin for (a) visible reflectance (0.6 μm) for 15 h LT and (b) daily infrared brightness temperature (11.0 μm) in K. Green and red rectangles show domains for internal and external cloudiness evaluation.

The long-term mean differences of temperature and reflectance after and before the reservoir construction is presented in Figure 11, respectively. The comparison analysis used local cloudiness data estimated from the GridSat-B1/ISCCP dataset in two distinct time frames, as described earlier in Section 2.2.

As a first approximation, both images suggest that the cloudiness over the external areas exceeds cloudiness over the water surface of the reservoir. Therefore, Figure 12 endorses our and other recent results [5], pointing out that the presence of the reservoir induces a lake breeze system that inhibits the cloud formation because of dry air subsidence from the upper atmospheric layers to near-surface levels above the water surface. However, Evan et al. [50] suggest that the difference in cloudiness frequency may be due to satellite viewing geometries affecting the cloudiness data acquisition on a long-term basis.

If the changes in the reflectance (R) (Figure 12a) and in the infrared brightness temperature (BT) (Figure 12b) were associated with satellite viewing geometry artifacts, it would affect the overall sampled area in a similar way. Both images in Figure 12 show that it is not the case, as the cloudiness reduction over the widest area of the water surface of the reservoir exceeds cloudiness reduction over the external areas. The differences in the external area present smaller magnitudes and may be attributed to surface heterogeneity and limitations in the sampling process due to interannual variability.

Figure 13a,b shows the histograms of the mean visible reflectance data over the water surface area in the same timeframes used earlier "before" and "after" the flooding of the hydropower reservoir.

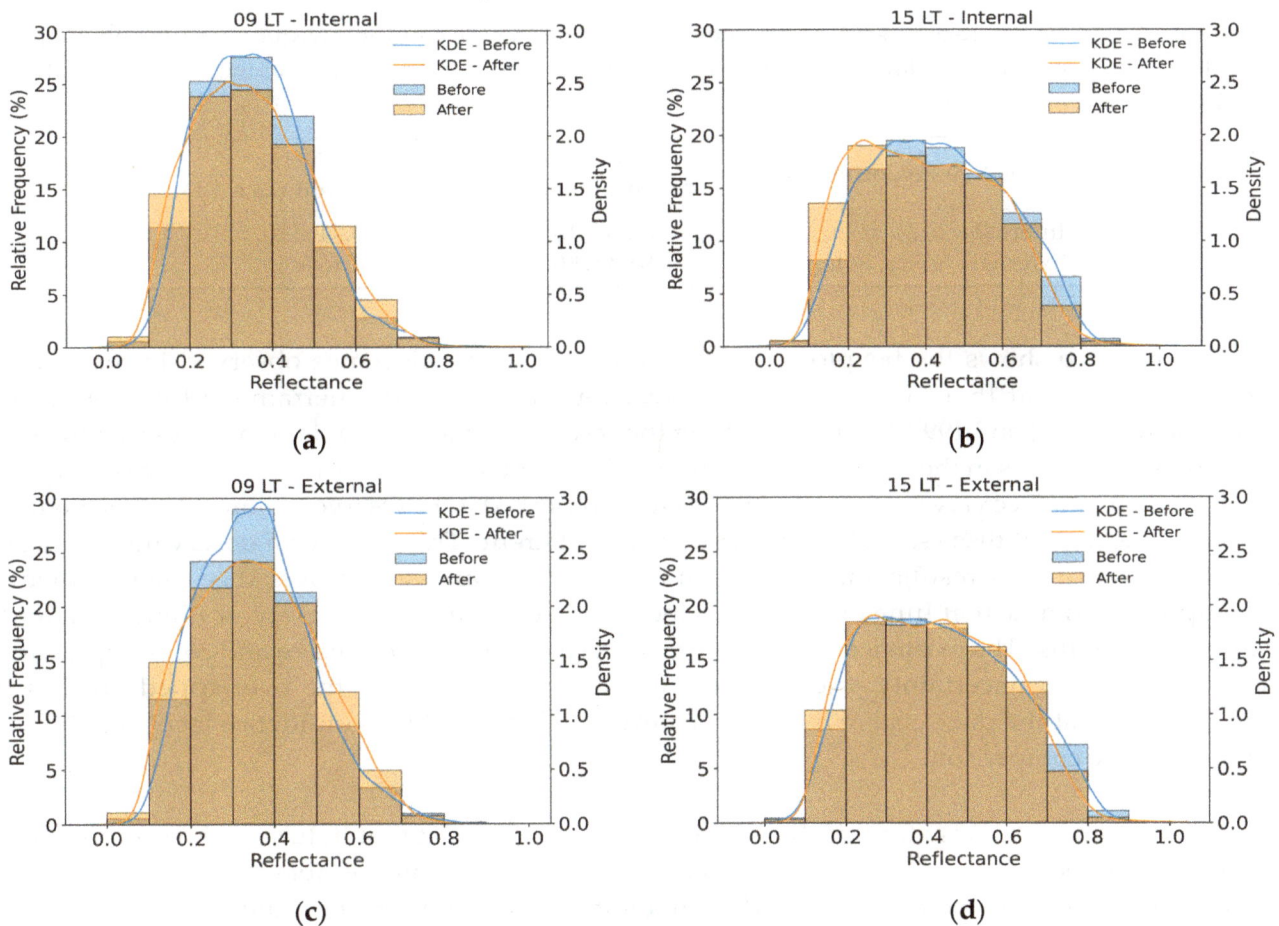

Figure 13. Relative frequency histogram (%) and kernel density estimates (KDE), right axis, of visible reflectance (R) over the water surface area based on GridSat-B1/ISCCPP data acquired before (blue boxes) and after (brown boxes) filling the Serra da Mesa Reservoir: (**a**) 09 h LT (morning period) in the internal area; (**b**) 15 h LT (afternoon period) in the internal area; (**c**) 09 h LT (morning period) in the external area, and (**d**) 15 h LT (afternoon period) in the external area.

In the same way, Figure 13c,d shows histograms of cloudiness frequency over the reservoir's external areas.

Figure 12a,c shows the histograms obtained for 9 h LT in both timeframes (after and before the flooding). Both present very similar profiles with the maximum frequency occurring in quite the same visible reflectance. Nevertheless, the histograms obtained for 1997–2009 (the after flooding period) presents more spread than the one obtained for 1984–1996 (before filling). The histograms for the afternoon period (15h LT), shown in Figures 13b and 12d, also have very similar profiles, except for a left shift on the maximum frequency after filling the reservoir. Considering the typical reflectance values for surface (0.08 to 0 0.15) and cloud top (0.5 to 0.9), this shift suggests that the reduction in visible reflectance is linked to a decrease in the cloudiness frequency. In the afternoon, no discernible change of cloudiness frequency occurs in the visible reflectance for the external areas surrounding the reservoir. The K–S test results endorse this conclusion comparing the empirical frequency distributions of reflectance raw data (Table 5). The null hypothesis at a 1% significance level is rejected only for the distributions observed in the water surface area during the afternoon (p-value $= 3.74 \times 10^{-3}$). Such results point out that the statistically significant change in cloudiness induced by the hydropower lake occurred only over the water surface area during the afternoon. Moreover, the influence of the satellite viewing geometry was discarded once the shift occurs only for part of the Serra da Mesa hydropower area.

Table 5. p-values for K–S two-sided hypothesis test comparing the empirical frequency distribution of cloud reflectance (R) raw data in the internal and surroundings (external) areas of the reservoir. Null hypothesis $H0 : \mu_{after} = \mu_{before}$.

Alternative Hypothesis $H1$	p-Value Morning (9 h LT)	p-Value Afternoon (15 h LT)
Internal : $\mu_{after} \neq \mu_{before}$	5.03×10^{-1}	3.74×10^{-3}
External : $\mu_{after} \neq \mu_{before}$	8.94×10^{-1}	4.16×10^{-1}

Figure 14a,b shows the boxplot of the monthly mean of reflectance observed in the internal and external areas at 15 h LT throughout the year for the same two timeframes 1984–1996 (before filling the reservoir) and 1997–2009 (after filling the reservoir). Figure 14a demonstrates that the most substantial differences in the median of the observed reflectance occur during the winter season from July to September. Decreased reflectance values were observed after reservoir filling. The mismatched notches indicate that these samples are significantly different. This long-term observation confirms the GOES-16 previous results and suggests a reduced cloudiness area over the water. However, it is important to note that June presented an increased reflectance after reservoir filling, differently from other months. The existence of fewer data samples (553 samples before and 395 samples after) may have affected uncertainty, widening the notches. In this sense, the overlapped notches for June indicate that the datasets are not significantly different at a 95% confidence level, leading to a non-conclusive comparison.

Another possible cause for this distinct result relies on the fact that in June, the reservoir surface temperature exceeds surface air temperature and tends to produce an unstable atmospheric boundary layer (ABL), as anticipated by Garrat [51]. Water surface temperatures (T_w) are, on average, higher than air temperatures (T_a) (Figure 3a,b). The ABL over the Serra da Mesa Reservoir remains unstable ($T_w > T_a$) around 3/4 of the time on a monthly scale (Figure 4b). However, the stability of the ABL varied at both diurnal (Figure 4a) and seasonal timescales (Figure 4b). Unstable ABL conditions were prevalent in summer and autumn, but June is the month marked by the higher gradients between the reservoir surface and atmosphere (Figure 4b). The maximum decrease for the water column temperature until 20 m depth occurred for June (Figure 4c), resulting in a consistent and persistent gradient between the reservoir surface and atmosphere. The mean values of T_w and T_a presented in Figure 4 are based on 3-year data acquired in the Serra da Mesa reservoir. More investigation is required to clarify the reason for this behavior in June. Nevertheless, it happened consistently during the data acquisition period. The lower wind speeds for June (Figure 3a) resulted in less mechanical mixing at the lower ABL and, consequently, an enhanced unstable ABL. Possibly, the cloudiness may be increased due to enhanced evaporation from the water surface after reservoir flooding. Meanwhile, Figure 14b shows slighter differences in reflectances between the two periods most of the year for the external area.

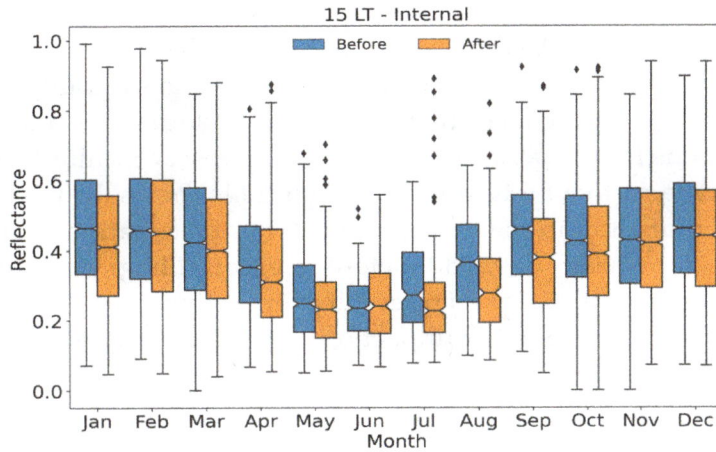

(a)

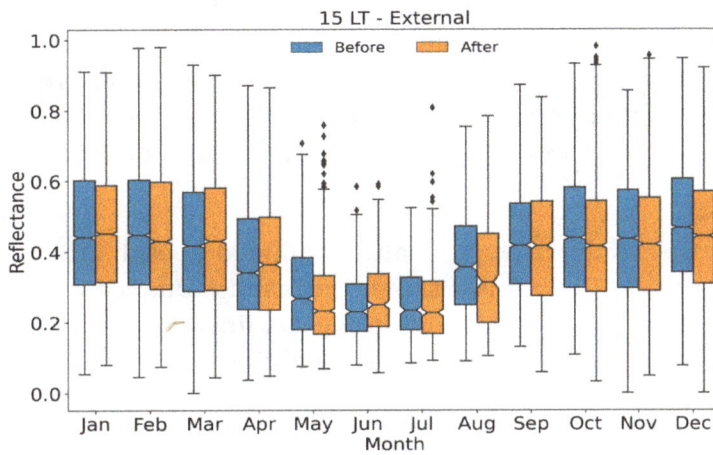

(b)

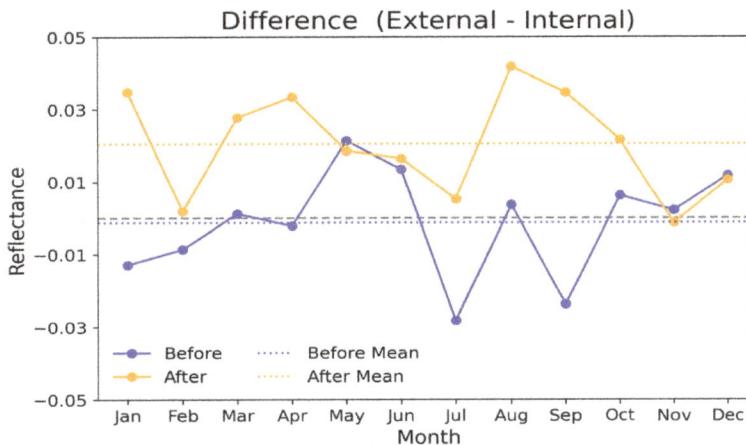

(c)

Figure 14. Boxplot for the monthly visible reflectance observed in the internal (**a**) and external (**b**) reservoir area at 15h LT before (blue boxes) and after (orange boxes) construction of the Serra da Mesa hydropower plant. Panel (**c**) shows the difference between external and internal areas for the period before and after lake filling. The box represents the interquartile range (IQR) (25–75%), and the whiskers are the +/− 1.5 × IQR bounds of data values. The notch (horizontal line) indicates the median (95% confidence). Overlapped notches mean the datasets are not significantly different at 95% confidence level [52].

An irregular behavior is noted in the dry months (May to August) possibly affected by fewer data samples as mentioned before. A small reduction is also noted at the end of the year (October to December). This result reveals the uncertainty inherent to the sampling method. Factors such as interannual variability and land cover change may affect cloudiness unevenly along the years, so that identical distributions for the periods before and after reservoir filling are unlikely to occur at monthly scales. To overcome partly of this limitation, Figure 14c shows the relative difference between the reflectance from external and internal areas for each period mentioned above. The reflectance differences oscillate near zero (mean = −0.0013) before reservoir filling, while the differences oscillate around positive values (mean = 0.0204) after reservoir filling. This result supports previous findings that suggest the existence of a considerable impact on cloudiness frequency caused by the Serra da Mesa Reservoir.

3.5. The Lake Breeze Influence on Surface Incoming Solar Irradiation

An estimate of the typical k_{0t_clear} value for the Serra da Mesa reservoir was obtained by fitting Equation (7) to the global solar irradiance data acquired in three ground measurement stations near the hydropower dam. Cloudy samples were manually screened and data were converted to observed clearness index (k_{t_clear}). The fitting procedure was performed for 5-degree intervals of the solar zenith angle in order to reduce uncertainties in k_{0t_clear} estimates. Ground data acquired at a solar elevation lower than 20° were discarded.

Table 6 summarizes the results for k_{t_clear} values obtained in each ground measurement site for two solar zenith angle intervals. The last line shows the results obtained by fitting the model for data from the three sites altogether. The k_{t_clear} listed in the last line was assumed as the regional values. The difference between the regional k_{t_clear} and the site-specific k_{t_clear} ranges from 1.3% for low zenith angles to ~3.5% for high zenith angles.

Table 6. List of ground measurement sites near the Serra da Mesa Reservoir used to estimate the cloudless atmospheric transmittances k_{t_clear} values. The simple model, described in Equation (7), provided the k_{t_clear} values for 5-degree intervals of solar zenith angles. The table presents the results for two of the intervals. The last line exhibits the results obtained by using data from the three measurement sites altogether to estimate regional k_{t_clear} values.

Met. Station	k_{t_Clear} ($\theta z < 5°$)	k_{t_Clear} ($55° < \theta z < 60°$)
BRB	0.760	0.710
A22	0.750	0.670
A24	0.770	0.680
Regional k_{t_clear}	**0.759**	**0.685**

The clearness index k_t was calculated using Equation (6) for C_{eff} values provided by GOES-16 as a function of day and time and then averaged over the year for three areas: internal (water surface), reservoir borders, and external areas. The calculation of GHI was performed by multiplying hourly clearness index (k_t) by the irradiance at the top of the atmosphere according to Equation (5). Table provides information on the enhanced incoming solar irradiance at the water surface of the Serra da Mesa Reservoir compared to the surrounding external areas as reference. Table results demonstrate that reduced local cloudiness increases the annual average of incoming solar irradiance around 1.73% from 7 h to 17 h LT. The solar irradiance enhancement reaches up to 4.51% for the afternoon (from 14 h to 16 h LT).

Table 7. Comparison of annual mean effective cloud cover index, C_{eff}, and respective estimates of mean surface incoming solar irradiance (GHI_{est}) at the water surface and external areas of the Serra da Mesa Hydropower Reservoir. The incoming global solar irradiance in external areas was the reference value to calculate the irradiance enhancement. Morning differences were not significant (not shown).

Region	Daily (7–17 h)			Afternoon (14–16 h)		
	C_{eff} avg.	GHI_{est} avg.	GHI_{est} enhanc.	C_{eff} avg.	GHI_{est} avg.	GHI_{est} enhanc.
Internal	0.187	523.8	1.73%	0.196	549.0	4.51%
Border	0.188	521.6	1.30%	0.207	542.1	3.20%
External Combined	0.198	514.9	-	0.234	525.3	-

4. Conclusions

This work evaluated the formation of the lake breeze in the Serra da Mesa Reservoir, located in the Central region of Brazil, and its impacts on the regional cloudiness climate. The study evaluated the spatial and time distributions of cloudiness based on in situ measurements and two satellite-derived datasets.

In situ data covering a 5-year period taken from an offshore buoy system confirmed a prevailing breeze mechanism in the reservoir that superimposes large-scale atmospheric flow and generates a modified regional wind climatology. The difference between the typical heat fluxes in the internal (water surface) and external areas combined with locally calm winds explains the development of the lake breeze most of the day.

High-resolution satellite imagery was used for assessment and seasonal analysis of the spatial distribution of cloudiness, which identified the lake contour signature on afternoon cloud fields all year long, suggesting a persistent pattern of lake breeze. Moreover, the spatial analysis allowed identifying a statistically significant reduction of ~5.7% in the effective cloud cover index (C_{eff}) over the water surface area compared to the surrounding areas of the reservoir. A more detailed hourly basis analysis showed that the largest cloudiness differences occur from 12 h to 16 h (Local Time).

In addition, two sets of 15 years of satellite images, one before and the other after the reservoir filling, allowed the assessment of the gross impact of the hydropower plant construction on the regional cloud regime. A change in cloudiness frequency distribution was detected after lake filling. Such a difference in cloudiness frequency between before and after was statistically significant, evidencing reservoir impact on regional cloud regime over the flooded area. On the other hand, the external areas surrounding the formed lake did not reveal a significant change in the cloudiness pattern, dismissing any data trend.

In summary, the study strongly suggests that induced lake breeze circulation enhances cloudless skies over the flooded area during daytime at Serra da Mesa. This conclusion was supported by remote and in situ measurements, but some limitations should be mentioned: This is a study case for a specific reservoir and, despite their potential for extrapolation, these conclusions should not be assumed for other tropical reservoirs. Further assessments are needed to evaluate the extent of the phenomenon for other locations. Another limitation concerns the quantitative analysis presented herein. Uncertainty related to the cloudiness sampling method and interannual variability may affect the comparisons, requiring a longer-term analysis to provide climatologically consistent measures.

Furthermore, a preliminary assessment of the incoming solar irradiance at the lake surface indicated an increment of 1.73% on the daily average and up to 4.51% increase for the afternoon timeframe. These are substantial values from the perspective of solar energy resource assessment. The enhancement in horizontal surface solar irradiance over the hydroelectric lake compared to the surrounding areas corresponds to a relative increment in the annual photovoltaic yield from the reported average ranging from 1622 kWh/KWp [53] up to 1696 kWh/kWp. Furthermore, as floating photovoltaic (FPV) plants typically perform better at lower temperatures, an additional gain ranging from 10% to 15% can be expected according to Rosa-Clot and Tina [54], leading to a net photovoltaic yield for the water surface of up to 1949 kWh/Kwp. Our research team is already working on the next step, i.e., evaluating the solar irradiance enhancement and FPV yield in several hydropower plants operating in different climate regions of the Brazilian territory to provide a solar energy assessment that supports the FPV technology deployment. In the near future, solar-hydro hybrid plants can become a great alternative to integrate both energy resources due to the breeze mechanism in several tropical hydropower dams combined with the high solar irradiance in the Brazilian territory.

Author Contributions: Conceptualization, A.R.G., A.T.A., F.R.M., R.S.C., M.S.G.C. and E.B.P.; methodology, R.S.C., M.S.G.C., E.V.M. and S.V.P.; software, S.V.P. and R.B.P.; formal analysis, A.R.G., A.T.A., M.P.P., F.J.L.L. and R.B.P.; investigation, A.R.G., A.T.A. and E.V.M; resources, M.P.P. and F.J.L.L.; data curation, M.P.P., F.J.L.L. and R.B.P.; writing—original draft preparation, E.B.P., A.R.G., F.R.M. and A.T.A.; writing—review and editing, all authors; visualization, S.V.P. and R.B.P.; supervision, F.R.M.; project administration, E.B.P.; funding acquisition, E.B.P. All authors have read and agreed to the published version of the manuscript.

Acknowledgments: The authors thank NOAA for providing GOES-16 satellite imagery and hosting the International Satellite Cloud Climatology Project (ISCCP) website (https://www.ncdc.noaa.gov/isccp) maintained by the ISCCP research group which provided GridSat B1 product. Thanks also to National Meteorology Institute (INMET) and to SONDA project for providing quality surface data. Finally, we acknowledge the National Institute for Space Research (INPE) for supporting the research team and the essential contribution of the Earth System Science Postgraduate Program (PG-CST) to this research.

Appendix A

The assessment of lake breeze occurrence according to Equation (4) requires an estimate of ambient wind speed (u_a), defined as the component unaffected by lake breeze. In order to differentiate the reservoir-modified wind, we considered the ambient wind speed as measured by the ridge-top site (A24 in Figure 1).

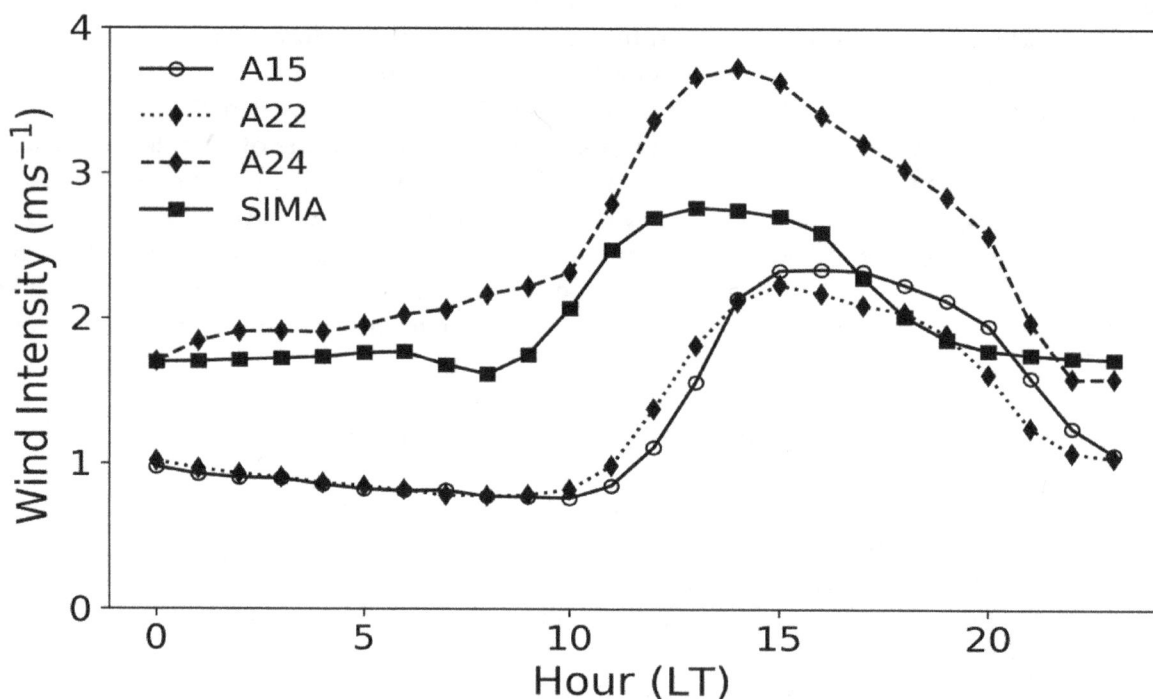

Figure A1. Daily cycle of mean hourly wind speed at three sites nearby reservoir (A15, A22, and A24) and inside the reservoir (SIMA) calculated over the whole period shown in Table 2.

Figure A1 shows the mean daily cycle of wind speed at four sites: SIMA (inside the reservoir), A15, A22, and A24 (surroundings the reservoir). All the sites show a diurnal cycle with a maximum in the early afternoon. The main aim was to examine how wind intensity on reservoir valley and surroundings (SIMA, A15 e A22) compared with the prevailing wind direction as measured by the ridge-top site (A24) (Figure 1 and Table 2). Figure A1shows that the ambient winds (A24) are more intense than the wind-terrain system (SIMA, A15 e A22), as expected.

Appendix B

The sun glint phenomenon affects satellite visible reflectance measurements due to sunlight specular reflection over water surfaces. It is geometrically dependent and thus presents a recurrent characteristic for geostationary satellites lasting several days at specific hours due to slow change on Earth axis declination. The abnormally high reflectance observed over the water in the austral summer maps shown in Figure 14 caught attention on the possible occurrence of sun glint over the Serra da Mesa reservoir. For its detection, a comparative analysis of surface reflectance for consecutive hours was performed, as shown in Figure A2. The detached pattern of the 10 h LT curve is observed from 5 December to 29 January. The differences from cloudiness prior (9 h) and after (11 h) suggest that clouds did not cause the brightness observed at 10 h.

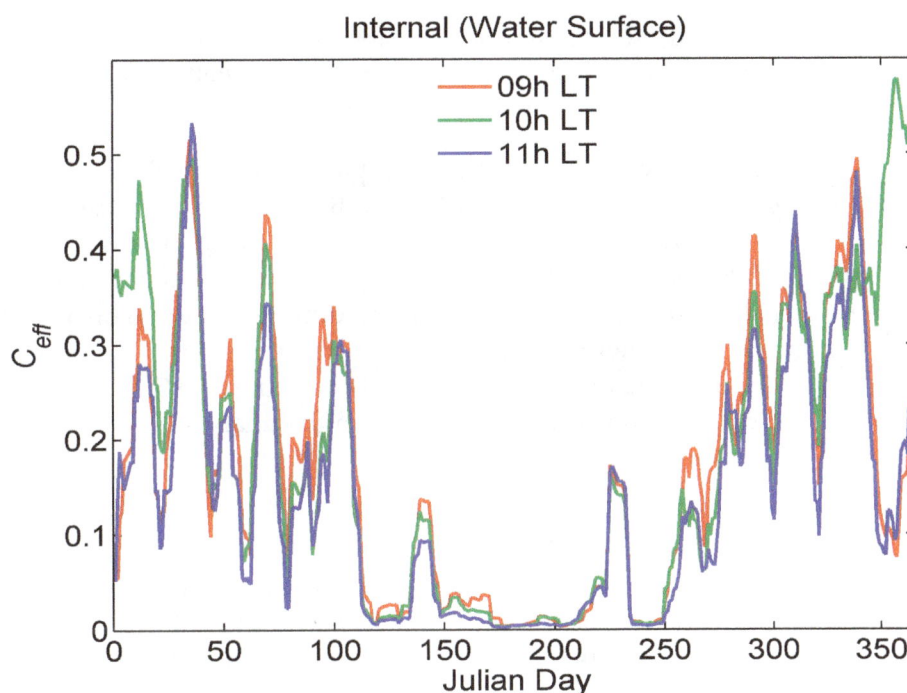

Figure A2. Annual cycle of hourly effective cloudiness index (C_{eff}) for 9 h, 10 h and 11 h LT for the Serra da Mesa reservoir internal area.

Despite the empirical evidence, an analytical evaluation of sun–satellite geometry was also performed to confirm the sun glint occurrence. It is known from basic optics that specular reflection implies identical incidence and reflected zenith angles along the plane that contains the incident vector and surface normal vector. Incidence vector is given by solar geometry while the reflected is given by satellite view geometry, which is fixed for a geostationary satellite for any point in Earth's surface. For a combined assessment of these geometries, we plotted the analemma for 9 h, 10 h, and 11 h LT for the SIMA buoy coordinates (Table 2) at Serra da Mesa reservoir in Figure A3. Satellite viewing geometry (zenith and azimuth angles) was extracted from GOES-16 raw navigation files and plotted over. The analemma curve shows the sun position (elevation and azimuth) at the same hour throughout the year.

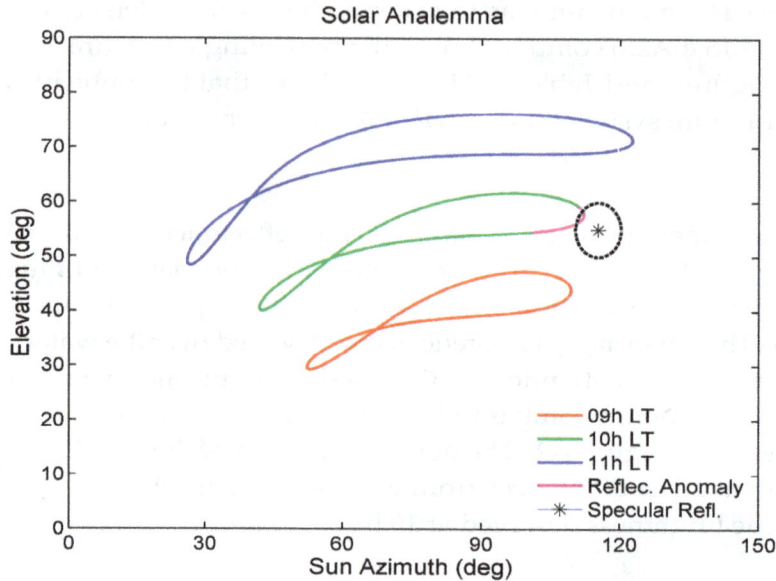

Figure A3. Analemma for 9 h (red), 10 h (green), and 11 h (blue) LT at Serra da Mesa reservoir (SIMA buoy coordinates) and corresponding specular reflection opposed to satellite viewing angles (black). Dash-dotted black line indicates a 5° margin from the specular position.

From Figure A3, it is possible to confirm that the reflectance anomaly observed during December and January at 10 h LT was caused by sun glint. Solar position cannot be treated as a point source. A margin of 5° from specular reflection position was included to account for angular variations due to sun disc solid angle (0.53°), lake extent, and mostly due to wavy water surface. This margin seems fairly conservative since any slope on the water facet causes a double deviation on a specular angle due to reflection symmetry. For oceans, where the water is far wavier, the sun glint is of greater concern and there is a well-established literature on its detection and correction [55,56]. For the work sequence, the 10 h LT samples affected by sun glint were excluded from the analysis.

References

1.   Biggs, W.G.; Graves, M.E. A Lake Breeze Index. *J. Appl. Meteorol.* **1962**, *1*, 474–480. [CrossRef]
2.   Rabin, R.M.; Stadler, S.; Wetzel, P.J.; Stensrud, D.J.; Gregory, M. Observed effects of landscape variability on convective clouds. *Bull. Am. Meteorol. Soc.* **1990**, *71*, 272–280. [CrossRef]
3.   Segal, M.; Arritt, R.W.; Shen, J.; Anderson, C.; Leuthold, M. On the clearing of cumulus clouds downwind from lakes. *Mon. Weather Rev.* **1997**, *125*, 639–646. [CrossRef]
4.   Asefi-Najafabady, S.; Knupp, K.; Mecikalski, J.R.; Welch, R.M.; Phillips, D. Ground-based measurements and dual-Doppler analysis of 3-D wind fields and atmospheric circulations induced by a meso-γ-scale inland lake. *J. Geophys. Res. Atmos.* **2010**, *115*. [CrossRef]
5.   Iakunin, M.; Salgado, R.; Potes, M. Breeze effects at a large artificial lake: Summer case study. *Hydrol. Earth Syst. Sci.* **2018**, *22*, 5191–5210. [CrossRef]
6.   Crosman, E.T.; Horel, J.D. Sea and Lake Breezes: A Review of Numerical Studies. *Bound. Layer Meteorol.* **2010**, *137*, 1–29. [CrossRef]
7.   Maeda, E.E.; Ma, X.; Wagner, F.H.; Kim, H.; Oki, T.; Eamus, D.; Huete, A. Evapotranspiration seasonality across the Amazon Basin. *Earth Syst. Dyn.* **2017**, *8*, 439–454. [CrossRef]
8.   Silva Dias, M.A.F.; Silva Dias, P.L.; Longo, M.; Fitzjarrald, D.R.; Denning, A.S. River breeze circulation in eastern Amazonia: Observations and modelling results. *Theor. Appl. Climatol.* **2004**, *78*, 111–121. [CrossRef]
9.   Yin, X.; Nicholson, S.E.; Ba, M.B. On the diurnal cycle of cloudiness over Lake Victoria and its influence on evaporation from the lake. *Hydrol. Sci. J.* **2000**, *45*, 407–424. [CrossRef]
10. Stivari, S.M.S.; De Oliveira, A.P.; Karam, H.A.; Soares, J. Patterns of local circulation in the Itaipu Lake Area: Numerical simulations of lake breeze. *J. Appl. Meteorol.* **2003**, *42*, 37–50. [CrossRef]

11. Moura, M.A.L.; Meixner, F.X.; Trebs, I.; Lyra, R.F.D.F.; Andreae, M.O.; Nascimento Filho, M.F.D. Do Evidência observacional das brisas do lago de Balbina (Amazonas) e seus efeitos sobre a concentração do ozônio. *Acta Amaz.* **2004**, *34*, 605–611. [CrossRef]

12. Ackerman, S.A.; Heidinger, A.; Foster, M.J.; Maddux, B. Satellite Regional Cloud Climatology over the Great Lakes. *Remote Sens.* **2013**, *5*, 6223–6240. [CrossRef]

13. Assireu, A.T.; Pimenta, F.M.; Freitas, R.M. De Observações e modelagem da camada limite interna no entorno de extensos sistemas aquáticos. *Ciência Nat.* **2016**, *38*, 305. [CrossRef]

14. Ekhtiari, N.; Grossman-Clarke, S.; Koch, H.; de Souza, W.M.; Donner, R.V.; Volkholz, J. Effects of the Lake Sobradinho reservoir (Northeastern Brazil) on the regional climate. *Climate* **2017**, *5*, 50. [CrossRef]

15. Stivari, S.M.S.; De Oliveira, A.P.; Soares, J. On the climate impact of the local circulation in the Itaipu Lake area. *Clim. Chang.* **2005**, *72*, 103–121. [CrossRef]

16. Peel, M.C.; Finlayson, B.L.; McMahon, T.A. Updated world map of the Köppen-Geiger climate classification. *Hydrol. Earth Syst. Sci.* **2007**, *11*, 1633–1644. [CrossRef]

17. Diniz, F.D.A.; Ramos, A.M.; Rebello, E.R.G. Brazilian climate normals for 1981–2010. *Pesqui. Agropecuária Bras.* **2018**, *53*, 131–143. [CrossRef]

18. Vera, C.; Higgins, W.; Amador, J.; Ambrizzi, T.; Garreaud, R.; Gochis, D.; Gutzler, D.; Lettenmaier, D.; Marengo, J.; Mechoso, C.R.; et al. Toward a unified view of the American monsoon systems. *J. Clim.* **2006**, *19*, 4977–5000. [CrossRef]

19. Stech, J.L.; Lima, I.B.T.; Novo, E.M.L.M.; Silva, C.M.; Assireu, A.T.; Lorenzzetti, J.A.; Carvalho, J.C.; Barbosa, C.C.; Rosa, R.R. Telemetric monitoring system for meteorological and limnological data acquisition. *SIL Proc. 1922–2010* **2006**, *29*, 1747–1750. [CrossRef]

20. OSCAR—Observing Systems Capability Analysis and Review Tool. Available online: https://www.wmo-sat. info/oscar/satellites (accessed on 10 January 2020).

21. Moser, W.; Raschke, E. Incident Solar Radiation over Europe estimated from METEOSAT data. *J. Clim. Appl. Meteorol.* **1984**, *26*, 166–170. [CrossRef]

22. Stuhlmann, R.; Rieland, M.; Raschke, E. An improvement of the IGMK model to derive total and diffuse solar radiation at the surface from satellite data. *J. Appl. Meteorol.* **1990**, *29*, 586–603. [CrossRef]

23. Knapp, K.R. Scientific data stewardship of international satellite cloud climatology project B1 global geostationary observations. *J. Appl. Remote Sens.* **2008**, *2*, 23548. [CrossRef]

24. Rossow, W.B.; Walker, A.W.; Beushebel, D.; Roiter, M. *International Satellite Cloud Climatology Project (ISCCP): Description of New Cloud Datasets*; WMO/TD737; World Climate Research Programme (ICSU and WMO): Geneva, Switzerland, 1996; Volume 115.

25. Kidder, S.Q.; Haar, T.H.V. Clouds and aerosols. In *Satellite Meteorology*; Gulf Professional Publishing: Houston, TX, USA, 1995; pp. 259–305. [CrossRef]

26. Rossow, W.B.; Garder, L.C. Cloud detection using satellite measurements of infrared and visible radiances for ISCCP. *J. Clim.* **1993**, *6*, 2341–2369. [CrossRef]

27. Bottino, M.J.; Ceballos, J.C. Daytime cloud classification over South American region using multispectral GOES-8 imagery. *Int. J. Remote Sens.* **2015**, *36*, 1–19. [CrossRef]

28. Espinar, B.; Ramírez, L.; Drews, A.; Beyer, H.G.; Zarzalejo, L.F.; Polo, J.; Martín, L. Analysis of different comparison parameters applied to solar radiation data from satellite and German radiometric stations. *Sol. Energy* **2009**, *83*, 118–125. [CrossRef]

29. Samuelsson, P.; Tjernström, M. Mesoscale flow modification induced by land-lake surface temperature and roughness differences. *J. Geophys. Res. Atmos.* **2001**, *106*, 12419–12435. [CrossRef]

30. Segal, M.; Arritt, R.W. Nonclassical mesoscale circulations caused by surface sensible heat- flux gradients. *Bull. Am. Meteorol. Soc.* **1992**, *73*, 1593–1604. [CrossRef]

31. Doran, J.C.; Shaw, W.J.; Hubbe, J.M. Boundary layer characteristics over areas of inhomogeneous surface fluxes. *J. Appl. Meteorol.* **1995**, *34*, 559–571. [CrossRef]

32. Beyer, H.G.; Pereira, E.B.; Martins, F.R.; Abreu, S.L. Assessing satellite derived irradiance information for South America within the UNEP resource assessment project SWERA. In Proceedings of the 5th EuroSun, Freiburg, Germany, 20–23 June 2004; pp. 3–771.

33. Martins, F.R.; Pereira, E.B.; Silva, S.A.B.; Abreu, S.L.; Colle, S. Solar energy scenarios in Brazil, Part one: Resource assessment. *Energy Policy* **2008**, *36*, 2853–2864. [CrossRef]

34.   Cano, D.; Monget, J.M.; Albuisson, M.; Guillard, H.; Regas, N.; Wald, L. A method for the determination of the global solar radiation from meteorological satellite data. *Sol. Energy* **1986**, *37*, 31–39. [CrossRef]

35.   Dagestad, K.F.; Olseth, J.A. A modified algorithm for calculating the cloud index. *Sol. Energy* **2007**, *81*, 280–289. [CrossRef]

36.   Bischoff-Gauß, I.; Kalthoff, N.; Fiebig-Wittmaack, M. The influence of a storage lake in the Arid Elqui Valley in Chile on local climate. *Theor. Appl. Climatol.* **2006**, *85*, 227–241. [CrossRef]

37.   Thiery, W.; Martynov, A.; Darchambeau, F.; Descy, J.P.; Plisnier, P.D.; Sushama, L.; Van Lipzig, N.P.M. Understanding the performance of the FLake model over two African Great Lakes. *Geosci. Model. Dev.* **2014**, *7*, 317–337. [CrossRef]

38.   Potes, M.; Salgado, R.; Costa, M.J.; Morais, M.; Bortoli, D.; Kostadinov, I.; Mammarella, I. Lake-atmosphere interactions at Alqueva reservoir: A case study in the summer of 2014. *Tellus Ser. A Dyn. Meteorol. Oceanogr.* **2017**, *69*, 1272787. [CrossRef]

39.   Sills, D.M.L.; Brook, J.R.; Levy, I.; Makar, P.A.; Zhang, J.; Taylor, P.A. Lake breezes in the southern Great Lakes region and their influence during BAQS-Met 2007. *Atmos. Chem. Phys.* **2011**, *11*, 7955–7973. [CrossRef]

40.   Physick, W. Numerical model of the sea-breeze phenomenon over a lake or gulf. *J. Atmos. Sci.* **1976**, *33*, 2107–2135. [CrossRef]

41.   Estoque, M.A. Further Studies of a Lake Breeze Part 1: Observational Studies. *Mon. Weather Rev.* **1980**, *109*, 611–618. [CrossRef]

42.   Corner, N.T.; McKendry, I.G. Observations and numerical modelling of lake Ontario breezes. *Atmos. Ocean* **1993**, *31*, 481–499. [CrossRef]

43.   Miranda, A.C.; Miranda, H.S.; Lloyd, J.; Grace, J.; Francey, R.J.; Mcintyre, J.A.; Meir, P.; Riggan, P.; Lockwood, R.; Brass, J. Fluxes of carbon, water and energy over Brazilian cerrado: An analysis using eddy covariance and stable isotopes. *Plant Cell Environ.* **1997**, *20*, 315–328. [CrossRef]

44.   Lorenzzetti, J.A.; Araújo, C.A.S.; Curtarelli, M.P. Mean diel variability of surface energy fluxes over Manso Reservoir. *Inland Waters* **2015**, *5*, 155–172. [CrossRef]

45.   Bechtold, P.; Pinty, J.P.; Mascart, P. A numerical investigation of the influence of large-scale winds on sea-breeze- and inland-breeze-type circulations. *J. Appl. Meteorol.* **1991**, *30*, 1268–1279. [CrossRef]

46.   Arritt, R.W. Effects of the large-scale flow on characteristic features of the sea breeze. *J. Appl. Meteorol.* **1993**, *32*, 116–125. [CrossRef]

47.   Simpson, J.E. *Sea Breeze and Local Wind*; Cambridge University Press: New York, NY, USA, 1994; p. 252.

48.   King, P.W.S.; Leduc, M.J.; Sills, D.M.L.; Donaldson, N.R.; Hudak, D.R.; Joe, P.; Murphy, B.P. Lake breezes in southern Ontario and their relation to tornado climatology. *Weather Forecast.* **2003**, *18*, 795–807. [CrossRef]

49.   Physick, W.L. Numerical experiments on the inland penetration of the sea breeze. *Q. J. R. Meteorol. Soc.* **1980**, *106*, 735–746. [CrossRef]

50.   Evan, A.T.; Heidinger, A.K.; Vimont, D.J. Arguments against a physical long-term trend in global ISCCP cloud amounts. *Geophys. Res. Lett.* **2007**, *34*, L04701. [CrossRef]

51.   Garratt, J.R. Review: The atmospheric boundary layer. *Earth Sci. Rev.* **1994**, *37*, 89–134. [CrossRef]

52.   McGill, R.; Tukey, J.W.; Larsen, W.A. Variations of box plots. *Am. Stat.* **1978**, *32*, 12–16. [CrossRef]

53.   Pereira, E.B.; Martins, F.R.; Gonçalves, A.R.; Costa, R.S.; Abreu, S.L.; Ruther, R.; Lima, F.J.L.; Pereira, S.V.; Souza, J.G. *Atlas Brasileiro de Energia Solar*, 2nd ed.; INPE: Sao Jose dos Campos, Brazil, 2017; ISBN 978-85-17-00090-4.

54.   Rosa-Clot, M.; Tina, G.M. *Submerged and Floating Photovoltaic Systems: Modelling, Design and Case Studies*; Academic Press: Cambridge, MA, USA, 2017; ISBN 9780128123232.

55.   Kay, S.; Hedley, J.D.; Lavender, S. Sun glint correction of high and low spatial resolution images of aquatic scenes: A review of methods for visible and near-infrared wavelengths. *Remote Sens.* **2009**, *1*, 697–730. [CrossRef]

56.   Zhang, H.; Wang, M. Evaluation of sun glint models using MODIS measurements. *J. Quant. Spectrosc. Radiat. Transf.* **2010**, *111*, 492–506. [CrossRef]

5

Combined Multi-Layer Feature Fusion and Edge Detection Method for Distributed Photovoltaic Power Station Identification

Yongshi Jie [1,2,3], Xianhua Ji [4], Anzhi Yue [1,3,5,*], Jingbo Chen [1,3], Yupeng Deng [1,2], Jing Chen [1,2] and Yi Zhang [1,2]

[1] Aerospace Information Research Institute, Chinese Academy of Sciences, Beijing 100094, China; jieys@radi.ac.cn (Y.J.); chenjb@aircas.ac.cn (J.C.); dengyp@radi.ac.cn (Y.D.); chenjing185@mails.ucas.ac.cn (J.C.); zhangyi@radi.ac.cn (Y.Z.)

[2] University of Chinese Academy of Sciences, Beijing 100049, China

[3] National Engineering Laboratory for Integrated Aero-Space-Ground-Ocean Big Data Application Technology, Xi'an 710129, China

[4] Engineering Quality Supervision Center of Logistics Support Department of the Military Commission, Beijing 100142, China; jixianhua1001@163.com

[5] Huizhou Academy of Space Information Technology, Institute of Remote Sensing and Digital Earth, Chinese Academy of Sciences, Huizhou 516006, China

* Correspondence: yueaz@aircas.ac.cn

Abstract: Distributed photovoltaic power stations are an effective way to develop and utilize solar energy resources. Using high-resolution remote sensing images to obtain the locations, distribution, and areas of distributed photovoltaic power stations over a large region is important to energy companies, government departments, and investors. In this paper, a deep convolutional neural network was used to extract distributed photovoltaic power stations from high-resolution remote sensing images automatically, accurately, and efficiently. Based on a semantic segmentation model with an encoder-decoder structure, a gated fusion module was introduced to address the problem that small photovoltaic panels are difficult to identify. Further, to solve the problems of blurred edges in the segmentation results and that adjacent photovoltaic panels can easily be adhered, this work combines an edge detection network and a semantic segmentation network for multi-task learning to extract the boundaries of photovoltaic panels in a refined manner. Comparative experiments conducted on the Duke California Solar Array data set and a self-constructed Shanghai Distributed Photovoltaic Power Station data set show that, compared with SegNet, LinkNet, UNet, and FPN, the proposed method obtained the highest identification accuracy on both data sets, and its F1-scores reached 84.79% and 94.03%, respectively. These results indicate that effectively combining multi-layer features with a gated fusion module and introducing an edge detection network to refine the segmentation improves the accuracy of distributed photovoltaic power station identification.

Keywords: distributed photovoltaic power stations; remote sensing images; convolutional neural network; multi-layer features; edge

1. Introduction

Renewable energy is a sustainable and inexhaustible energy, including biomass energy, wind energy, solar energy, etc., which plays an important role in solving the energy crisis. Biomass energy can be converted into Eco-fuels, and it has been found that Eco-fuels are a sustainable energy scenario at the local scale [1]. The main use of wind energy is to convert energy into electricity through wind turbines. Solar energy is a clean and safe renewable energy source (RES) with strong development potential and application value [2]. Photovoltaic power generation is an effective way to use solar energy [3], of which there are two main forms: Centralized photovoltaic power generation and distributed photovoltaic power generation [4,5]. Centralized photovoltaic power stations are installed primarily in the desert and other ground areas and the generated electricity is usually incorporated into the national public power grid [6], while distributed photovoltaic power stations are generally installed on tops of buildings and the generated electricity is mainly for the inhabitants' own use [7]. Distributed photovoltaic power stations have advantages such as unlimited installed capacity, no occupation of land resources [8], and no pollution. Thus, exploitation of distributed photovoltaic power generation is an important solar energy development mode that has entered a stage of rapid development and is supported by Chinese policy [9,10]. The International Energy Agency predicts that the world's total renewable energy generation will grow by 50% between 2019 and 2024, with solar photovoltaic generation alone accounting for nearly 60% of the prospective growth. Distributed photovoltaic generation is expected to account for approximately half of the growth in total photovoltaic power generation [11]. The installed capacity of distributed photovoltaic power stations is currently growing rapidly. Consequently, the ability to accurately and efficiently acquire the installation locations, distribution, and total area of distributed photovoltaic power stations over a wide range is of importance to energy companies, governmental departments, and investors. For example, obtaining information of distributed photovoltaic power stations can help optimize power system planning [12]. The information of distributed photovoltaic power stations and solar irradiance data of building surfaces can be combined to predict the power generation potential [13]. Moreover, it can also support the development of open data and energy systems and facilitate the development of the energy field [14]. However, due to the spontaneity and randomness of distributed photovoltaic power station construction, it is difficult to obtain accurate information regarding the quantity and distribution of distributed photovoltaic power stations solely from governmental department planning information. In addition, distributed photovoltaic power stations are generally installed on the tops of buildings, making it difficult to investigate their distribution and area manually. High-resolution remote sensing imagery has the characteristics of high spatial resolution, high efficiency, and wide coverage. Thus, it provides the possibility for automatic identification of large-scale distributed photovoltaic power stations.

Traditional distributed photovoltaic power station identification methods rely mainly on manually designed features, and it is difficult to accurately obtain the location and area of photovoltaic power stations. Malof [15] pioneered the use of manual features for extracting distributed photovoltaic power stations and proposed a method that first obtains all the maximally stable extreme regions (MSERs) [16] from an image and then filters out the areas with low confidence. Then, color features and shape features in the remaining candidate area are extracted for classification by a support vector machine (SVM) [17]. However, this method does not obtain photovoltaic panel areas accurately. Later, Malof [18] used color, texture, and other features in the neighborhood of each pixel to represent the pixel, and then used a random forest [19] to predict the category of each pixel. However, this method also has difficulty accurately obtaining the location and area information of photovoltaic panels. On the basis of the research conducted by the authors of [18], Malof [20] cascaded the random forest and convolutional neural network [21] to identify distributed photovoltaic power stations. However, this method still relies on feature information designed by humans. In a later work, Malof [22] proposed a distributed photovoltaic power station identification model based on a VGG model [23]. However, its ability to accurately obtain the locations and shapes of photovoltaic panels is limited.

As deep learning technology has developed, a series of convolutional neural network (CNN) models have been proposed [23–30]. Semantic segmentation technology based on deep learning can use a CNN, which has strong feature-learning ability, to automatically learn object features from massive amounts of data. Compared with earlier machine learning methods, such as SVMs and random forests, CNNs significantly improved the object extraction accuracy. Semantic segmentation technology has been widely applied and developed rapidly in fields such as medical image segmentation, automatic driving, and video segmentation. Jiang [31] used a CNN model and small data sets to extract the heart and lungs. Zhou [32] proposed the UNet++ model that has achieved high accuracy in nodule, nuclei, and liver segmentation. In addition to 2D medical image segmentation, the 3D full convolutional neural network can be used to realize organ segmentation in CT images [33]. Deep learning has become a robust and effective method for medical image segmentation [34]. In the field of automatic driving, CCNet [35] and ACFNet [36], respectively, used spatial context information and class context information to achieve the segmentation of objects in the street scene. Gated-scnn [37] combined shape and semantic information to extract targets on the street. In addition, in order to improve the performance of target segmentation in automatic driving, the idea of knowledge distillation has been used to retain the model's high precision while reducing the computation [38]. For the video semantic segmentation task, Paul [39] proposed an efficient video segmentation method that combines a convolutional neural network running on the GPU with an optical stream running on the CPU. Pfeuffer [40] added recurrent neural network into the video segmentation model to make full use of the time information of video sequence and improved the accuracy of video segmentation. Jain [41] proposed a video segmentation model with two input branches, which made use of the feature information of the current frame and the context information of the previous frame. Nekrasov [42] proposed a video segmentation algorithm without reliance on the optical flow, which further improved the efficiency of video segmentation. In addition to the natural image domain, semantic segmentation methods based on fully convolutional neural networks (FCN) [43] models have been widely used for object identification from remote sensing imagery, including road extraction, building extraction, and water extraction. For example, Zhou [44] proposed a road extraction method based on encoder-decoder structure and series-parallel dilated convolutions. Wu [45] added attention mechanism to the model [44], which further improved the accuracy of road extraction. Xu [46] designed a road extraction model based on DenseNet [30] and attracted local and global attention. Gao [47] used the refined residual convolutional neural network to extraction road in high-resolution remote sensing images. Xu [48] used deep convolutional neural network to extract buildings and optimized the results with guided filters. Yang [49] used DenseNet [30] and the spatial attention module to extract buildings. Huang [50] presented a residual refinement network for building extraction that fused aerial images and LiDAR point cloud data. Sun [51] proposed a building extraction method combining multi-scale convolutional neural network and SVM. Yu [52] proposed a water body extraction method based on convolutional neural networks, which used both spectral and spatial information from Landsat images Chen [53] proposed a cascade hyperpixel segmentation and convolutional neural network classification method to extract urban water bodies. Li [54] used fully convolutional network to extract water bodies from GeoFen-2 images with limited training data. Some previous deep learning-based semantic segmentation methods have been applied to the identification of distributed photovoltaic power stations. Yuan [55] was the first to introduce an FCN model for distributed photovoltaic power station identification. However, the adopted FCN model requires up-sampling by a large multiple, which may cause the loss of feature information. Subsequently, SegNet [56] and UNet [57] were used to identify distributed photovoltaic power stations [58,59]. Although the identification results of those models are superior to the results of traditional methods, they still do not solve the problem that photovoltaic panels with small areas are easily missed and densely installed photovoltaic panels are easily adhered.

To solve the above problems, this paper proposes a distributed photovoltaic power station identification method that combines multi-layer features and edge detection. The main contributions aims of this paper are as follows:

- To address the problem that small photovoltaic panels are difficult to recognize, a gated fusion module is introduced into the encoder-decoder model to effectively fuse multi-layer features, which improves the model's ability to identify small photovoltaic panels.

- To address the problem of edge blurring, a multi-task learning model that combines edge detection and semantic segmentation is proposed to refine the edges of the segmentation results using feature information of the target edge.
- Comparative experiments are conducted on the Duke California Solar Array data set [60] and the Shanghai Distributed Photovoltaic Power Station data set, and the results verify the effectiveness of the proposed method.

The remainder of this article is organized as follows. Section 2 introduces the distributed photovoltaic power station identification model designed in this paper, including the encoder-decoder architecture, gated fusion module, and edge detection network. Section 3 presents the experiments and results analysis on the two data sets, including the experimental data, evaluation metrics, experimental settings, and the experimental results. The results are analyzed and compared with those of other methods. Finally, Section 4 concludes this paper.

2. Model Architecture and Design

The model proposed in this paper was composed of a semantic segmentation network and an edge detection network. These 2 networks were trained in parallel for multi-task learning, as shown in Figure 1. The semantic segmentation network was used to extract the semantic features of photovoltaic panels, and its architecture included an encoder-decoder structure based on UNet. The encoder was Efficientnet-B1 [61]. In the semantic segmentation network, a gated fusion module was introduced to control the transmission of valuable information, effectively fuse multi-layer features, and improve the ability to identify small photovoltaic panels. The edge detection network was used to extract the edge features of the photovoltaic panels and guide the semantic segmentation network to produce segmentation results with more refined edges to alleviate the problem of blurred and unrefined edges in segmentation results.

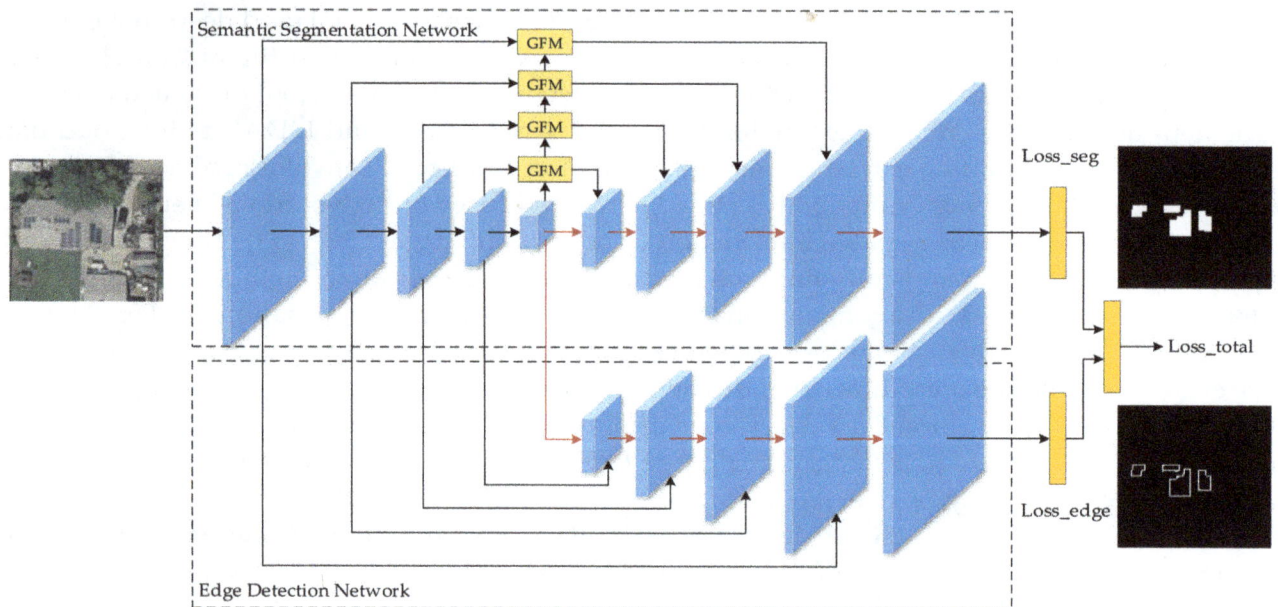

Figure 1. Structure of the proposed model.

2.1. Semantic Segmentation Network with Gated Fusion Multi-Layer Features

A semantic segmentation network was used to extract the semantic features of photovoltaic panels. Efficientnet-B1 uses an encoder, and a gated fusion module was introduced to effectively fuse multi-layer features.

2.1.1. Encoder and Decoder

This study adopted EfficientNet-B1, which has strong feature representation capabilities, as the encoder for feature extraction. This decoder is the same as that used in the original UNet. The Efficientnet-B1 network structure is shown in Figure 2. The basic component of Efficientnet-B1 is the MBConv module. In the MBConv module, a 1×1 convolution is first used to change the channels of the input features, followed by a depth-wise convolution. Then, the channel attention mechanism of SENet [62] is introduced, and finally, a 1×1 convolution is used to reduce the channels of the feature maps.

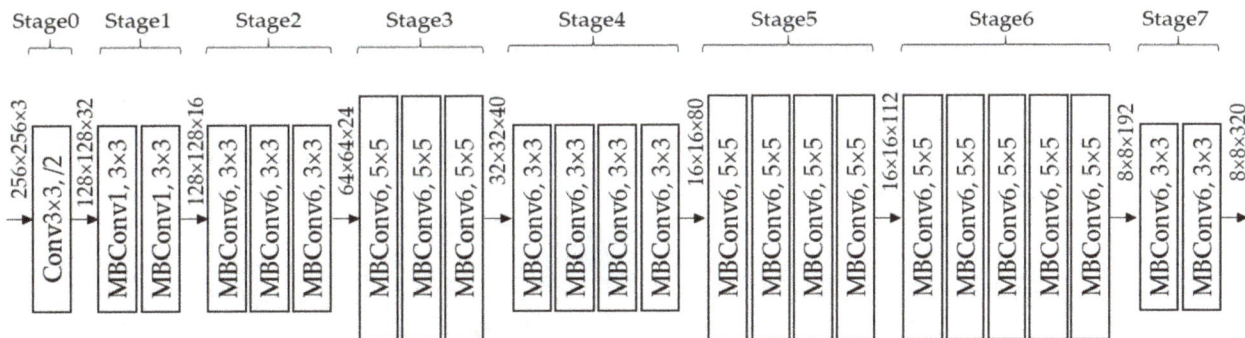

Figure 2. Structure of EfficientNet-B1.

The original UNet encoder structure consists of 5 stages. The feature resolution at each stage is successively changed to half of that of the previous stage through down-sampling, and the features of each stage are fused with the corresponding decoder features through skip connections. Based on the UNet structure, this paper adopted the output features of Stages 0, 2, 3, 5, and 7 of Efficientnet-B1 as the 5 encoder blocks used in the encoder of our model, as shown in Figure 3, which assumes that the size of the input image is $256 \times 256 \times 3$.

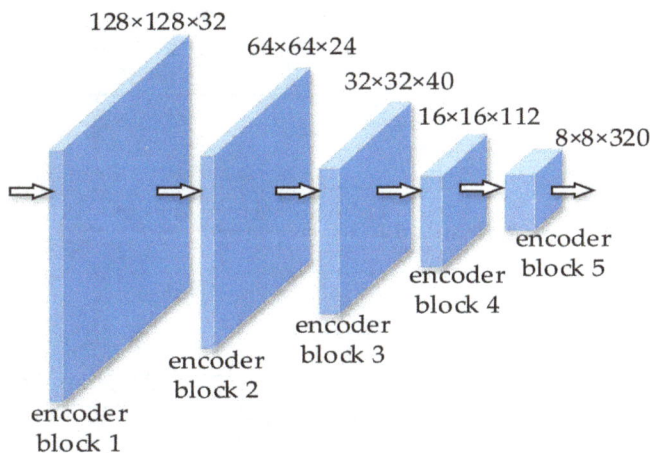

Figure 3. Encoder structure in the proposed model.

The decoder is mainly used to gradually up-sample the low-resolution high-level features to restore the original size of the input image. During the up-sampling process, the corresponding features of the encoder and decoder are concatenated through skip connections. The decoder structure block is shown in Figure 4. The decoding features represent the output feature of the previous decoder block, and the encoding features represent the features passed to the corresponding decoder block through the skip connections. First, the decoding features are up-sampled twice and then concatenated with the encoding features on the channel dimension. The number of channels of the concatenated features is the sum of the number of channels of the two features. After the concatenation and two

3×3 convolutional layers, the output features of the decoder block are obtained. The output features of the current decoder block are the input decoding features for the next decoder block.

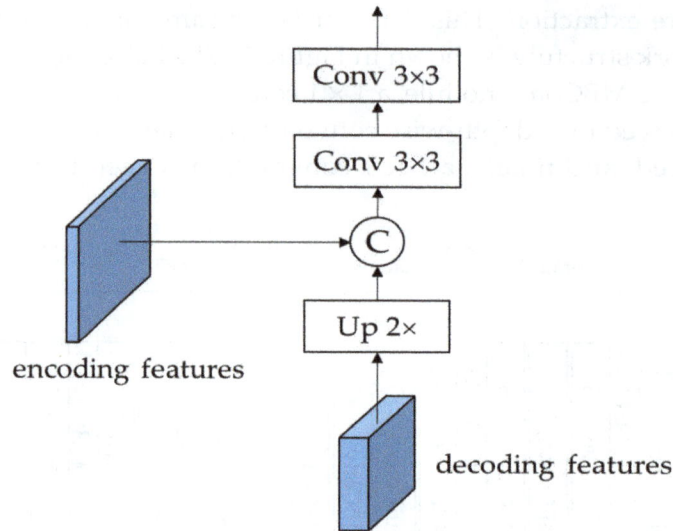

Figure 4. Structure of the decoder blocks.

2.1.2. Gated Fusion Module

Inspired by the research conducted by the authors of [63], a gating fusion module was introduced to effectively fuse the multi-layer features to improve the ability to identify small photovoltaic panels. The gating fusion module structure is shown in Figure 5. The input is the feature of the adjacent layer of the encoder, and the features generated by the gating unit are used to measure the usefulness of the feature at each position in the spatial dimension. This arrangement controls the transmission of useful information and suppresses the transmission of useless information.

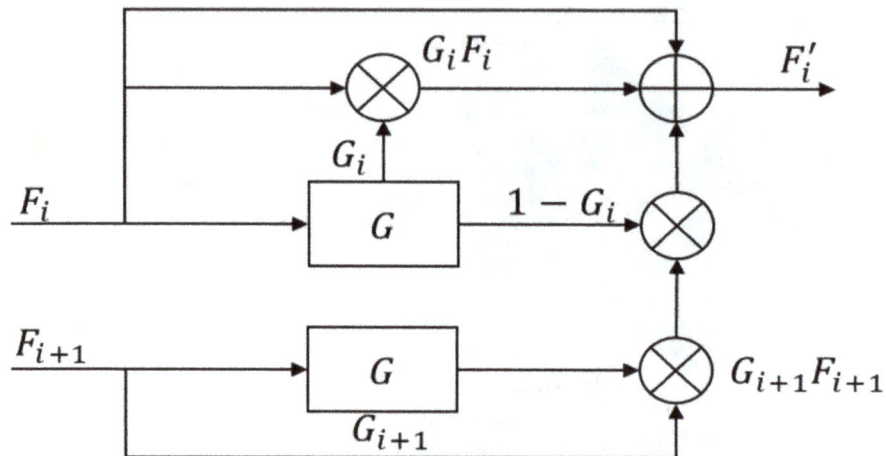

Figure 5. Structure of the gated fusion module.

The input to the gated fusion module consists of the features F_i from layer i and the features F_{i+1} from the adjacent layer $i + 1$. Due to the differences in the feature sizes and the channel numbers, F_{i+1} is first up-sampled twice, and the number of channels in F_{i+1} is converted to be the same as that in F_i. Then, F_{i+1} is input into the gating unit G. The output of gated fusion module is F_i'.

The purpose of gating unit G feeds the input features into a 1×1 convolution and then obtains the gated features G_i through the sigmoid function, as shown in Equation (1). The gated feature graph is used to judge the usefulness of the spatial position features of the input features. The range of the gated feature values is [0, 1]. A value less than 0.5 (approximately 0) corresponds to useless feature information, whereas a value greater than 0.5 (approximately 1) corresponds to useful feature information. The transfer of useful information and useless information is controlled by element-by-element multiplication between the gated features and the input features of the gating unit:

$$G_i = \sigma(w_i * F_i), \tag{1}$$

where σ is the sigmoid function, the asterisk ('$*$') represents the convolution operation, and w_i is the weight parameter of the convolution.

The entire gated fusion module process can be defined as shown in Equation (2). For a position (x, y), when $G_{i+1}(x, y)$ is larger and $G_i(x, y)$ is smaller, F_{i+1} transmits useful information to F_i that F_i lacks at this position. When $G_{i+1}(x, y)$ is smaller or $G_i(x, y)$ is larger, this useless information is suppressed to reduce information redundancy:

$$F'_i = (1 + G_i) \odot F_i + (1 - G_i) \odot G_{i+1} \odot F_{i+1}, \tag{2}$$

where $\odot$ denotes element-by-element multiplication.

2.2. Combining Edge Detection for Multi-Task Learning

The edge detection network was used to extract the edge features of photovoltaic panels. The semantic segmentation network was trained using multi-task learning so that the network model produced segmentation results with refined edges.

2.2.1. Edge Detection Network

Distributed photovoltaic stations have dense distribution characteristics, and the identified results of adjacent photovoltaic panels are prone to adhesion. In this paper, edge information extracted by the edge detection network was combined with the semantic segmentation network to ameliorate the problem of edge blurring.

In this paper, an encoder-decoder structure was adopted in the edge detection network, as shown in Figure 6. This is the same encoder used in semantic segmentation network for feature extraction and feature sharing. The decoder structure of the edge detection network is also the same as that of the semantic segmentation network. The object edge feature information is gradually obtained through multiple up-sampling operations, and the edge feature extracted by the encoder is fused by skip connections during the up-sampling process.

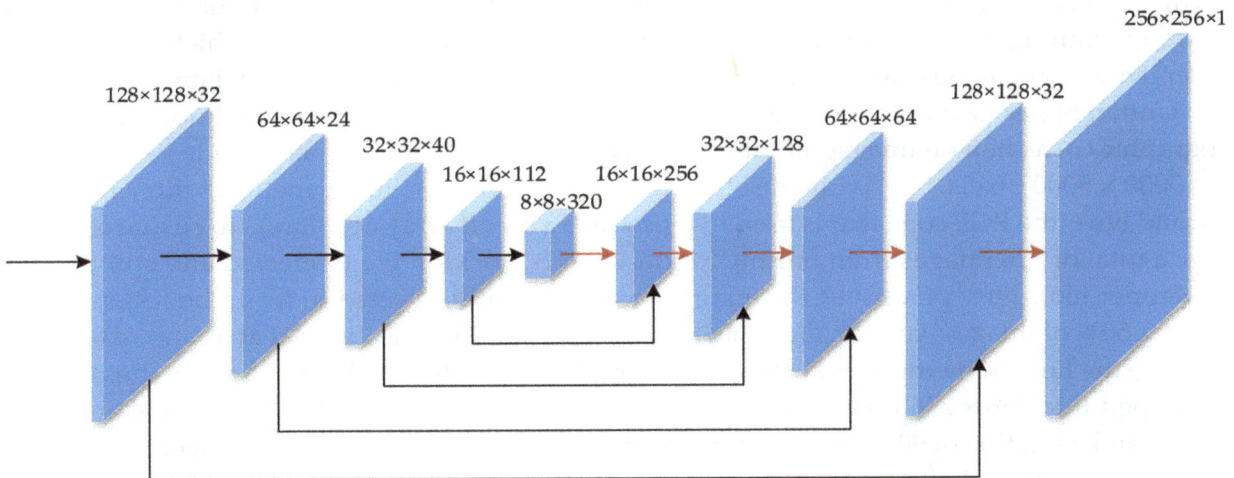

Figure 6. Structure of the edge detection network.

2.2.2. Loss Function

In the parallel training of 2 networks, a semantic segmentation loss function and an edge detection loss function are used to supervise the learning process for the semantic and edge features of photovoltaic panels, respectively. The semantic segmentation network loss function is calculated from the segmentation predictions and segmentation labels, while the edge detection loss function is calculated from the edge predictions and edge labels. Both the semantic segmentation and edge detection of photovoltaic power stations are binary classification tasks. In addition, compared with the background, the segmentation labels and edge labels account for only a small proportion. To avoid sample imbalance problems, a loss function composed of binary cross entropy (BCE) and the Dice loss function (Dice), namely, BCE + Dice [64,65], is used in both the semantic segmentation network and edge detection network. During training, the 2 loss functions are summed to obtain the total model loss, as shown in the following equation:

$$Loss\_total = Loss\_seg + Loss\_edge, \tag{3}$$

where $Loss\_total$ is the total loss function of our proposed model, $Loss\_seg$ is the loss function of the semantic segmentation network and $Loss\_edge$ is the loss function of the edge detection network.

The BCE loss function is shown in Equation (4). The Dice loss function is given by Equation (5).

$$BCE = -\frac{1}{n} \sum_{i=1}^{n} (g_i \times log(p_i) + (1 - g_i) \times log(1 - p_i)), \tag{4}$$

$$Dice = 1 - \frac{2|G \cap P|}{|G| + |P|} = 1 - \frac{2 \sum_{i=1}^{n} (g_i \times p_i)}{\sum_{i=1}^{n} g_i^2 + \sum_{i=1}^{n} p_i^2}, \tag{5}$$

where n represents the number of pixels in the image, g_i represents the value of the i-th pixel in the label, p_i denotes the value of the i-th pixel in the prediction result map, and G and P denote the label and prediction result map, respectively.

3. Experimental and Result Analysis

3.1. Experimental Data

The experimental data in this study consisted of the Duke California Solar Array and Shanghai Distributed Photovoltaic Power Station data sets.

1. Duke California Solar Array data set

This data set is currently the largest manually labelled distributed photovoltaic power station data set, containing images and coordinate information of object boundary which can be used to train semantic segmentation and object detection algorithms. The images in the data set are collected by the United States Geological Survey (USGS), which uses remote sensing technology to perform orthographic correction on images, eliminating distortions caused by camera and terrain. The image size is 5000 × 5000 pixels, the spatial resolution is 0.3 m, and each image includes three bands: Red, green, and blue (the RGB code that is used to reproduce a broad array of colors). To ensure comparable results, a total of 526 images from Fresno, Modesto, and Stockton were selected and split following SolarMapper [66]. Fifty percent of the images were randomly selected to form the test set, and the remaining 50% of images were divided into a training set and verification set at a ratio of 8:2.

Given the limited memory available on the graphics card, the original images in the training set were clipped into 256 × 256 image blocks and the data were augmented by horizontal and vertical mirroring and a rotation of 90 degrees. Finally, a total of 85,448 image blocks were collected for training. During the training of the edge detection network, photovoltaic panel edge labels are needed. In this study, the edge labels were obtained based on the semantic segmentation labels. Some sample images, segmentation labels, and edge labels from this data set are shown in Figure 7.

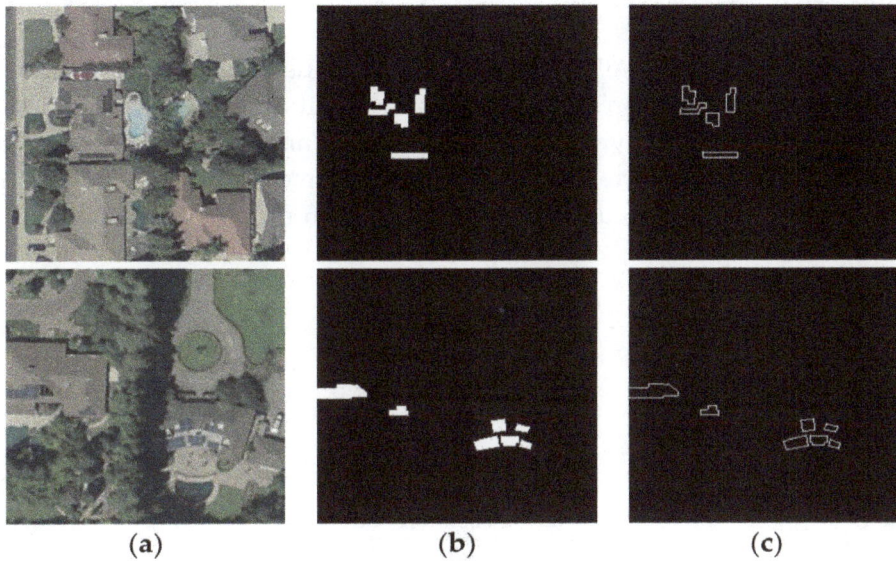

Figure 7. Samples from the Duke California Solar Array data set: (**a**) Image; (**b**) segmentation label; (**c**) edge label.

2. Shanghai Distributed Photovoltaic Power Station Data Set

To verify the effectiveness of the proposed method in this paper for identifying domestically distributed photovoltaic power stations, the Shanghai Distributed Photovoltaic Power Station data set was constructed. The images were collected from the Songjiang and Pudong New districts in Shanghai. The data set contains 1000 aerial images with a size of 2048 × 2048 and a spatial resolution of 0.1 m and the images include three bands: Red, green, and blue. The data set images were randomly divided into a training set, a validation set, and a test set at a ratio of 7:1:2. The training set data were clipped into 256 × 256 image blocks. Then, the data were augmented by horizontal and vertical mirroring and rotations of 90, 180 and 270 degrees. Contrast transformation and brightness transformation was carried out. Finally, a total of 55,560 image blocks were collected for training. Some sample images, segmentation labels, and edge labels for this data set are shown in Figure 8.

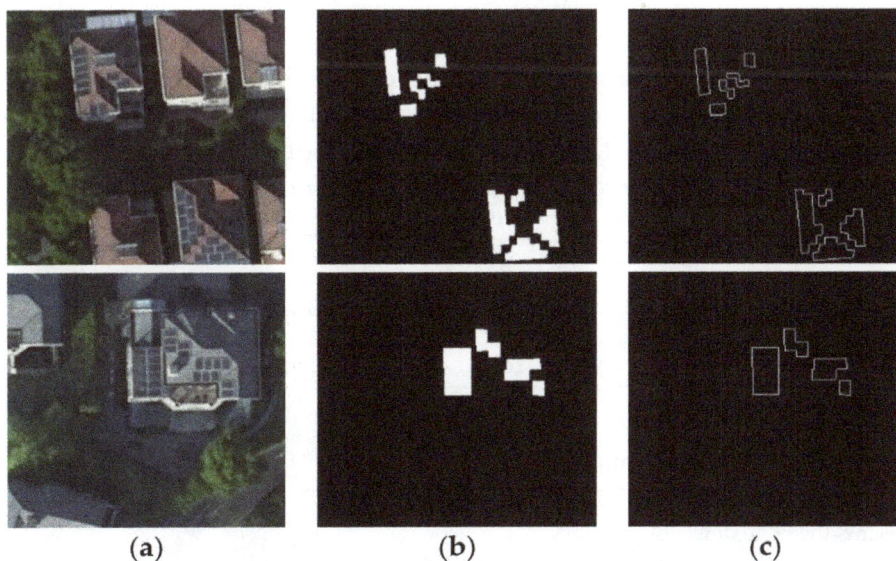

Figure 8. Samples from the Shanghai Distributed Photovoltaic Power Station data set: (**a**) Image; (**b**) segmentation label; (**c**) edge label.

3.2. Evaluation Metrics

In this study, IoU, precision, recall, and F1-scores were used as evaluation metrics. The IoU is the ratio of the intersection and union of the predicted result area and the labelled area. Precision represents the ratio of pixels correctly predicted as positive among all pixels predicted as positive. Recall represents the ratio of pixels correctly predicted as positive among all positive pixels. The F1 is a metric that combines precision and recall. The four evaluation metrics are calculated as shown in the following equations:

$$IoU = \frac{TP}{TP + FP + FN},\tag{6}$$

$$Precision = \frac{TP}{TP + FP},\tag{7}$$

$$Recall = \frac{TP}{TP + FN},\tag{8}$$

$$F1 = \frac{2 \times Precision \times Recall}{Precision + Recall},\tag{9}$$

where TP (true positive) represents the number of pixels that are both predicted and labelled as positive FP (false positive) represents the number of pixels that are predicted as positive but labelled as negative, and FN (false negative) represents the number of pixels that are predicted as negative but labelled as positive.

3.3. Experimental Setting

1. Experimental environment

The computer used in the experiments was equipped with an Ubuntu 16.04.5 LTS operating system, an Intel (R) Xeon (R) E5-2678 v3 CPU, and two NVIDIA TITAN XP graphics cards, each with 12 GB of memory. PyTorch was used to build all the semantic segmentation models.

2. Training strategy and hyperparameter settings

All the models were trained using the Adam optimizer to help ensure a fast convergence speed. The batch size of the input images in each training epoch was 64. The initial learning rate was 1×10^{-3} and the learning rate decay adopted the cosine annealing learning rate decline strategy. The cycle was 10, and the minimum learning rate was 1×10^{-5}.

3.4. Experimental Results

To verify the effectiveness of the proposed method, EfficientNet-B1-UNet was considered as the baseline network. Then, the gated fusion module and edge detection network were added successively. The experiments used the Duke California Solar Array data set and the Shanghai Distributed Photovoltaic Power Station data set. The experimental results on the Duke California Solar Array data set are shown in Table 1.

Table 1. Experimental results of each improved module on the Duke California Solar Array data set (%).

Methods	IoU	Precision	Recall	F1
Effi-UNet	72.41	85.40	82.64	84.00
Effi-UNet + GFM	73.33	86.03	83.24	84.61
Effi-UNet + GFM + EDN	73.60	86.17	83.45	84.79

Effi-UNet represents UNet, which uses EfficientNet-B1 as the encoder; GFM represents the gated fusion module, and EDN represents the edge detection network.

On the Duke California Solar Array data set, by adding the gated fusion module, the IoU of the test set was increased from 72.41% to 73.33%, F1 was increased from 84.00% to 84.61%, and recall was increased from 82.64% to 83.24%. By adding the edge detection network, the IoU of the network model was further improved from 73.33% to 73.60% and F1 was improved from 84.61% to 84.79%.

The experimental results of the Shanghai Distributed Photovoltaic Power Station data set are shown in Table 2.

Table 2. Experimental results from successively improved models on the Shanghai distributed photovoltaic power station data set (%).

Methods	IoU	Precision	Recall	F1
Effi-UNet	87.40	93.08	93.47	93.27
Effi-UNet + GFM	88.34	93.54	94.08	93.81
Effi-UNet + GFM + EDN	88.74	93.88	94.19	94.03

On the Shanghai Distributed Photovoltaic Power Station data set, adding the gating fusion module increased the IoU of the test set from 87.40% to 88.34%, the F1-score from 93.27% to 93.81%, and the recall from 93.47% to 94.08%. After adding the edge detection network, the IoU of the network model was further improved to 88.74% and the F1-score improved to 94.03%.

The added modules improved all four evaluation metrics. This shows that the gated fusion module and edge detection network proposed in this paper can improve the accuracy of distributed photovoltaic panel identification tasks.

3.5. Results Analysis

1. The influence of the gating fusion module on the segmentation results

Figure 9 shows a sample image, and its segmentation results are shown both before and after adding the gated fusion module. The first two rows of images are from the Duke California Solar Array data set and the second two rows of images are from the Shanghai Distributed Photovoltaic Power Station data set. The first column is the sample image, the second column is the labelled image, and the third column shows the segmentation results of Effi-UNet. Compared with the labelled image, the Effi-UNet results failed to detect of some small photovoltaic panels. The fourth column shows the segmentation results of Effi-UNet + GFM, revealing that, with the help of the GFM module, the network's ability to identify small photovoltaic panels was improved, which verifies the effectiveness of the module.

2. The influence of the edge detection network on the segmentation results

By extracting edge information and conducting multi-task learning of the edge detection and segmentation networks, more refined segmentation results can be generated. In Figure 10, the first two rows of sample images come were sourced from the Duke California Solar Array data set, while the second two rows of sample images are were sourced from the Shanghai Distributed Photovoltaic Power Station data set. The first column is the sample image, and the second column is the segmentation label. The third column is the Effi-UNet + GFM segmentation results. Compared with the segmentation label, the segmentation results of adjacent photovoltaic panels were adhered. The fourth column and the fifth column, respectively, represent the semantic segmentation results and edge detection results of Effi-UNet + GFM + EDN, and the sixth column is the label of edge detection. With the help of the edge detection network, fine edge results were obtained, distinguishing adjacent photovoltaic panels insofar as possible and alleviating the adhesion problem.

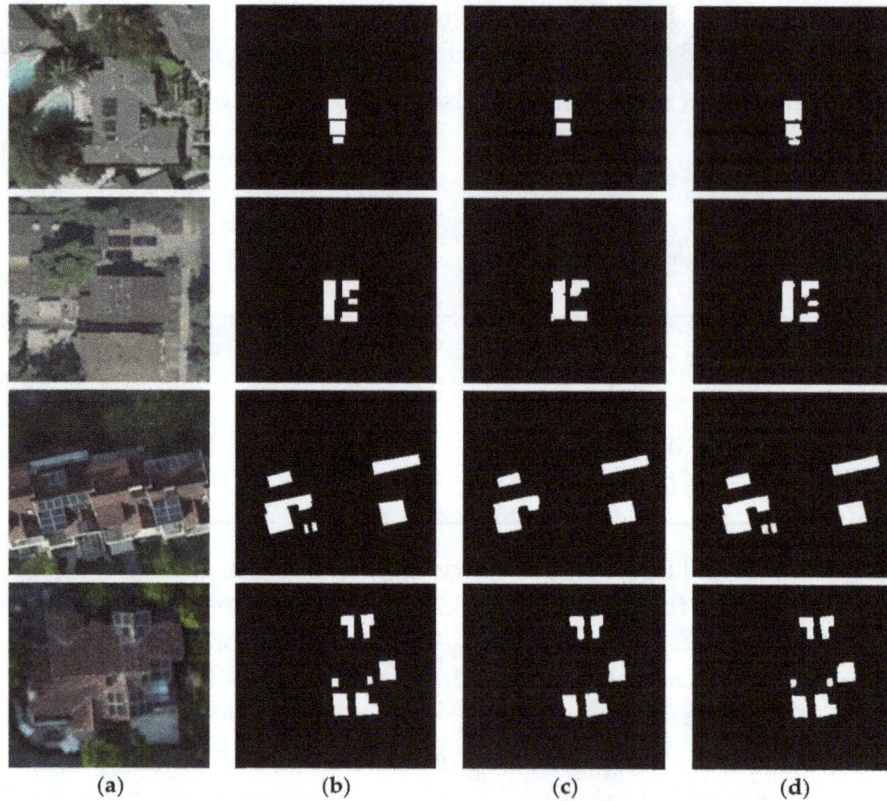

Figure 9. Result samples before and after adding GFM: (**a**) Image; (**b**) label; (**c**) Effi-UNet results; (**d**) Effi-UNet + GFM results.

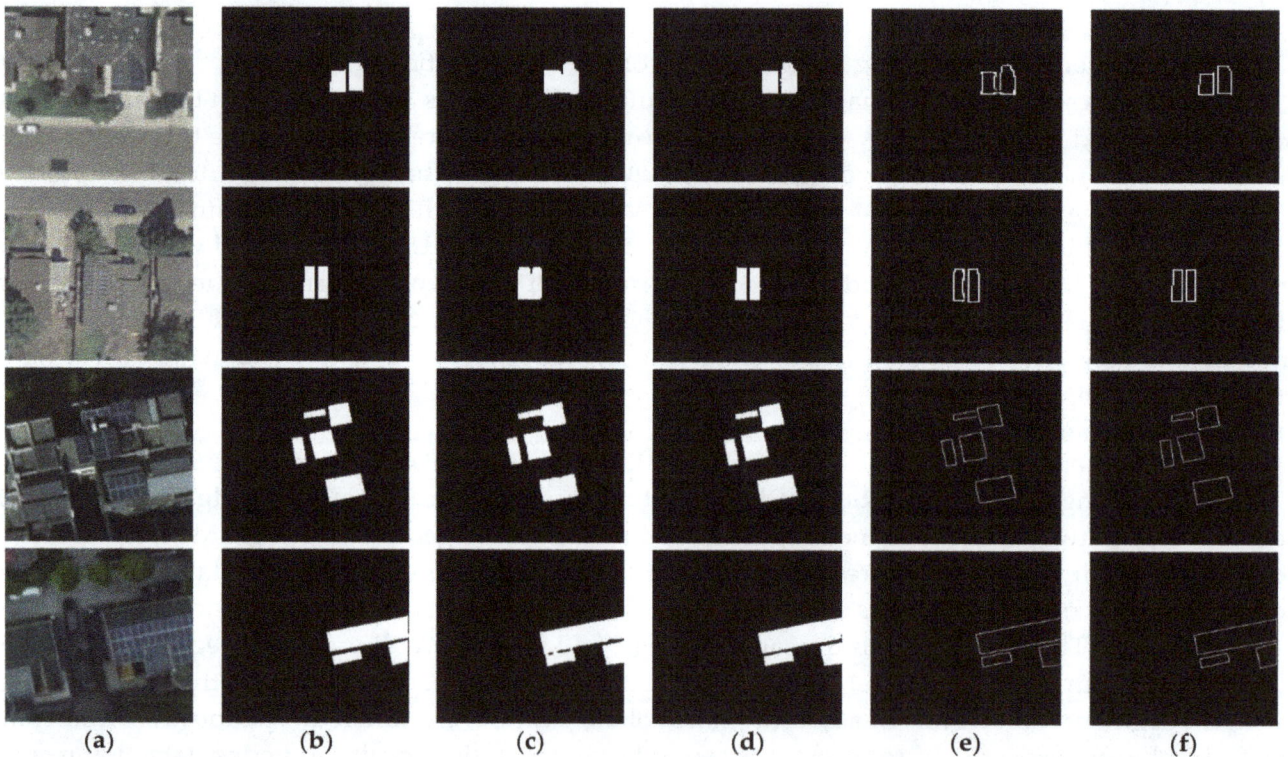

Figure 10. Result samples before and after adding the edge detection network: (**a**) Image; (**b**) segmentation label; (**c**) Effi-UNet + GFM results; (**d**) Effi-UNet + GFM + EDN segmentation results; (**e**) Effi-UNet + GFM + EDN edge detection results; (**f**) edge label.

3.6. Comparisons with Other Methods

To further verify the effectiveness of the proposed method, the identification method proposed in this paper was compared with SegNet, LinkNet [67], UNet, and FPN [68] on the adopted two data sets. The results and analysis are as follows.

3.6.1. Results on the Duke California Solar Array Data Set

The experimental results of each method on the test set of the Duke California Solar Array data set are shown in Table 3. The results show that the proposed method outperformed the other methods on all the evaluation metrics. The IoU of the proposed method in this paper reached 73.60%, and its F1-score reached 84.79%. Moreover, the IoU of the proposed method was 6.6% better than the IoU of SolarMapper [66]. The analysis of the results is as follows: (1) Although LinkNet, UNet, and FPN combine features from different layers, they do not consider the differences between the high-level and low-level features, nor do they make full use of object edge information. (2) In this paper, based on the encoder and decoder structure network, the multi-layer features were fused effectively by the gated fusion module, and the useful information was transferred by the gated mechanism improving the ability to identify small photovoltaic panels. (3) Based on the semantic segmentation network, the method in this paper combined an edge detection network for multi-task learning to ameliorate the edge-blurring problem.

Figure 11 shows some of the experimental results of each method on the Duke California Solar Array data set. The segmentation results in the first and second rows show that the method proposed in this paper was better at identifying small photovoltaic panels compared with the other methods. In the segmentation results shown in the third and fourth rows, although each method identified the photovoltaic panel in the image, the method in this paper obtained more refined edges.

Table 3. Accuracy of each method on the Duke California Solar Array data set (%).

Methods	IoU	Precision	Recall	F1
SegNet	66.97	83.48	77.20	80.22
SolarMapper	67.00	—	—	—
LinkNet	69.23	83.60	80.11	81.82
UNet	70.28	83.83	81.30	82.54
FPN	71.11	84.79	81.50	83.11
Our method	73.60	86.17	83.45	84.79

3.6.2. Results on the Shanghai Distributed Photovoltaic Power Station Data Set

Table 4 shows the evaluation results of each model on the Shanghai Distributed Photovoltaic Power Station data set, revealing that the method proposed in this paper outperformed all the other methods on all the evaluation metrics. The IoU of the method in this paper reached 88.74%, and its F1-score reached 94.03% Due to the encoder-decoder structure, the method proposed in this paper effectively fused features from multiple layers, improved the ability to identify small photovoltaic panels, and refined the segmentation edge results using the edge detection network. Therefore, compared with the other methods, the method in this paper achieved higher accuracy.

Figure 12 shows an example of the experimental results of the proposed method and the compared methods on the Shanghai Distributed Photovoltaic Power Station data set. As seen from the results in the first row, the method proposed in this paper was better at identifying small photovoltaic panels, and the identification results were more complete. In the second row, the two separate photovoltaic panels were difficult to identify due to their small sizes. Compared with the other methods, the proposed method not only recognized them but also obtained more refined edges in the identification results. In the third row, multiple photovoltaic panels were close to each other, which was likely to cause adhesion problems in the identification process. Compared with the other methods, with the help of the edge detection network, the identification results of the method proposed in this paper had more refined edges and alleviated the adhesion problem. A comparison of the results in the fourth row shows that the identification results of the proposed method had more refined edges.

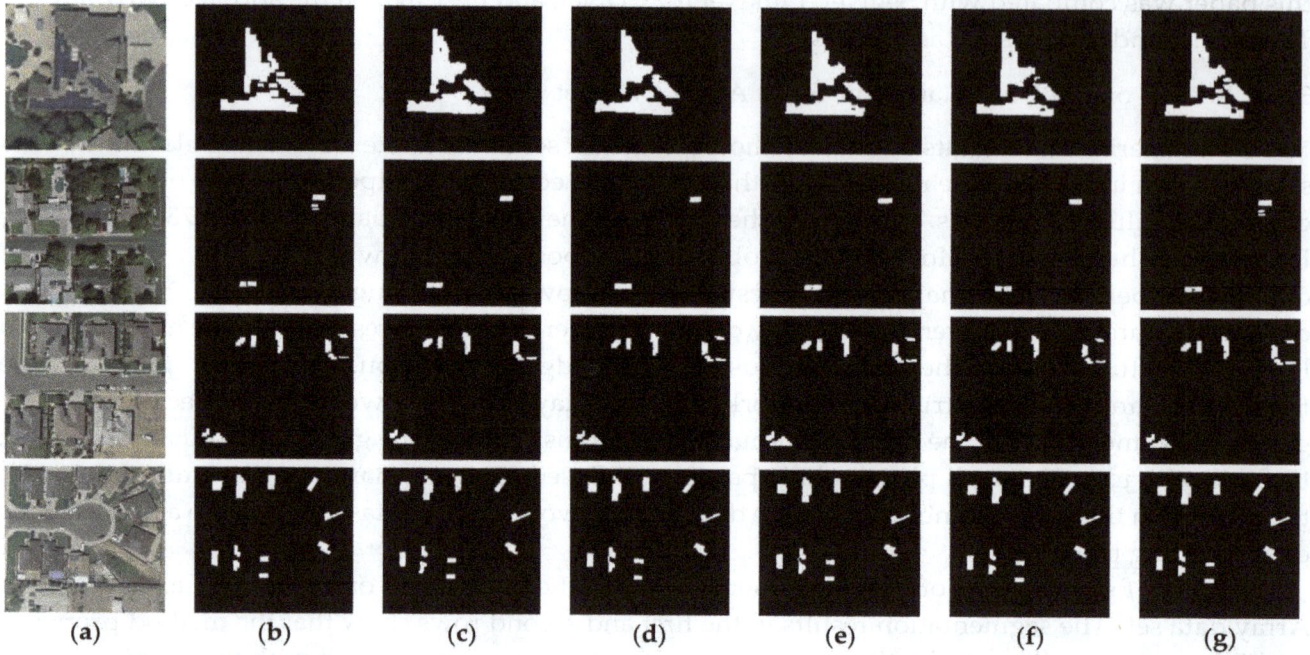

Figure 11. Sample results of each method on the Duke California Solar Array data set: (**a**) Image; (**b**) label; (**c**) SegNet; (**d**) LinkNet; (**e**) UNet; (**f**) FPN; (**g**) our method.

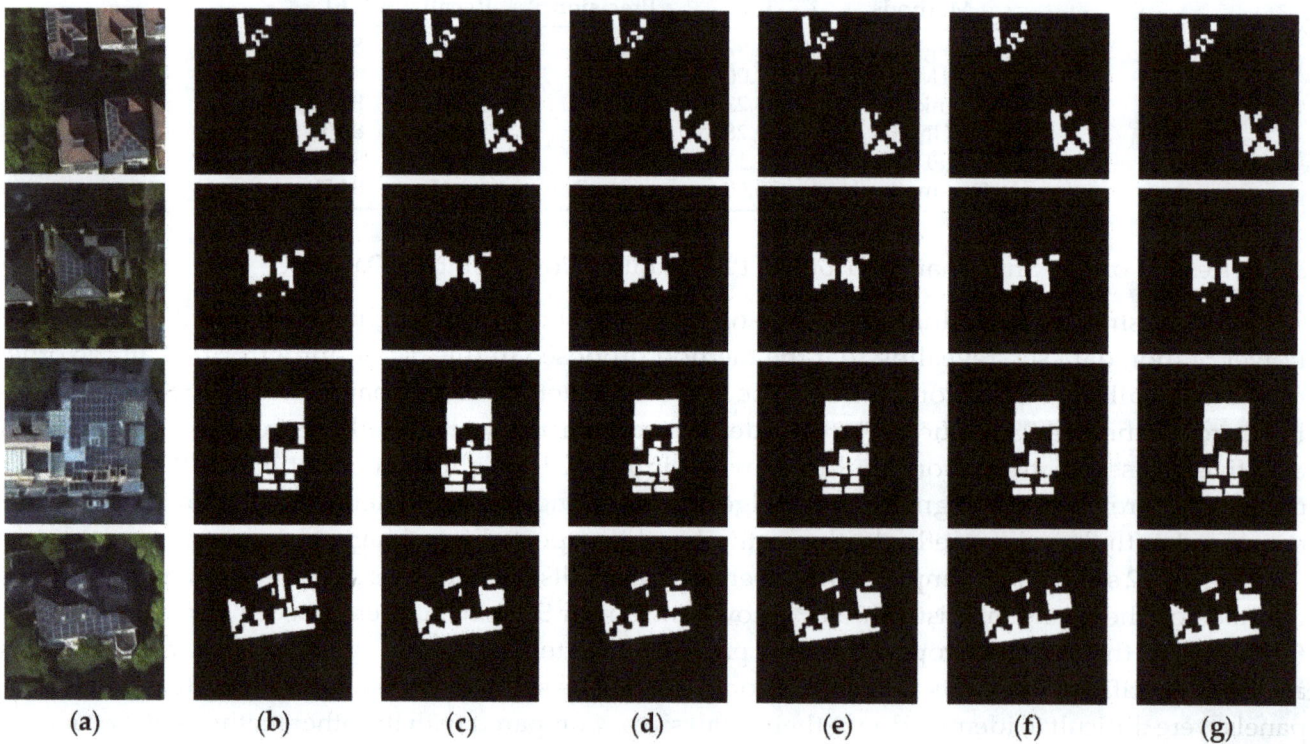

Figure 12. Sample results of each method on the Shanghai Distributed Photovoltaic Power Station data set: (**a**) Image; (**b**) label; (**c**) SegNet; (**d**) LinkNet; (**e**) UNet; (**f**) FPN; (**g**) our method.

Table 4. Accuracy of each method on the Shanghai Distributed Photovoltaic Power Station data set (%).

Methods	IoU	Precision	Recall	F1
SegNet	85.32	91.97	92.19	92.08
LinkNet	85.96	92.29	92.62	92.45
UNet	86.32	92.43	92.89	92.66
FPN	86.77	92.70	93.14	92.92
Our method	88.74	93.88	94.19	94.03

4. Conclusions

This paper presented a novel fully connected convolutional neural network model that can automatically extract distributed photovoltaic power stations from remote sensing imagery. A distributed photovoltaic power station identification method that combines multi-layer features and edge detection was proposed to solve two problems: That small photovoltaic panels are difficult to identify and that adjacent photovoltaic panels can easily adhere. The model structure was composed of a semantic segmentation network and an edge detection network. A gated fusion module was introduced into the semantic segmentation network to conduct effective multi-layer feature fusion, and an edge detection network was used to guide the production of segmentation results with refined edges. Experiments on the Duke California Solar Array data set and the Shanghai Distributed Photovoltaic Power Station data set showed that the problem of missed small photovoltaic panels was improved and that the identification accuracy was enhanced by introducing a gating fusion module. By combining the edge detection network and semantic segmentation network for multi-task learning, the edge information of the photovoltaic panel was used to constrain the segmentation results, resulting in the extraction of photovoltaic panels with finer edges, which further improved the identification accuracy. Compared with SegNet, LinkNet, UNet and FPN, the method proposed in this paper achieved the highest identification accuracy on both data sets, and its F1-scores reached 84.79% and 94.03%, respectively.

However, there are also some limitations in this study: (1) In terms of data source, due to the limitations of the current data set, the trained model is only applicable to RGB optical images and cannot be directly used to images containing more bands. (2) In terms of the spatial resolution of the image, the training and testing of the method in this paper were carried out on the images with the same spatial resolution. Due to the differences of solar panels in images with different resolutions, the accuracy may be uncertain when the trained model is directly used to predict images with different resolutions. (3) Since the training data only includes distributed photovoltaic power stations, the trained model cannot be used to identify centralized photovoltaic power stations. The future work will be carried out from the following aspects: (1) Explore the application of our method in multi-spectral images and further improve the segmentation performance with more spectral information. (2) Multiple images of different spatial resolutions will be collected to train our method so that our method can identify distributed photovoltaic power stations in images with different resolutions. (3) A centralized photovoltaic power station data set will be constructed, and our method will be extended to the identification of centralized photovoltaic power stations. (4) In addition, the extracted results of distributed photovoltaic power stations will be combined with solar radiation data to assess the power generation potential.

Author Contributions: Y.J., A.Y. and X.J. designed the network architecture. Y.J. performed the experiments and wrote the paper. X.J. and J.C. (Jingbo Chen) revised the paper. Y.D., J.C. (Jing Chen) and Y.Z. built the data set. All authors have read and agreed to the published version of the manuscript.

Acknowledgments: The authors sincerely thank the editors and reviewers. We also sincerely thank the authors of the Duke California Solar Array data set.

References

1.   Nastasi, B.; de Santoli, L.; Albo, A.; Bruschi, D.; Basso, G.L. RES (Renewable Energy Sources) availability assessments for Eco-fuels production at local scale: Carbon avoidance costs associated to a hybrid biomass/H2NG-based energy scenario. *Energy Procedia* **2015**, *81*, 1069–1076. [CrossRef]

2.   Moriarty, P.; Honnery, D. Feasibility of a 100% Global Renewable Energy System. *Energies* **2020**, *13*, 5543. [CrossRef]

3.   Li, W.; Ren, H.; Chen, P.; Wang, Y.; Qi, H. Key Operational Issues on the Integration of Large-Scale Solar Power Generation—A Literature Review. *Energies* **2020**, *13*, 5951. [CrossRef]

4.   Li, H.; Lin, H.; Tan, Q.; Wu, P.; Wang, C.; De, G.; Huang, L. Research on the policy route of China's distributed photovoltaic power generation. *Energy Rep.* **2020**, *6*, 254–263.

5.   Xin-gang, Z.; Zhen, W. Technology, cost, economic performance of distributed photovoltaic industry in China. *Renew. Sustain. Energy Rev.* **2019**, *110*, 53–64. [CrossRef]

6.   Yi, T.; Tong, L.; Qiu, M.; Liu, J. Analysis of Driving Factors of Photovoltaic Power Generation Efficiency: A Case Study in China. *Energies* **2019**, *12*, 355. [CrossRef]

7.   Ahmed, R.; Sreeram, V.; Mishra, Y.; Arif, M.D. A review and evaluation of the state-of-the-art in PV solar power forecasting: Techniques and optimization. *Renew. Sustain. Energy Rev.* **2020**, *124*, 109792. [CrossRef]

8.   Mancini, F.; Nastasi, B. Solar energy data analytics: PV deployment and land use. *Energies* **2020**, *13*, 417. [CrossRef]

9.   Han, M.; Xiong, J.; Wang, S.; Yang, Y. Chinese photovoltaic poverty alleviation: Geographic distribution, economic benefits and emission mitigation. *Energy Policy* **2020**, *144*, 111685. [CrossRef]

10.   Xu, M.; Xie, P.; Xie, B.C. Study of China's optimal solar photovoltaic power development path to 2050. *Resour. Policy* **2020**, *65*, 101541. [CrossRef]

11.   International Energy Agency. Renewables 2019. 2019. Available online: https://www.iea.org/reports/renewables-2019 (accessed on 6 February 2020).

12.   Lv, T.; Yang, Q.; Deng, X.; Xu, J.; Gao, J. Generation expansion planning considering the output and flexibility requirement of renewable energy: The case of Jiangsu Province. *Front. Energy Res.* **2020**, *8*, 39. [CrossRef]

13.   Nassar, Y.F.; Hafez, A.A.; Alsadi, S.Y. Multi-Factorial Comparison for 24 Distinct Transposition Models for Inclined Surface Solar Irradiance Computation in the State of Palestine: A Case Study. *Front. Energy Res* **2020**, *7*, 163. [CrossRef]

14.   Manfren, M.; Nastasi, B.; Groppi, D.; Garcia, D.A. Open data and energy analytics-An analysis of essential information for energy system planning, design and operation. *Energy* **2020**, *213*, 118803. [CrossRef]

15.   Malof, J.M.; Hou, R.; Collins, L.M.; Bradbury, K.; Newell, R. Automatic solar photovoltaic panel detection in satellite imagery. In Proceedings of the 2015 International Conference on Renewable Energy Research and Applications (ICRERA), Palermo, Italy, 22–25 November 2015; pp. 1428–1431.

16.   Matas, J.; Chum, O.; Urban, M.; Pajdla, T. Robust wide-baseline stereo from maximally stable extremal regions. *Image Vis. Comput.* **2004**, *22*, 761–767. [CrossRef]

17.   Cortes, C.; Vapnik, V. Support-vector networks. *Mach. Learn.* **1995**, *20*, 273–297. [CrossRef]

18.   Malof, J.M.; Bradbury, K.; Collins, L.M.; Newell, R.G. Automatic detection of solar photovoltaic arrays in high resolution aerial imagery. *Appl. Energy* **2016**, *183*, 229–240. [CrossRef]

19.   Breiman, L. Random forests. *Mach. Learn.* **2001**, *45*, 5–32. [CrossRef]

20.   Malof, J.M.; Collins, L.M.; Bradbury, K.; Newell, R.G. A deep convolutional neural network and a Random Forest classifier for solar photovoltaic array detection in aerial imagery. In Proceedings of the 2016 IEEE International Conference on Renewable Energy Research and Applications (ICRERA), Birmingham, UK, 20–23 November 2016; pp. 650–654.

21.   LeCun, Y.; Bengio, Y. Convolutional networks for images, speech, and time series. *Handb. Brain Theory Neural Netw.* **1995**, *3361*, 1995.

22.   Malof, J.M.; Collins, L.M.; Bradbury, K. A deep convolutional neural network, with pre-training, for solar photovoltaic array detection in aerial imagery. In Proceedings of the 2017 IEEE International Geoscience and Remote Sensing Symposium (IGARSS), Fort Worth, TX, USA, 22–28 July 2017; pp. 874–877.

23.   Simonyan, K.; Zisserman, A. Very deep convolutional networks for large-scale image recognition. *arXiv* **2014**, arXiv:1409.1556.

24. Krizhevsky, A.; Sutskever, I.; Hinton, G.E. Imagenet classification with deep convolutional neural networks. In Proceedings of the Advances in Neural Information Processing Systems, Lake Tahoe, NV, USA, 3–6 December 2012; pp. 1097–1105.

25. Szegedy, C.; Liu, W.; Jia, Y.; Sermanet, P.; Reed, S.; Anguelov, D.; Erhan, D.; Vanhoucke, V.; Rabinovich, A. Going deeper with convolutions. In Proceedings of the IEEE Conference on Computer Vision and Pattern Recognition, Boston, MA, USA, 7–12 June 2015; pp. 1–9.

26. Ioffe, S.; Szegedy, C. Batch normalization: Accelerating deep network training by reducing internal covariate shift. *arXiv* **2015**, arXiv:1502.03167.

27. Szegedy, C.; Vanhoucke, V.; Ioffe, S.; Shlens, J.; Wojna, Z. Rethinking the inception architecture for computer vision. In Proceedings of the IEEE Conference on Computer Vision and Pattern Recognition, Las Vegas, NV, USA, 27–30 June 2016; pp. 2818–2826.

28. Szegedy, C.; Ioffe, S.; Vanhoucke, V.; Alemi, A. Inception-v4, inception-resnet and the impact of residual connections on learning. In Proceedings of the Thirty-first AAAI Conference on Artificial Intelligence, Honolulu, HI, USA, 21–26 July 2017.

29. He, K.; Zhang, X.; Ren, S.; Sun, J. Deep residual learning for image recognition. In Proceedings of the IEEE Conference on Computer Vision and Pattern Recognition, San Francisco, CA, USA, 4–9 February 2016; pp. 770–778.

30. Huang, G.; Liu, Z.; Van Der Maaten, L.; Weinberger, K.Q. Densely connected convolutional networks. In Proceedings of the IEEE Conference on Computer Vision and Pattern Recognition, San Francisco, CA, USA, 4–9 February 2017; pp. 4700–4708.

31. Jiang, F.; Grigorev, A.; Rho, S.; Tian, Z.; Fu, Y.; Jifara, W.; Adil, K.; Liu, S. Medical image semantic segmentation based on deep learning. *Neural Comput. Appl.* **2018**, *29*, 1257–1265. [CrossRef]

32. Zhou, Z.; Siddiquee MM, R.; Tajbakhsh, N.; Liang, J. Unet++: A nested u-net architecture for medical image segmentation. In *Deep Learning in Medical Image Analysis and Multimodal Learning for Clinical Decision Support*; Springer: Cham, Switzerland, 2018; pp. 3–11.

33. Roth, H.R.; Oda, H.; Zhou, X.; Shimizu, N.; Yang, Y.; Hayashi, Y.; Oda, M.; Fujiwara, M.; Misawa, K.; Mori, K. An application of cascaded 3D fully convolutional networks for medical image segmentation. *Comput. Med. Imaging Graph.* **2018**, *66*, 90–99. [CrossRef] [PubMed]

34. Hesamian, M.H.; Jia, W.; He, X.; Kennedy, P. Deep learning techniques for medical image segmentation: Achievements and challenges. *J. Digit. Imaging* **2019**, *32*, 582–596. [CrossRef] [PubMed]

35. Huang, Z.; Wang, X.; Huang, L.; Huang, C.; Wei, Y.; Liu, W. Ccnet: Criss-cross attention for semantic segmentation. In Proceedings of the IEEE International Conference on Computer Vision, Seoul, Korea, 27 October–2 November 2019; pp. 603–612.

36. Zhang, F.; Chen, Y.; Li, Z.; Hong, Z.; Liu, J.; Ma, F.; Han, J.; Ding, E. Acfnet: Attentional class feature network for semantic segmentation. In Proceedings of the IEEE International Conference on Computer Vision, Seoul, Korea, 27 October–2 November 2019; pp. 6798–6807.

37. Takikawa, T.; Acuna, D.; Jampani, V.; Fidler, S. Gated-scnn: Gated shape cnns for semantic segmentation. In Proceedings of the IEEE International Conference on Computer Vision, Seoul, Korea, 27 October–2 November 2019; pp. 5229–5238.

38. Liu, Y.; Chen, K.; Liu, C.; Qin, Z.; Luo, Z.; Wang, J. Structured knowledge distillation for semantic segmentation. In Proceedings of the IEEE Conference on Computer Vision and Pattern Recognition, Long Beach, CA, USA, 15–20 June 2019; pp. 2604–2613.

39. Paul, M.; Mayer, C.; Gool, L.V.; Timofte, R. Efficient video semantic segmentation with labels propagation and refinement. In Proceedings of the IEEE Winter Conference on Applications of Computer Vision, Snowmass Village, CO, USA, 1–5 March 2020; pp. 2873–2882.

40. Pfeuffer, A.; Schulz, K.; Dietmayer, K. Semantic segmentation of video sequences with convolutional lstms. In Proceedings of the 2019 IEEE Intelligent Vehicles Symposium (IV), Paris, France, 9–12 June 2019; pp. 1441–1447.

41. Jain, S.; Wang, X.; Gonzalez, J.E. Accel: A corrective fusion network for efficient semantic segmentation on video. In Proceedings of the IEEE Conference on Computer Vision and Pattern Recognition, Long Beach, CA, USA, 16-20 June 2019; pp. 8866–8875.

42. Nekrasov, V.; Chen, H.; Shen, C.; Reid, I. Architecture search of dynamic cells for semantic video segmentation. In Proceedings of the IEEE Winter Conference on Applications of Computer Vision, Snowmass Village, CO, USA, 1–5 March 2020; pp. 1970–1979.

43. Long, J.; Shelhamer, E.; Darrell, T. Fully convolutional networks for semantic segmentation. In Proceedings of the IEEE Conference on Computer Vision and Pattern Recognition, Boston, MA, USA, 7–12 June 2015; pp. 3431–3440.

44. Zhou, L.; Zhang, C.; Wu, M. D-LinkNet: LinkNet With Pretrained Encoder and Dilated Convolution for High Resolution Satellite Imagery Road Extraction. In Proceedings of the CVPR Workshops, Salt Lake City, UT, USA, 19–21 June 2018; pp. 182–186.

45. Wu, M.; Zhang, C.; Liu, J.; Zhou, L.; Li, X. Towards accurate high resolution satellite image semantic segmentation. *IEEE Access* **2019**, *7*, 55609–55619. [CrossRef]

46. Xu, Y.; Xie, Z.; Feng, Y.; Chen, Z. Road extraction from high-resolution remote sensing imagery using deep learning. *Remote Sens.* **2018**, *10*, 1461. [CrossRef]

47. Gao, L.; Song, W.; Dai, J.; Chen, Y. Road extraction from high-resolution remote sensing imagery using refined deep residual convolutional neural network. *Remote Sens.* **2019**, *11*, 552. [CrossRef]

48. Xu, Y.; Wu, L.; Xie, Z.; Chen, Z. Building extraction in very high resolution remote sensing imagery using deep learning and guided filters. *Remote Sens.* **2018**, *10*, 144. [CrossRef]

49. Yang, H.; Wu, P.; Yao, X.; Wu, Y.; Wang, B.; Xu, Y. Building extraction in very high resolution imagery by dense-attention networks. *Remote Sens.* **2018**, *10*, 1768. [CrossRef]

50. Huang, J.; Zhang, X.; Xin, Q.; Sun, Y.; Zhang, P. Automatic building extraction from high-resolution aerial images and LiDAR data using gated residual refinement network. *ISPRS J. Photogramm. Remote Sens.* **2019**, *151*, 91–105. [CrossRef]

51. Sun, G.; Huang, H.; Zhang, A.; Li, F.; Zhao, H.; Fu, H. Fusion of multiscale convolutional neural networks for building extraction in very high-resolution images. *Remote Sens.* **2019**, *11*, 227. [CrossRef]

52. Yu, L.; Wang, Z.; Tian, S.; Ye, F.; Ding, J.; Kong, J. Convolutional neural networks for water body extraction from Landsat imagery. *Int. J. Comput. Intell. Appl.* **2017**, *16*, 1750001. [CrossRef]

53. Chen, Y.; Fan, R.; Yang, X.; Wang, J.; Latif, A. Extraction of urban water bodies from high-resolution remote-sensing imagery using deep learning. *Water* **2018**, *10*, 585. [CrossRef]

54. Li, L.; Yan, Z.; Shen, Q.; Cheng, G.; Gao, L.; Zhang, B. Water body extraction from very high spatial resolution remote sensing data based on fully convolutional networks. *Remote Sens.* **2019**, *11*, 1162. [CrossRef]

55. Yuan, J.; Yang, H.H.L.; Omitaomu, O.A.; Bhaduri, B.L. Large-scale solar panel mapping from aerial images using deep convolutional networks. In Proceedings of the 2016 IEEE International Conference on Big Data (Big Data), Washington, DC, USA, 5–8 December 2016; pp. 2703–2708.

56. Badrinarayanan, V.; Kendall, A.; Cipolla, R. SegNet: A Deep Convolutional Encoder-Decoder Architecture for Image Segmentation. *IEEE Trans. Pattern Anal. Mach. Intell.* **2017**, *39*, 2481–2495. [CrossRef]

57. Ronneberger, O.; Fischer, P.; Brox, T.U. Convolutional networks for biomedical image segmentation. In Proceedings of the International Conference on Medical Image Computing and Computer-Assisted Intervention, Munich, Germany, 5–9 October 2015.

58. Camilo, J.; Wang, R.; Collins, L.M.; Bradbury, K.; Malof, J.M. Application of a semantic segmentation convolutional neural network for accurate automatic detection and mapping of solar photovoltaic arrays in aerial imagery. *arXiv* **2018**, arXiv:1801.04018.

59. Castello, R.; Roquette, S.; Esguerra, M.; Guerra, A.; Scartezzini, J.L. Deep learning in the built environment: Automatic detection of rooftop solar panels using Convolutional Neural Networks. *J. Phys. Conf. Ser.* **2019**, *1343*, 012034. [CrossRef]

60. Bradbury, K.; Saboo, R.; Johnson, T.L.; Malof, J.M.; Devarajan, A.; Zhang, W.; Collins, L.M.; Newell, R.G. Distributed solar photovoltaic array location and extent dataset for remote sensing object identification. *Sci. Data* **2016**, *3*, 1–9. [CrossRef]

61. Tan, M.; Le, Q.V. Efficientnet: Rethinking model scaling for convolutional neural networks. *arXiv* **2019**, arXiv:1905.11946.

62. Hu, J.; Shen, L.; Sun, G. Squeeze-and-excitation networks. In Proceedings of the IEEE Conference on Computer Vision and Pattern Recognition, Salt Lake City, UT, USA, 18–22 June 2018; pp. 7132–7141.

63. Li, X.; Zhao, H.; Han, L.; Tong, Y.; Yang, K. Gff: Gated fully fusion for semantic segmentation. *arXiv* **2019**, arXiv:1904.01803.

64. Milletari, F.; Navab, N.; Ahmadi, S.A. V-net: Fully convolutional neural networks for volumetric medical image segmentation. In Proceedings of the 2016 Fourth International Conference on 3D Vision (3DV), Stanford, CA, USA, 25–28 October 2016; pp. 565–571.

65. Patravali, J.; Jain, S.; Chilamkurthy, S. 2D-3D fully convolutional neural networks for cardiac MR segmentation. In *International Workshop on Statistical Atlases and Computational Models of the Heart*; Springer: Cham, Switzerland, 2017; pp. 130–139.

66. Malof, J.M.; Li, B.; Huang, B.; Bradbury, K.; Stretslov, A. Mapping solar array location, size, and capacity using deep learning and overhead imagery. *arXiv* **2019**, arXiv:1902.10895.

67. Chaurasia, A.; Culurciello, E. Linknet: Exploiting encoder representations for efficient semantic segmentation. In Proceedings of the 2017 IEEE Visual Communications and Image Processing (VCIP), St. Petersburg, Russia, 10–13 December 2017; pp. 1–4.

68. Lin, T.Y.; Dollár, P.; Girshick, R.; He, K.; Hariharan, B.; Belongie, S. Feature pyramid networks for object detection. In Proceedings of the IEEE Conference on Computer Vision and Pattern Recognition, Honolulu, HI, USA, 21–26 July 2017; pp. 2117–2125.

6

The Global Wind Resource Observed by Scatterometer

Ian R. Young [1,*], Ebru Kirezci [1] and Agustinus Ribal [1,2]

[1] Department of Infrastructure Engineering, The University of Melbourne, Melbourne, VIC 3010, Australia; edemirci@student.unimelb.edu.au (E.K.); agustinus.ribal@unimelb.edu.au (A.R.)

[2] Department of Mathematics, Faculty of Mathematics and Natural Sciences, Hasanuddin University, Makassar 90245, Indonesia

* Correspondence: ian.young@unimelb.edu.au

Abstract: A 27-year-long calibrated multi-mission scatterometer data set is used to determine the global basin-scale and near-coastal wind resource. In addition to mean and percentile values, the analysis also determines the global values of both 50- and 100-year return period wind speeds. The analysis clearly shows the seasonal variability of wind speeds and the differing response of the two hemispheres. The maximum wind speeds in each hemisphere are comparable but there is a much larger seasonal cycle in the northern hemisphere. As a result, the southern hemisphere has a more consistent year-round wind climate. Hence, coastal regions of southern Africa, southern Australia, New Zealand and southern South America appear particularly suited to coastal and offshore wind energy projects. The extreme value analysis shows that the highest extreme wind speeds occur in the North Atlantic Ocean with extreme wind regions concentrated along the western boundaries of the North Atlantic and North Pacific Oceans and the Indian Ocean sector of the Southern Ocean. The signature of tropical cyclones is clearly observed in each of the well-known tropical cyclone basins.

Keywords: wind speed; extreme value analysis; scatterometer

1. Introduction

The determination of the global offshore wind energy resource and the design and operation of coastal and offshore wind energy projects require long-term measurements of wind conditions. There are a variety of systems which can be used to provide such information, including: in situ anemometer data, remote sensing applications, and numerical modelling. Anemometer records obviously provide a direct measurement at the location of interest, but they are limited to specific areas and long-term records are seldom available at locations of interest. Satellite systems such as altimeters, radiometers and scatterometers have the advantage that they provide global coverage and the data record for each of these systems is now approximately 30 years. Numerical models have clearly advanced in terms of their accuracy and resolution, but they still require validation against actual measurements.

A number of previous studies have used an altimeter, radiometer and scatterometer measurements to assess global climatology of wind speed and wave height [1–4]. These studies have generally used relatively short satellite records (decade) for these studies. The advent of long-term databases of multiple-mission satellite data for altimeter [4–6], radiometer [7] and scatterometer [8] provides the opportunity to examine both the climatology and extreme value global distribution of wind speed. Each of these satellite systems has similar error statistics [9], however, the sampling pattern of altimeters and the degradation of radiometer measurements in heavy rain mean that scatterometers are the preferred instrument for the measurement of global-scale wind speed.

The present paper uses a long-term consistently calibrated and validated archive of scatterometer data compiled over a 27-year period [8] to investigate global wind climatology and extreme value (1 in 50- and 100-year) wind speed estimates. In addition, the data are processed to obtain the distribution of wind speeds along coastlines, thus making it directly applicable as a resource for the offshore wind energy industry.

The arrangement of the paper is as follows. Following this introduction, Section 2 provides an overview of scatterometer measurement of wind speed, the datasets and the methods used to estimate climatologic values and extremes values. Section 3 describes both the ocean-scale basin and coastline analyses of the data. This is followed by a discussion of the results in Section 4 and conclusions in Section 5.

2. Materials and Methods

2.1. Scatterometer Wind Speed Measurement

Scatterometers (like altimeters) are "active" sensors in that they transmit a microwave radar signal to the water surface and monitor the received energy. In contrast, radiometers are "passive" sensors in that they monitor radiation emanating naturally from the water surface. The transmitted energy is reflected from the water surface as a result of Bragg scattering [10]. Bragg scattering is constructive interferences between the transmitted wave and the surface water waves. Water waves with a wavelength $\lambda_w = \lambda / (2 \cos \theta)$ satisfy the Bragg resonance conditions, where λ is the microwave scatterometer wavelength and θ is the incident angle of the scatterometer energy on the water surface. Scatterometers typically operate in either the C-band (5.255 GHz) or the Ku-band (13.4 GHz), for which the microwave wavelengths are 5 and 2 cm, respectively. Hence, the wavelengths of the Bragg scattering water waves, λ are of order centimetres. These short centimetre-scale waves respond almost instantaneously to the local wind and hence the Bragg scattering provides a means of indirectly sensing the wind speed.

The radar cross-section, σ_0 is defined as the ratio of the received power at the satellite antenna to the transmitted power. The radar cross-section, σ_0, can be related to the neutral stability wind speed, measured at an elevation of 10 m, U_{10} and the wind direction, ϕ through a Geophysical Model Function (GMF), $\sigma_0 = GMF(U_{10}, \theta, \phi, \rho, \lambda)$, where ρ is the radar polarization [11–14]. There are two unknowns in this relationship, the wind speed, U_{10} and wind direction, ϕ. In addition, it is necessary to resolve the 180-degree wind direction ambiguity. Hence, a minimum of three measurements of the radar cross-section is required at each location. Multiple measurements (often more than three) are obtained by having an antenna configuration which can image the same location at different angles. There is a wide range of both fan and pencil beam scatterometer antenna designs which achieve these requirements, whilst measuring over a broad swath [8].

As noted above, U_{10} is the wind velocity at a reference height of 10 m for a neutral stability boundary layer. That is, a marine boundary layer for which the air and water temperatures are equal. In such a case, the boundary layer follows a logarithmic form [15–17]. If the water temperature is greater than the air temperature, then there is cold air overlaying warm air and hence density differences result in a vertical circulation. In this case, the boundary layer is described as unstable and there is a more uniform distribution of wind speed with height than represented by the logarithmic form. Conversely, if the water temperature is less than the air temperature, there is warm air overlaying cold air and the boundary layer is described as stable. In this case, the velocity profile varies more rapidly with height than described by the logarithmic profile [7]. Such stability effects have an influence on wind speed measurements and predictions over the ocean, irrespective of the data source. For instance, estimations of wind speeds at a reference height (usually 10 m) are impacted irrespective of whether they come from models, anemometers or remote sensing systems.

In the case of remote sensing measurements, including scatterometers, as described above, short-wavelength waves are being used as a proxy for wind speed. These short waves respond to the wind speed at a height comparable to their wavelengths, which are very short [7]. The calibration process used for such instruments [8] relates the radar backscatter σ_0 to anemometer data at a reference height of 10 m. As these data very commonly come from buoy mounted instruments at heights less than 10 m, they are corrected to a height of 10 m using the assumption of a neutrally stable logarithmic boundary layer. As such neutral conditions occur rarely over the world's oceans [7], this process introduces some error. As the calibrations are, however, carried out across many different buoys and over many years, a broad range of conditions is encountered. Hence, it is commonly assumed that such stability effects are averaged out in the calibration. Because of its long duration and number of buoys far from land, the NDBC buoy network [18] around the United States coastline is typically used for such calibrations. As shown by Young and Donelan [7], in the context of altimeter and radiometer, when such calibrations are applied to other geographic locations, with larger stability effects, they can result in a bias in the data. Although such detailed comparisons have not been undertaken for scatterometer, as the frequency of operation is similar to the altimeter, the impacts should be comparable. The stability impact of the altimeter is a function of air–water temperature difference and wind speed [7]. For a temperature difference of 2 °C and a wind speed of 5 ms^{-1}, the error is approximately 10%. This percentage error increases as the wind speed decreases and the temperature difference increases.

Noting that the mechanism used by scatterometers, and other remote sensing systems, to measure wind speed relies on the use of short-scale waves on the water surface as a proxy for the wind speed, such systems only measure wind speeds over the oceans (not land).

2.2. Scatterometer Database

Ribal and Young [8] have compiled a long-term database of duration 27-years (1992 to 2019) of scatterometer data. The data includes the satellite missions ERS-1, ERS-2, QUIKSCAT, METOP-A, OCEANSAT-2, METOP-B and RAPIDSCAT. Each scatterometer was independently calibrated against NDBC buoy data [18] and validated against other scatterometers in orbit at the same time at cross-over locations. In addition, as the performance of buoy anemometers is questionable at high wind speeds due to ocean wave sheltering [19–23], Ribal and Young [8] calibrated the scatterometers for wind speeds above 25 ms^{-1} using platform data.

Scatterometers are typically placed in near-polar sun-synchronous orbits (with the exception of RAPIDSCAT). The instrument measures over a swath varying in width between 1000 and 1500 km and with a spatial resolution of 25 km in both along-track and cross-track directions. As a result, a single scatterometer will image almost the full surface of the earth (between latitudes of ±80°) twice per day. As shown in Figure 1, for much of the period of the available data, there were multiple scatterometers in orbit. As a result, each location around the globe is imaged multiple times per day. As such, there are generally sufficient observations to form accurate monthly statistics for mean and percentile wind speed [7] as well as defining the tail of the probability distribution function for extreme value analysis [24].

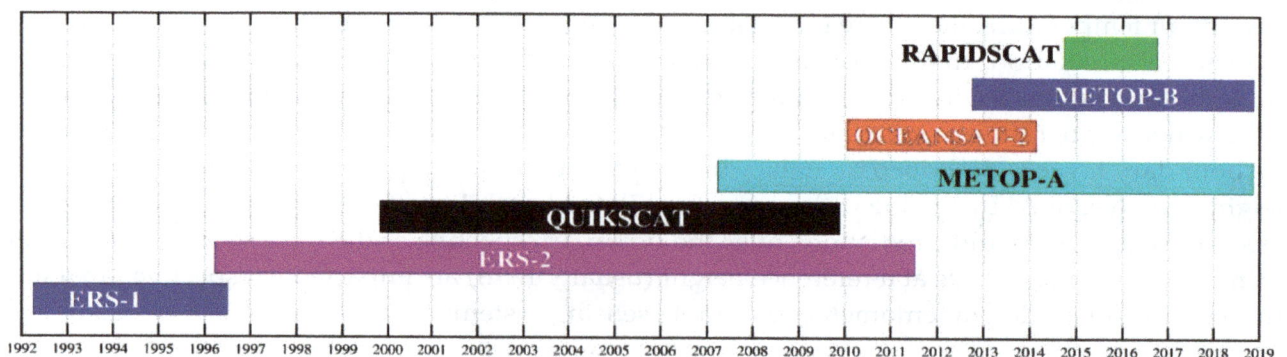

Figure 1. Scatterometers in the Ribal and Young [8] database and the duration of each mission.

The Ribal and Young [8] calibrations of the scatterometer systems used large co-location datasets (typically greater than 50,000 for each scatterometer) with buoy data. These calibrations produced typical error statistics between the scatterometer and buoy of the root mean square error, $RMSE \approx 1.0$ ms^{-1} and $|Bias| \approx 0.1$ ms^{-1}. In a triple co-location study involving scatterometer, altimeter, radiometer and buoys [9], scatterometers were shown to have error standard deviations of approximately 0.5 ms^{-1}. Interestingly, this was actually smaller than buoy mounted anemometers, 0.8 ms^{-1}. These error statistics are consistent with a range of other similar studies [25–30]. As a result of the consistent error statistics for scatterometers, they have largely become the reference for large-scale model validation of wind speeds over the oceans [31].

2.3. Climatology

In order to form stable monthly statistics, the data were binned into $2^\circ \times 2^\circ$ bins and the monthly mean ($\overline{U}_{10}$) and the monthly 10th ($U_{10}(10)$), 70th ($U_{10}(70)$), 90th ($U_{10}(90)$) and 99th ($U_{10}(99)$) percentiles determined for each bin [1,2,7], where U_{10} is the scatterometer calibrated wind speed at an elevation of 10 m.

2.4. Extreme Value Estimates

The aim of extreme value analysis (EVA) is to fit a theoretical probability distribution function (pdf) to a set of observations of events [32,33]. For the process to be valid, the data must satisfy two conditions. They must be independent and identically distributed (IID). As the focus of EVA is the distribution of the extreme tail of the distribution rather than the body of the pdf, independence can be interpreted as ensuring there are not multiple observations from the same storm. Unlike anemometer/buoy data, satellite data does not provide a time series at a fixed location where the passage of individual storms can be tracked. Hence, the requirement of independence is usually satisfied by ensuring a minimum time interval between observations. A value of 48 h has commonly been used [24,34–36] and is adopted here. The requirement that the data are identically distributed means that they should all be drawn from the same parent distribution. This requirement would not typically be satisfied where, for example, there are two or more independent meteorological systems which can generate extreme winds (e.g., tropical cyclones and trade winds). In such cases, the data should be partitioned, and the pdf of each partition determined separately [37]. The present dataset provides no mechanism to partition such data on a global scale. As a result, the present results may be limited in regions such as those where tropical cyclones dominate. Note such locations may also be impacted by the validity of the scatterometer data at high winds (see Section 4).

The application of EVA is typically aimed at determining the extreme wind speed associated with a defined probability of exceedance, P_r or the return period $T_r = 1/P_r$. As the length of the recorded time series of observations is very commonly shorter than the desired return period, EVA typically involves the fitting of a defined analytical pdf to the pdf of the observed extreme values and then extrapolating to the required probability level. Following Coles [33], there are two approaches which are commonly used to define the extreme value dataset and fit an appropriate analytical form: block maxima/annual maxima (AM) [33] and peaks over threshold (PoT) [34,38–41]. It can be shown [33,42] that for the AM method, in which the highest value in each year is selected, the resulting extreme value pdf will follow a generalized extreme value distribution. As only one value in each year is selected, the method ensures that the data are independent. The significant limitation of the method, however, is that, as only one value per year is selected, the pdf is typically constructed from only a limited number of data points (in the present case, 27 for the scatterometer data record). This results in considerable statistical sampling variability due to the small sample size. The PoT approach overcomes this limitation by selecting the peak value above a defined threshold, also ensuring there are no values separated by less than 48 h, as noted above. In such a case, it can be shown that the data will follow a generalized Pareto distribution (GPD) given by:

$$F(x) = 1 - \left[1 + k\left(\frac{x - A}{B}\right)^{-1/k}\right]$$ (1)

where $F(x)$ is the pdf and x is the variable under consideration (wind speed, U_{10}). In Equation (1), A defines the threshold value, B is a scale parameter and k is a shape parameter. It is common to select a high percentile for the threshold, in this case, the 90th percentile [24,34,36,41,43–45]. With the theoretical pdf defined in this manner, it can be fitted to the observed data determining the best-fit parameters B and k. The desired return period, in this case, the 100-year value is then defined as:

$$P_r(x < x^{100}) = 1 - N_y/(100 N_{PoT})$$ (2)

where N_{PoT} is the number of data points in the PoT analysis and N_y is the number of years spanned by the analysis.

As for the calculation of the climatology above, the data were binned into a 2° × 2° grid and both 50- and 100-year return period values of wind speed (U_{10}^{50}, U_{10}^{100}) determined for each grid square. In order to validate the EVA described above, scatterometer values of U_{10}^{100} were compared with deep water NDBC [18] buoy calculations using the same PoT analysis described above. The differences between buoy and scatterometer were quantified in terms of the relative error:

$$\Delta r = \left[U_{10}^{100}(Scat.) - U_{10}^{100}(Buoy)\right]/U_{10}^{100}(Buoy)$$ (3)

These values were then averaged across all buoys to obtain a mean error, r. Table 1 shows a comparison across the NDBC buoys. The scatterometer values of extreme wind speed are in reasonable agreement with the buoy data with a mean error of 15.2%. This value is quite similar to the results obtained by Takbash et al. [24] using altimeter derived wind speed data. It is interesting that the values of the scatterometer U_{10}^{100} are consistently higher than the buoy data. This may be because extreme values calculated over a 2° × 2° region are not the same as a time series from a point location. That is, over the spatial region more storm peaks occur than at a point. This difference is considered further in Section 4.

Table 1. Comparison of values of 100-year return period wind speed calculated using a PoT analysis for both scatterometer and NDBC deep-water buoys.

Buoy No.	Lat (°N), Lon(°E)	U_{10}^{100}(Buoy) (ms^{-1})	U_{10}^{100}(Scat.) (ms^{-1})
46001	56.23, 212.05	24.9	29.9
46002	42.61, 229.46	24.4	26.9
46003	51.33, 204.15	26.1	30.5
46005	46.14, 228.93	25.3	29.1
46006	40.78, 222.60	27.2	31.0
51005	24.42, 197.90	18.9	26.0
44004	38.48, 289.57	27.3	34.1
41002	31.76, 285.16	25.9	34.0
42001	25.90, 270.33	28.1	28.0
42002	26.09, 266.24	26.3	29.0
Error, r	-	-	15.2%

2.5. Coastal Wind Speeds

As noted above, both the climatological and extreme values wind speeds were evaluated globally on a 2° × 2° grid. As our aim is to obtain representative values for coastlines, these values need to be associated with coastal segments. In order to do this, we have adopted the Dynamic Interactive Vulnerability Assessment database (DIVA) [46,47]. DIVA is a database for the assessment of coastal vulnerability at the global scale. The database, as such, is not used in this analysis, rather, we have adopted the 9866 DIVA coastal locations as appropriate points to define the near-coastal wind climate. Using a Geographical Information System (GIS) model, representative wind speed values (climatological and extreme) at each DIVA point were associated with the closest 2° grid point [47,48]. To ensure scatterometer microwave radar returns are not corrupted by the proximity to land, all data closer than 25 km from land were excluded from the analysis. As such, the coastal values reported are characteristic of the open ocean conditions offshore from the coast.

3. Results

3.1. Ocean Basin Analysis

Figures 2 and 3 show the mean monthly ($\overline{U}_{10}$) and monthly 99th percentile ($U_{10}(99)$) wind speeds, respectively, colour contoured globally. The major climatological features previously reported from model and altimeter wind speed climatology [1,7] are apparent. The zonal variation of wind speeds is clear with the strongest wind speeds evident at high latitudes in winter in both hemispheres. The magnitudes of the mean winter wind speeds (Figure 2, January, July) at high latitudes ($\sim$12 ms^{-1}) in both hemispheres are comparable. The strongest mean winds are concentrated in the North Atlantic (Figure 2, January) and the Indian Ocean basin of the Southern Ocean (Figure 2, July). The summer comparison between the hemispheres at high latitudes are quite different from the North Atlantic decreasing significantly to $\sim$8.5 ms^{-1}, whilst the Southern Ocean decreases to only $\sim$10.5 ms^{-1}. That is, the Northern hemisphere has a much larger seasonal variation whilst the Southern Ocean experiences strong mean winds year-round. The North Atlantic experiences slightly stronger mean wind speeds than the North Pacific in both summer and winter.

The other striking feature of the distribution of ocean basin-scale wind speeds are the strong trade wind belts in the mid-latitudes of both hemispheres, which exist across all major ocean basins (Figure 2). Compared to higher latitudes, the trade wind belts are persistent for most of the year at mean wind speeds of $\sim$8 ms^{-1}. There are also a number of local features such as the strong localized mean wind speeds associated with the Somali jet in the Arabian Sea in June, July, August (Figure 2), associated with the onset of the Monsoon. Clear wind shadows are also seen east of New Zealand and South America (April to November Figure 2), where the strong westerly winds are blocked by the high mountain ranges.

The 99th percentile wind speeds, ($U_{10}(99)$) shown in Figure 3 indicate similar seasonal variations as the means. However, the trade winds are no longer apparent at these more extreme percentiles. That is, the trade winds are persistent year-round without being extreme. There is also a large triangularly shaped region at low latitudes in the southeastern Pacific where $U_{10}(99)$ is noticeably low, $\sim$10 ms^{-1}. The Somali jet which was prominent for $\overline{U}_{10}$ from June to August is no longer apparent for $U_{10}(99)$, indicating it is associated with consistent moderate winds but not extremes.

At these higher percentiles, the maximum wind speeds again occur in the respective winters of both hemispheres at high latitudes (Figure 3, January, July). Again, the southern hemisphere has a much smaller seasonal variation (extreme wind all year) than the northern hemisphere. Whereas the mean values were a maximum in the North Atlantic, the $U_{10}(99)$ maximum values are comparable across the North Atlantic, North Pacific and the Indian Ocean sector of the Southern Ocean ($\sim$22 ms^{-1}) (Figure 3, January, July).

Figures showing the distributions for the other percentiles are shown in the Supplementary Material (SM) ($U_{10}(10)$—Figure S1, $U_{10}(70)$—Figure S2, $U_{10}(90)$—Figure S3). The higher percentile cases are very similar to $U_{10}(99)$ (Figure 3). However, the spatial distributions of $U_{10}(10)$ (Figure S1) indicates that at this lower percentile the trade winds are now much more dominant. For significant, periods of the year the trade wind belts show the strongest winds globally at this lower percentile. This clearly demonstrates the differing structures of the wind speed pdfs in the trade winds compared to higher latitudes.

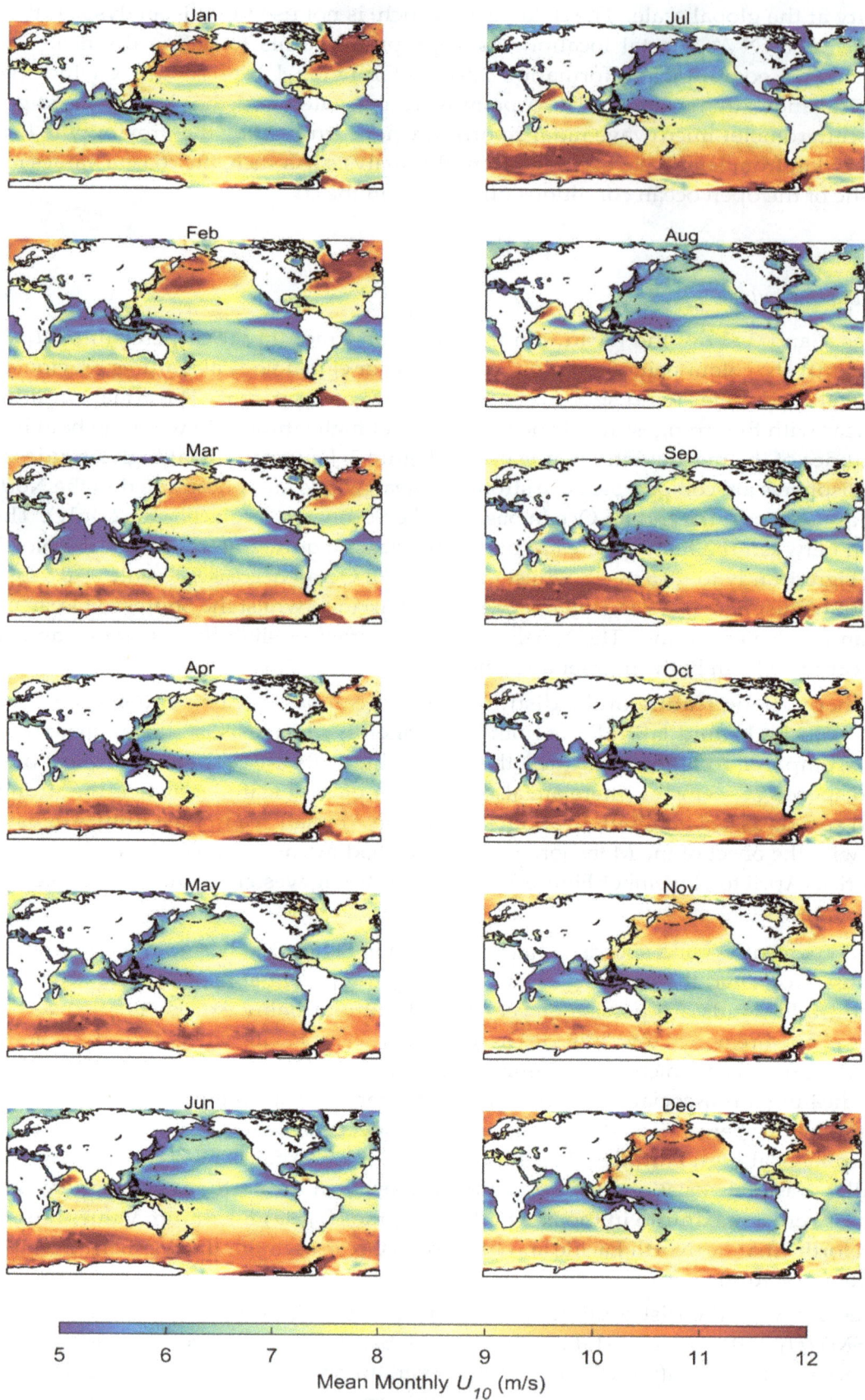

Figure 2. Mean global monthly wind speed at a reference elevation of 10 m (U_{10}) from scatterometer data, $\overline{U}_{10}$. The data were gridded at 2° resolution.

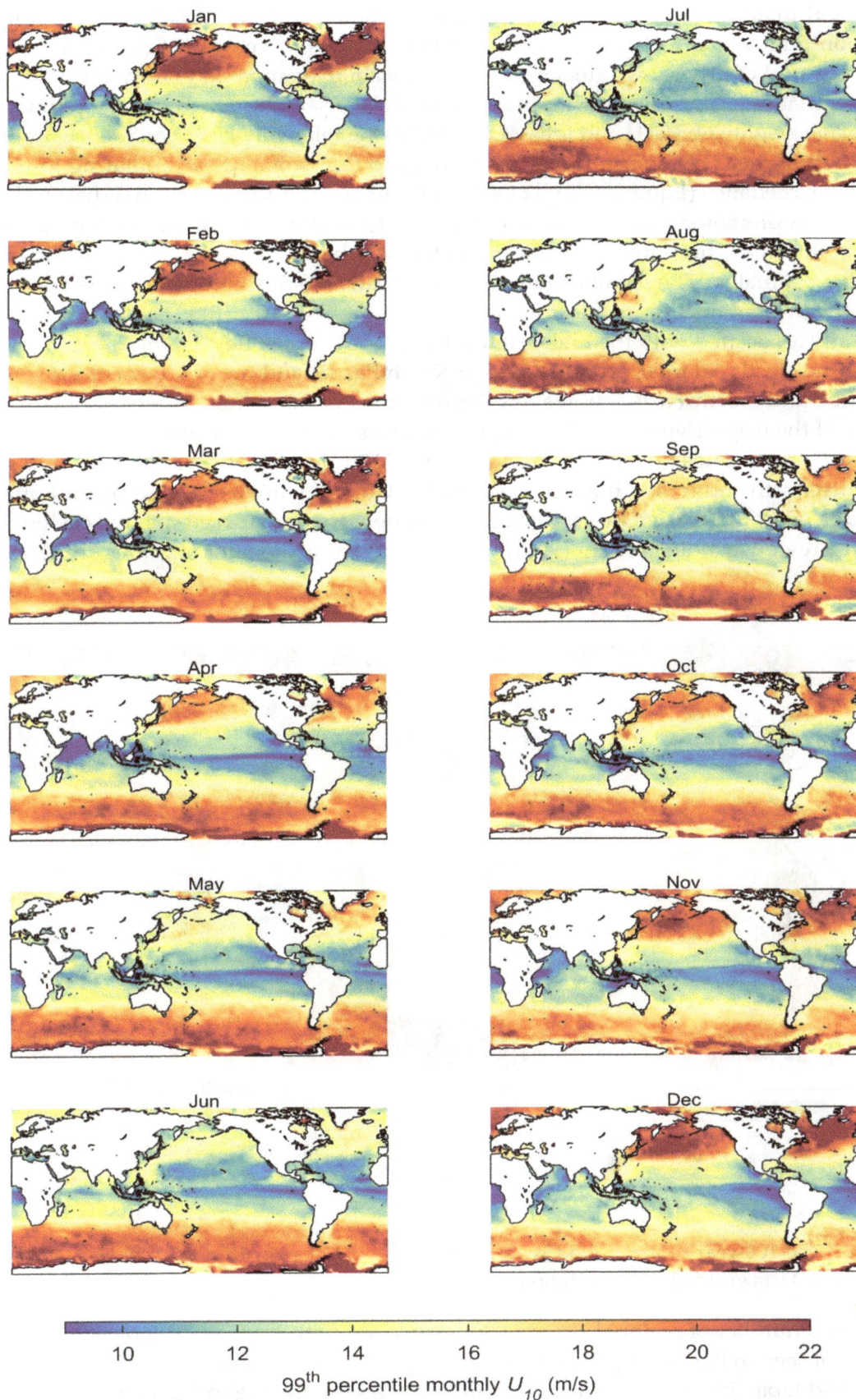

Figure 3. The 99th percentile global monthly wind speed at a reference elevation of 10 m (U_{10}) from scatterometer data, $U_{10}(99)$. The data were gridded at 2° resolution.

The global distribution of the 100-year return period wind speed U_{10}^{100} is shown in Figure 4. The corresponding results for the 50-year return period are shown in the Supplementary Material (Figure S7). Figure 4 clearly shows greater statistical variability compared to the percentile plots above. This level of variability is consistent with previous studies of satellite extreme value estimates of wind speed and wave height [24,34,41] and is associated with the relatively large confidence limits resulting from the fitting of the GPD distribution Equation (1) to the data and extrapolating to the required probability of exceedance (Equation (2)) [24,43]. Both the magnitude and spatial distribution of U_{10}^{100} are similar to that presented using a similar PoT analysis for altimeter data by Takbash et al. [24]. It is believed that this is the first time that scatterometer data has been processed to obtain such global extreme value estimates. The results have some similarities to the 99th percentile values (Figure 3), noting that the present results are effectivel for an even higher percentile. The maximum values are again generally associated with the higher latitudes of both hemispheres, with the maximum values occurring in the North Atlantic, followed by the North Pacific and the Southern Oceans. In both the North Atlantic and North Pacific, however, these regions of maximum values are displaced to the western side of the respective oceanic basins. This mirrors the tracks taken by storms in each of these basins. Atlantic hurricanes typically track across the Atlantic and either propagate into the Gulf of Mexico or turn north and follow the East Coast of the United States as they decay and often become higher latitude extra-tropical cyclones. The approximately triangular region in the eastern equatorial south Pacific previously noted for $U_{10}(99)$ (Figure 3) is again present for U_{10}^{100} (Figure 4).

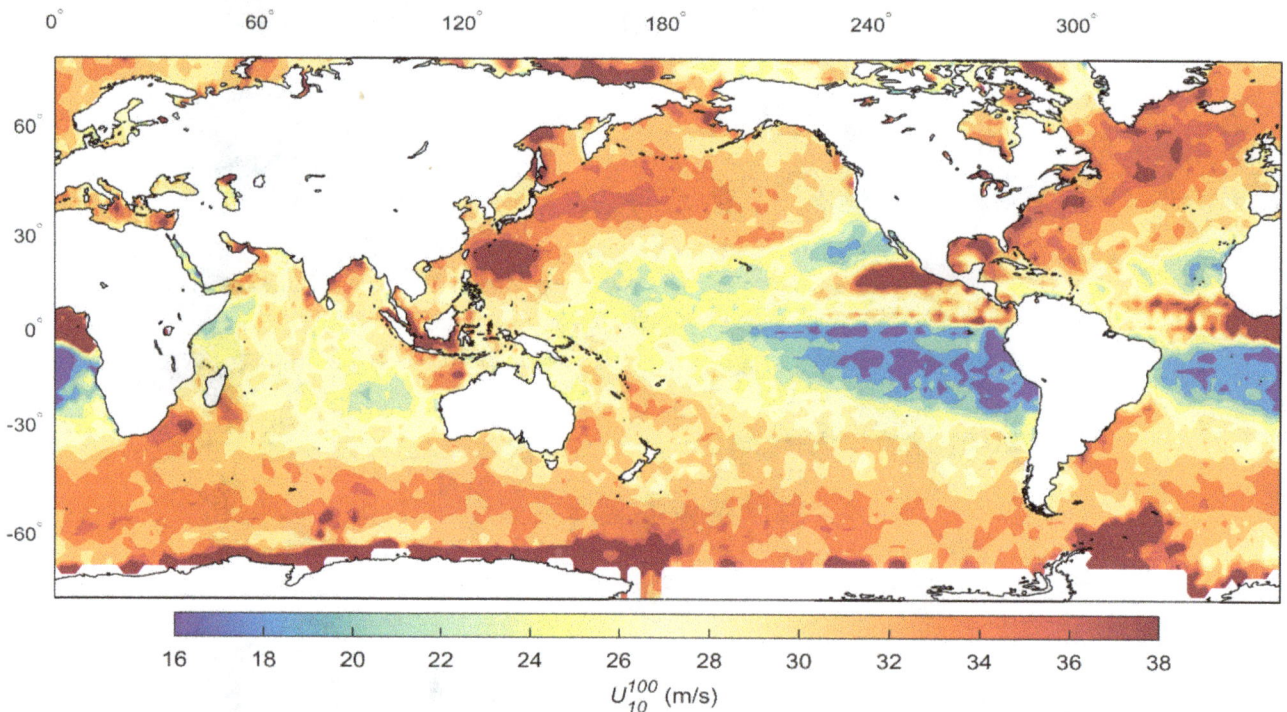

Figure 4. The 100-year return period wind speed, U_{10}^{100} obtained with a PoT analysis and a GPD distribution. Data gridded at 2° resolution.

There are a number of regional maxima, all associated with tropical cyclone activity in Figure 4, which were not seen in the lower percentile estimates in Figure 3. These include the area between the Philippines and Japan in the South China Sea, associated with the highest density of tropical cyclone occurrence globally (typhoons) [24]. The signature of North American hurricanes is also seen east of Mexico and in the Gulf of Mexico. Further tropical cyclone related maxima are also present northwest of Australia, the Bay of Bengal and Madagascar.

An area of lower values of U_{10}^{100} is located in the Indian Ocean west of Australia. This is an area where there is an unusually low density of tropical cyclone tracks [24]. There is also a region of large values of U_{10}^{100} off the southeast coast of Australia. It is believed that this is associated with the occurrence of so-called "East Coast lows" in this region which results in intense storm conditions [49]. The enhanced spatial resolution of the scatterometer compared to altimeter means that these localized regions of extreme wind speeds are visible in the present EVA but have not been resolved in previous studies.

One area of note is the region of high winds identified in the Gulf of Guinea (West Africa). It is believed that these values are spurious and associated with a poor fit to the extreme value pdf (see Section 4).

3.2. Coastal Wind Climate

The coastal distributions of the mean month wind speed, $\overline{U}_{10}$, and 99th percentile monthly wind speed, $U_{10}(99)$ are shown in Figures 5 and 6, respectively. Again, the 10th, 70th and 90th percentile values are shown in the SM ($U_{10}(10)$—Figure S4, $U_{10}(70)$—Figure S5, $U_{10}(90)$—Figure S6). The largest values of mean month wind speed ($\overline{U}_{10} > 10$ ms^{-1}) occur in the northern hemisphere winter along the East and West Coasts of Canada, the northeast coast of the United States, northern Europe and the coast of northern China and the Pacific coast of Russia. These results are consistent with the global distributions shown in Figure 2. With the exception of the southern tip of South America, southern hemisphere land masses are at lower latitudes than these regions in the northern Hemisphere, and hence experience lower mean monthly wind speeds than in the northern hemisphere.

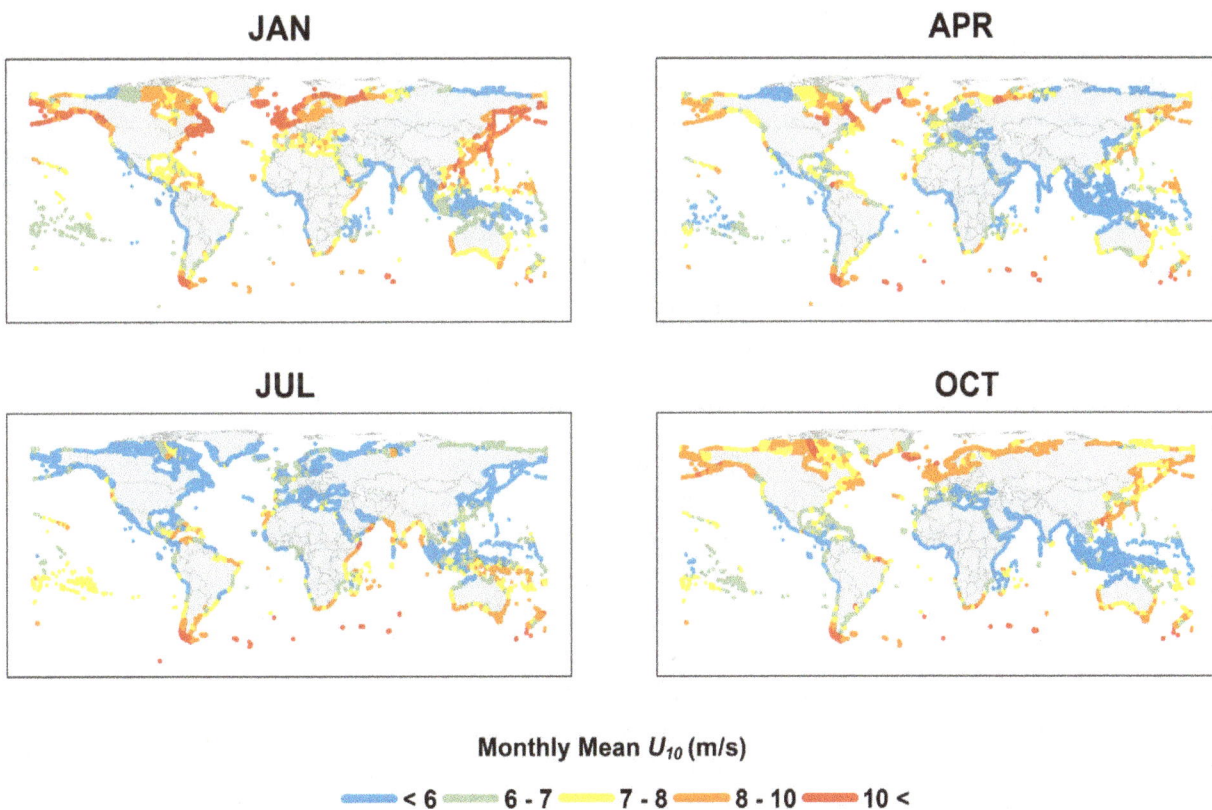

Figure 5. Mean monthly wind speed, at a reference elevation of 10 m (U_{10}), $\overline{U}_{10}$ at near-coastal locations defined by the DIVA dataset.

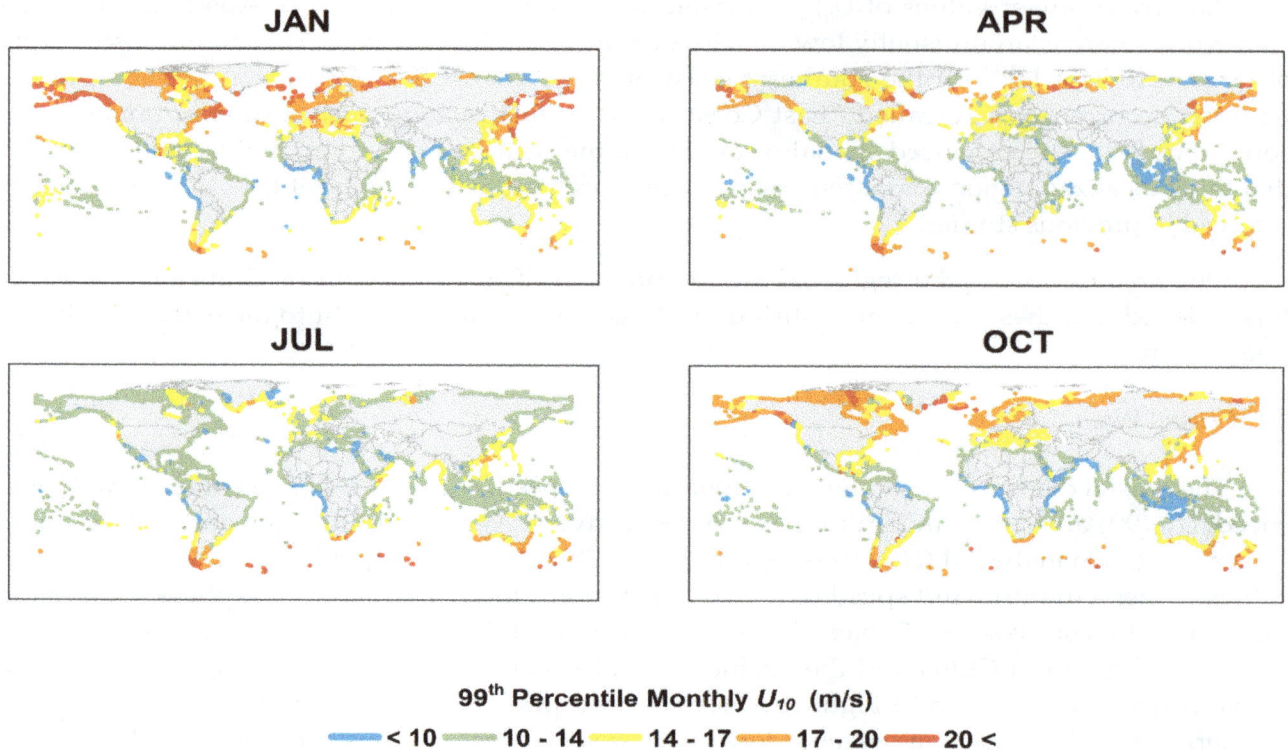

Figure 6. The 99th percentile monthly wind speed at a reference elevation of 10 m (U_{10}), $U_{10}(99)$ at near-coastal locations defined by the DIVA dataset.

As there is a much smaller seasonal variability in the southern hemisphere (Figure 2), however, the coastlines of southern Africa, southern Australia, New Zealand and southern South America experience values of $\overline{U}_{10}$ between 7 and 8 ms^{-1} year-round. In contrast, the northern hemisphere locations mentioned above vary between less than 6 ms^{-1} in summer and greater than 10 ms^{-1} in winter. Hence, the southern hemisphere provides a potentially more consistent wind climate for offshore wind energy generation.

The trade wind belts which were so prominent in the global distributions (Figure 2) do not have a significant impact on mean monthly near-coastal wind speeds. Rather, there is a low latitude/high latitude distribution. In contrast to the higher latitudes, described above, lower latitudes ($\sim \pm 20°$) have values of $\overline{U}_{10} < 6$ ms^{-1}, year-round.

The 99th percentile wind speeds, $U_{10}(99)$ (Figure 6) show a very similar spatial distribution as for $\overline{U}_{10}$. The high latitude northern hemisphere regions (Canada, northeast US, northern Europe, China, Pacific coast of Russia) show $U_{10}(99)$ varying between 10 and 20 ms^{-1} (summer to winter). In contrast, the high latitude southern hemisphere regions (southern Africa, southern Australia, New Zealand, southern South America) vary (summer to winter) between 14 to 17 ms^{-1} for $U_{10}(99)$. The low latitude regions ($\sim \pm 20°$) have values less than 10 ms^{-1}, year-round.

The 100-year return period near-coastal values, U_{10}^{100} (Figure 7) show regions in excess 32 ms^{-1} along the east coast of the United States, around Greenland, the Gulf of Mexico, the Bay of Bengal and South Africa. These same areas are also clear in the spatial distributions of Figure 4. Although Figure 4 shows very high values of U_{10}^{100} in the South China Sea, the values near the coast in this region are lower, ranging from 24 to 28 ms^{-1}. As also seen in Figure 4, there is a relatively large percentage of the oceanic basins with values of U_{10}^{100} between 20 to 28 ms^{-1}. In Figure 7, these areas occur down the east coast of the United States, the southern parts of South America, northern Europe, most of Asia, and most of Australia.

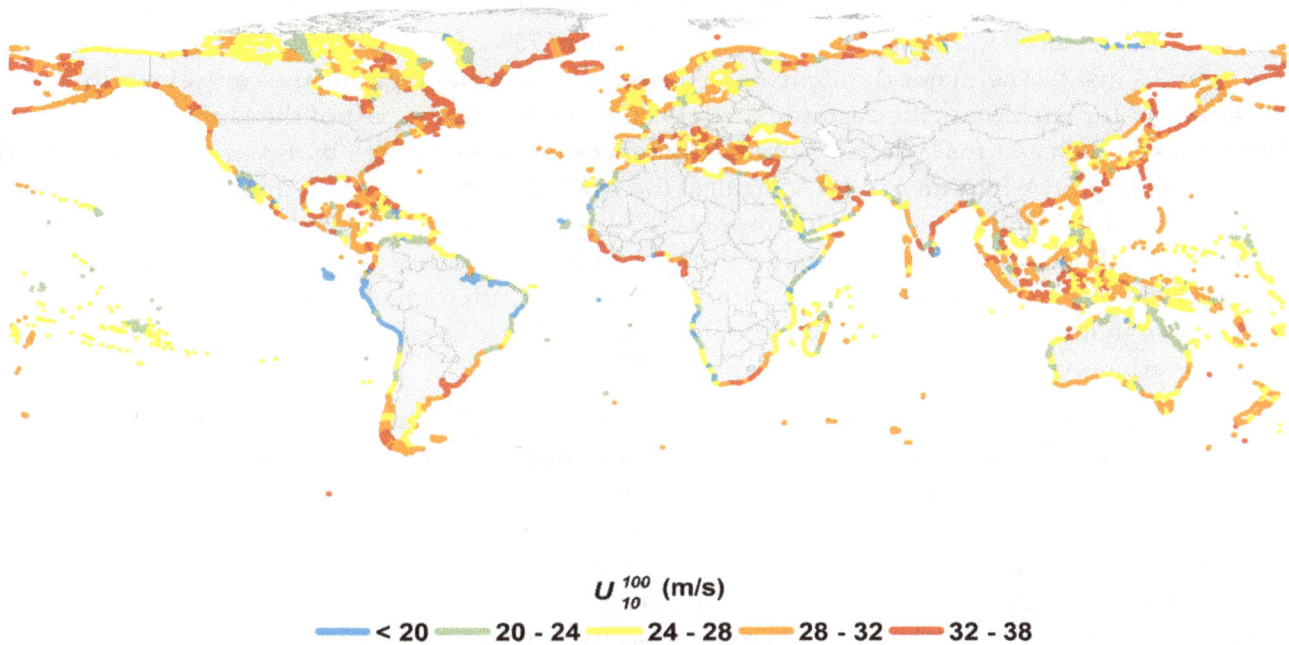

$$U_{10}^{100} \text{ (m/s)}$$

━━━ < 20 ━━━ 20 - 24 ━━━ 24 - 28 ━━━ 28 - 32 ━━━ 32 - 38

Figure 7. The 100-year return period wind speed, U_{10}^{100} at near-coastal locations obtained with a PoT analysis and a GPD distribution.

4. Discussion

The scatterometer data used in this study were selected for a number of reasons. Firstly, our desire is for a global assessment of the near-coastal wind resource, limiting consideration to either satellite or model data sources. Global wind models are now of high quality and certainly offer a valuable data resource. However, models are still limited by our understanding of the underlying physics and our ability to represent sub-grid scale influences. For this reason, we have opted for a satellite data source. Indeed, the present dataset represents a useful validation source for future model studies. Of the three potential global satellite data sources: altimeter, radiometer, scatterometer, all have some limitations. Altimeters measure only along a narrow beam at the satellite nadir. This means that, although the altimeter provides global coverage, the spatial resolution is limited, with altimeter tracks typically being separated by hundreds of kilometres [2,24,41]. Radiometers measure over a broad swath, eliminating the sampling constraint of altimeters. However, radiometer measurements are seriously degraded during rain events. As heavy rain events are often associated with strong winds, this constraint results in a "fair weather" bias for radiometer data. This limitation has been shown to make such data unusable for EVA [24]. In contrast, scatterometer data address both of these constraints, measuring over a broad swath and providing acceptable data in most rain events. Note, scatterometer data is degraded in heavy rain but not to the same magnitude as the radiometer. In addition, in triple collocation studies, it has been shown that the random errors in all three instruments are comparable and similar to buoys [9].

As noted above, global wind models have become common analysis and design tools and long-duration global reanalyses, such as ERA-Interim [50] and ERA5 [51] are commonly available. In order to assess the present composite scatterometer database against such reanalysis data, a comparison was made with the ERA5 dataset. ERA5 wind data is available at 0.25° resolution. These data were regretted to 2° resolution to make a comparison with the present scatterometer data. Rather than undertaking comparisons for every month as in Figure 2, both ERA5 and scatterometer data were averaged across the time period 1992 to 2019 and all months, to form annual average distributions of the 10 m elevation wind speed, U_{10}. Figure 8a,b shows the global distributions of these annual average values for scatterometer and ERA5, respectively. The spatial distributions of the major elements of the wind climatology are very similar between the two products. The strong zonal winds of the Southern Ocean, the strong winds of the North Atlantic and North Pacific and the distinct trade wind belts are all reproduced in both datasets. Figure 8c shows the difference between these results $(\overline{U}_{10}(Scatt.) - \overline{U}_{10}(ERA5))$. A total of 75% of locations have an absolute difference less than 0.5 ms^{-1}. The major differences occur in the equatorial regions, the central South Pacific, the southern Indian Ocean, the eastern North Pacific and the East Coast of the United States where differences as high as 1 ms^{-1} occur. Similar differences have been noted by Rivas and Stoffelen [31] when validating ERA-Interim and ERA5 against the ASCAT scatterometer. In this case, the differences were assumed to be limitations in the models' ability to represent the impacts of ocean currents. However, Quilfen et al. [52] noted similar differences between buoy and ERS scatterometer data in the equatorial Pacific, indicating the issue may be regional scatterometer bias. As noted earlier, it has been shown that for the altimeter and radiometer, atmospheric boundary layer stability can result in regional variations in wind speed measurements [7].

To investigate the potential impact of atmospheric stability, the air–water temperature difference $(\Delta T = T_a - T_w)$ is shown in Figure 8d. The values of T_a and T_w were obtained from the ERA-Interim archive. There is a remarkable degree of similarity between the wind speed difference (Figure 8c) and ΔT (Figure 8d). Many of the major ocean currents are clearly seen in this figure, including the Kuroshio current (Japan), the Gulf stream (East coast N. America) and the East Australian current. The warm waters across the equatorial regions are also clear. The high level of similarity between Figure 8c,d suggests that stability effects are important in describing the wind speed differences [7]. However, it is not possible to determine whether the scatterometer or ERA5 reanalysis is in error. It is likely that both datasets are impacted by these regional influences. Figure 8 provides a clear indication of the potential variability of boundary layer shape as a result of such global variations in air–water temperature differences. As pointed out by Young and Donelan [7], the vast majority of the world's oceans experience unstable atmospheric conditions for most of the year.

There are also other limitations of the present data which are important to consider. Although scatterometer passes generally have a spatial resolution of 25 km, the present data have been binned at 2° resolution to form stable statistics (monthly means and percentiles and extreme value estimates). This is consistent with a range of previous studies [1,2,7,24,34,41]. In the present application, however, it means that the near-coastal values are representative of wind speeds, of order, 100 km offshore. Therefore, the results do not account for local influences which may change wind speeds closer to shore. Also, this resolution will not resolve local orographic effects.

The present analysis uses a multi-mission scatterometer database which has been extensively calibrated and validated against buoy and platform anemometer measurements. These calibrations are, however, limited to wind speeds up to 25 ms^{-1}. At higher wind speeds, the radar return for scatterometers begins to saturate, limiting accuracy and the validity of the calibration used [53–55]. As a result, there is reduced confidence in tropical cyclone regions and particularly, the EVA analysis for these locations. Despite this, the present results (Figure 4) yield EVA estimates in tropical cyclone regions which are consistent with design practice for such areas [56]. In addition, the present EVA results also show a range of local responses to tropical cyclones not previously apparent in either the lower resolution altimeter studies [24] or model estimates of such extremes [43].

Extreme value analyses such as that undertaken here are sensitive to the detailed shape of the measured pdf. In the present case, a GPD has been used to fit the observed data (Equation (1)). This functional form has three free parameters giving maximum flexibility in the fit. However, this can sometimes give rise to unrealistic fits to the data. For instance, the shape parameter k can be either negative, resulting in a distribution with an upper bound, or positive resulting in an unbounded distribution. An example of such a poor fit to the data occurs in the Gulf of Guinea, where the GPD results in unrealistically large values of 100-year return period wind speeds, U_{10}^{100}. An alternative is to use a two-parameter exponential distribution in place of the GPD. However, the exponential distribution yields significantly higher values of U_{10}^{100} which, unlike the GPD, are not in agreement with buoy data (Table 1).

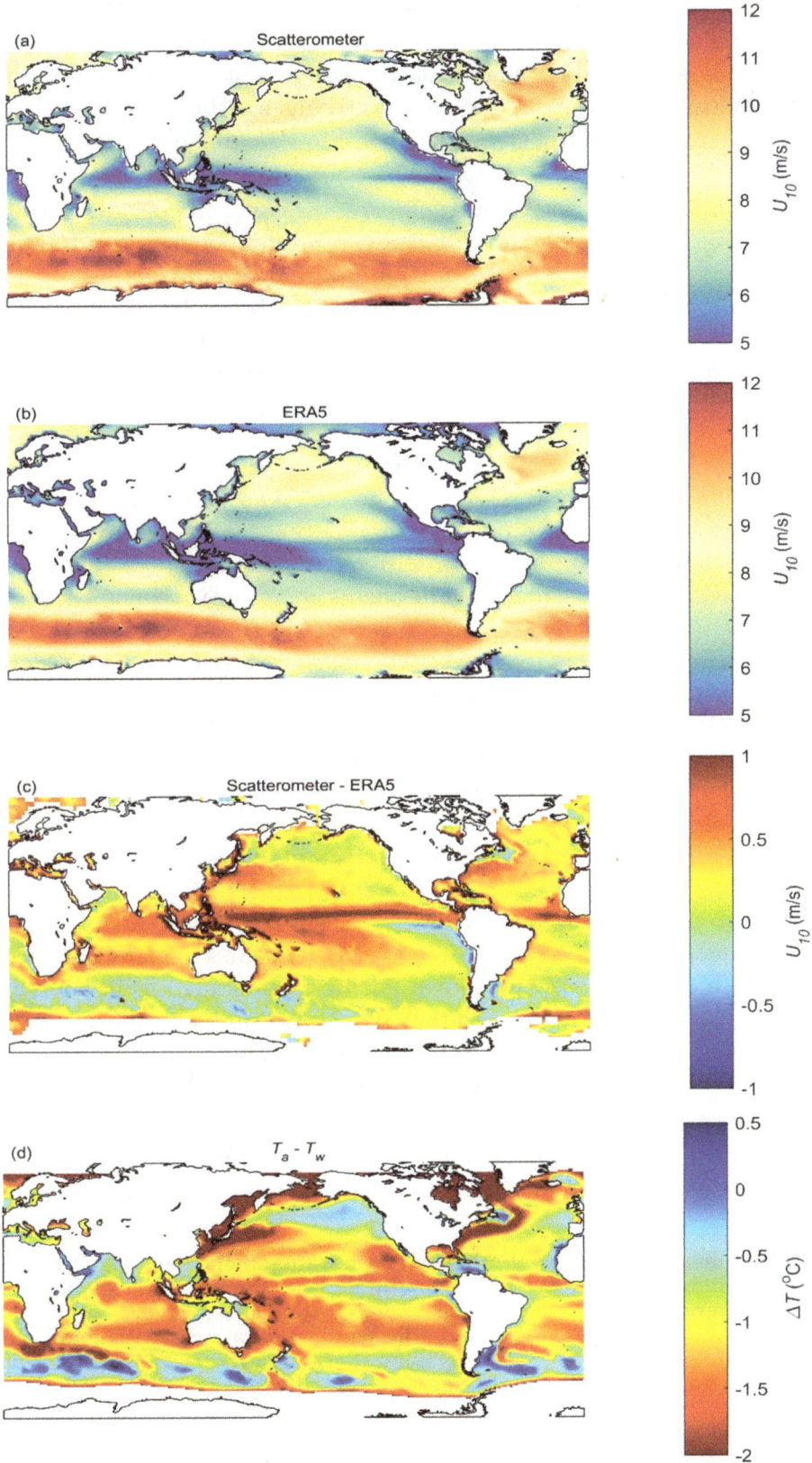

Figure 8. Comparison of mean annual wind speed at a reference height of 10 m. (**a**) Scatterometer wind speed, (**b**) ERA5 wind speed, (**c**) difference between scatterometer and ERA5 $\overline{U}_{10}(Scatt.) - \overline{U}_{10}(ERA5)$, (**d**) air–water temperature difference, $\Delta T = T_a - T_w$.

The present EVA for the scatterometer data is consistent with buoy data (Table 1). However, the mean error across all locations is approximately 15%. In addition, at all but one location, the scatterometer yields larger extreme values. This result is very similar to that of Takbash et al. [24] for altimeter data. This suggests that the $2° \times 2°$ binning used for these satellite datasets results in different extreme value statistics to a point buoy location. This is not surprising and suggests that more extreme events are identified by pooling the data over the spatial domain.

The climatologies in Figures 2 and 3 represent averages over the period 1992 to 2019. Multiple studies have shown that over this period global mean wind speeds have changed as a result of multi-decadal oscillations and possibly anthropogenic global climate change. In a range of studies using both remote sensing and reanalysis data [57–59], it has been shown that the largest increases have occurred in the Southern Ocean. Over the period from 1985 to 2018 the average rate of wind speed increase in the Southern Ocean is approximately $2 \text{ cm}^{-1} \text{ yr}^{-1}$.

5. Conclusions

The present analysis uses a calibrated scatterometer multi-mission database of 27-years duration to examine the global near-coastal wind resource. The data were processed to determine the global climatology of monthly mean and percentile values and to present these both on an ocean-basin scale basis and around global coastlines. In addition, an extreme value analysis was undertaken to determine 100-year return period wind speeds across oceanic basins and at near-coastal sites.

The results clearly show the seasonal variations of wind speed across the globe. Of particular note is the different seasonal impacts in the different hemispheres. Both hemispheres have a maximum mean and percentile wind speeds in their respective winters at high latitudes. However, the northern hemisphere has a much larger variation between summer and winter than the southern hemisphere. In their respective winters, the maximum values of mean and percentile conditions are comparable for both hemispheres. However, northern hemisphere summers are relatively calm, whereas the southern hemisphere (and particularly the Southern Ocean) is relatively windy year-round. This means that a range of southern hemisphere locations seem particularly suitable for wind energy generation, with a consistent wind climate year-round.

The 100-year return period wind speeds have some similarities to the upper percentile climate analysis. However, in the northern hemisphere, the maximum values are now displaced to the western boundaries of both the North Atlantic and North Pacific Oceans. This mirrors the locations of the major storm tracks in these regions. Although the analysis is limited by the accuracy of wind speed measurements from scatterometers above 30 ms^{-1}, the extreme value analysis clearly shows the regional impacts of all the major tropical cyclone basins.

The results presented in the paper and the data which it has been based upon have been archived for open use (see Data Availability below), providing a resource for the assessment of potential wind energy projects at locations around the world.

Supplementary Materials
Figure S1: The 10th percentile global monthly wind speed from scatterometer data; Figure S2: The 70th percentile global monthly wind speed from scatterometer data; Figure S3: The 90th percentile global monthly wind speed from scatterometer data; Figure S4: The 10th percentile monthly wind speed at near-coastal locations defined by the DIVA dataset; Figure S5: The 70th percentile monthly wind speed at near-coastal locations defined by the DIVA dataset; Figure S6: The 90th percentile monthly wind speed at near-coastal locations defined by the DIVA dataset; Figure S7: The 50-year return period wind speed, obtained with a PoT analysis and a GPD distribution.

Author Contributions: The project was conceptualized by I.R.Y. and the bulk of the analysis was undertaken by I.R.Y. The scatterometer dataset was developed by A.R. and the near-coastal analysis was undertaken by E.K. All authors contributed to the development and editing of the manuscript. All authors have read and agreed to the published version of the manuscript.

Data Access: The scatterometer data used for the analysis are available from the Australian Ocean Data Network (AODN) https://portal.aodn.org.au/. The ocean basin scale and near-coastal processed monthly mean and percentile data, as well as the 100-year return period data, are archived as NetCDF files at https://figshare.com/s/6e7a977914d2762ef9d9.

References

1. Young, I.R. Global ocean wave statistics obtained from satellite observations. *Appl. Ocean Res.* **1994**, *16*, 235–248. [CrossRef]
2. Young, I.R. Seasonal Variability of the global ocean wind and wave climate. *Int. J. Climatol.* **1999**, *19*, 931–950. [CrossRef]
3. Risien, C.M.; Chelton, D.B. A Global climatology of surface wind and wind stress fields from eight years of QuikSCAT scatterometer data. *J. Phys. Oceangr.* **2008**, *38*, 2379–2413. [CrossRef]
4. Young, I.R.; Sanina, E.; Babanin, A.V. Calibration and cross-validation of a global wind and wave database of altimeter, radiometer and scatterometer measurements. *J. Atmos. Ocean Techol.* **2017**, *34*, 1285–1306. [CrossRef]
5. Ribal, A.; Young, I.R. 33 years of globally calibrated wave height and wind speed data based on altimeter observations. *Sci. Data* **2019**, *6*, 77. [CrossRef] [PubMed]
6. Dodet, G.; Piolle, J.F.; Quilfen, Y.; Abdalla, S.; Accensi, M.; Ardhuin, F.; Passaro, M. The sea state CCI dataset v1: Towards a sea state climate data record based on satellite observations. *Earth Sys. Data Sci.* **2020**, *12*, 1929–1951. [CrossRef]
7. Young, I.R.; Donelan, M.A. On the determination of global ocean wind and wave climate from satellite observations. *Remote Sens. Environ.* **2018**, *215*, 228–241. [CrossRef]
8. Ribal, A.; Young, I.R. Calibration and cross-validation of global ocean wind speed based on scatterometer observations. *J. Atmos. Ocean. Tech.* **2020**, *37*, 279–297. [CrossRef]
9. Ribal, A.; Young, I.R. Global calibration and error estimation of altimeter, scatterometer, and radiometer wind speed using triple collocation. *Rem. Sens.* **2020**, *12*, 1997. [CrossRef]
10. Donelan, M.A.; Pierson, W.J. Radar scattering and equilibrium ranges in wind-generated waves with application to scatterometry. *J. Geophys. Res.* **1987**, *92*, 4977–5029. [CrossRef]
11. Freilich, M.H.; Dunbar, R.S. A Preliminary C-band Scatterometer Model Function for the ERS-1 AMI Instrument. In Proceedings of the First ERS-1 Symposium on Space at the Service of Our Environment, Cannes, France, 4–6 November 1993; European Space Agency: Paris, France, 1993.
12. Wentz, F.J.; Smith, D.K. A model function for the ocean-normalized radar cross section at 14 GHz derived from NSCAT observations. *J. Geophys. Res.* **1999**, *104*, 499–511. [CrossRef]
13. Hoffman, R.N.; Leidner, S.M. An introduction to the near-real-time QuikSCAT data. *Weather Forecast.* **2005**, *20*, 476–493. [CrossRef]
14. Ricciardulli, L.; Wentz, F.J. A scatterometer geophysical model function for climate-quality winds: QuikSCAT Ku-2011. *J. Atmos. Ocean. Tech.* **2015**, *32*, 1829–1846. [CrossRef]
15. Priestly, C.H.B. *Turbulent Transfer in the Lower Atmosphere*; University of Chicago Press: Chicago, IL, USA, 1959.
16. Lumley, J.L.; Panofsky, H.A. *The Structure of Atmospheric Turbulence*; Interscience: Woburn, MA, USA, 1964.
17. Web, E.K. Profile relationships: The log-linear range, and the extension to strong stability. *Q.J.R. Meteorol. Soc.* **1970**, *96*, 67–90. [CrossRef]
18. Evans, D.; Conrad, C.; Paul, F. *Handbook of Automated Data Quality Control Checks and Procedures of the National Data Buoy Center*; NOAA/National Data Buoy Centre: Stennis Space Center, MS, USA, 2003; p. 44.
19. Bender, L.C.; Guinasso, N.J.; Walpert, J.N.; Howden, S.D. A comparison of methods for determining significant wave heights—Applied to a 3-m discus buoy during Hurricane Katrina. *J. Atmos. Oceanic Technol.* **2010**, *27*, 1012–1028. [CrossRef]
20. Jensen, R.; Swail, V.R.; Bouchard, R.H.; Riley, R.E.; Hesser, T.J.; Blaseckie, M.; MacIsaac, C. Field Laboratory for Ocean. Sea State Investigation and Experimentation: FLOSSIE: Intra-measurement evaluation of 6N wave buoy systems. In Proceedings of the 14th Intenational Workshop on Wave Hindcasting and Forecasting/Fifth Coastal Hazard Symposium, Key West, FL, USA, 8–13 November 2015.
21. Large, W.; Morzel, G.J.; Crawford, G.B. Accounting for surface wave distortion of the marine wind profile in low-level ocean storms wind measurements. *J. Phys. Oceanogr.* **1995**, *25*, 2959–2971. [CrossRef]
22. Taylor, P.K.; Yelland, M.J. On the effect of ocean waves on the kinetic energy balance and consequences for the inertial dissipation technique. *J. Phys. Oceangr.* **2001**, *31*, 2532–2536. [CrossRef]
23. Zeng, L.; Brown, R.A. Scatterometer observations at high wind speeds. *J. Appl. Meteor.* **1998**, *37*, 1412–1420. [CrossRef]
24. Takbash, A.; Young, I.R.; Breivik, O. Global wind speed and wave height extremes derived from satellite records. *J. Climate* **2019**, *32*, 109–126. [CrossRef]

25. Accadia, C.; Zecchetto, S.; Lavagnini, A.; Speranza, A. Comparison of 10-m wind forecasts from a regional area model and QuikSCAT scatterometer wind observations over the Mediterranean Sea. *Mon. Wea. Rev.* **2007**, *135*, 1945–1960. [CrossRef]

26. Bentamy, A. Characterization of ASCAT measurements based on buoy and QuikSCAT wind vector observations. *Ocean Sci.* **2008**, *5*, 77–101.

27. Bentamy, A.; Quilfen, Y.; Flament, P. Scatterometer wind fields: A new release over the decade 1991–2001. *Can. J. Remote Sens.* **2002**, *28*, 431–449. [CrossRef]

28. Pickett, M.H.; Tang, W.; Rosenfeld, L.K.; Wash, C.H. QuikSCAT satellite comparisons with nearshore buoy wind data off the U.S. West Coast. *J. Atmos. Ocean. Techol.* **2003**, *20*, 1869–1879. [CrossRef]

29. Satheesan, K.; Sarkar, A.; Parekh, A.; Kumar, M.R.R.; Kuroda, Y. Comparison of wind data from QuikSCAT and buoys in the Indian Ocean. *Int. J. Remote Sens.* **2007**, *28*, 2375–2382. [CrossRef]

30. Verspeek, J.; Stoffelen, A.; Portabella, M.; Verhoef, A.; Vogelzang, J. ASCAT scatterometer ocean calibration. In Proceedings of the International Geoscience and Remote Sensing Symposium, Boston, MA, USA, 7–11 July 2008.

31. Rivas, M.B.; Stoffelen, A. Characterizing ERA-Interim and ERA5 surface wind biases using ASCAT. *Ocean. Sci.* **2019**, *15*, 831–852. [CrossRef]

32. Goda, Y. On the methodology of selecting design wave height. In Proceedings of the 21st International Conference on Coastal England, Malaga, Spain, 20–25 June 1988.

33. Coles, S. *An Introduction to Statistical Modelling of Extremes*; Springer: Berlin, Germany, 2001.

34. Vinoth, J.; Young, I.R. Global estimates of extreme wind speed and wave height. *J. Climate* **2011**, *24*, 1647–1665. [CrossRef]

35. Lopatoukhin, L.J.; Boukhanovsky, A.V. Estimates of Extreme wind Wave Heights Report No. WMO/TD-1041. Available online: https://www.wmo.int/pages/prog/amp/mmop/documents/JCOMM-TR/J-TR-9-ExtWaveHeight/JCOMM-TR-9-Extr-Wave-Height-Full.pdf (accessed on 13 December 2019).

36. Caires, S.; Sterl, A. 100-year return value estimates for ocean wind speed and significant wave height from the ERA-40 data. *J. Climate* **2005**, *18*, 1032–1048. [CrossRef]

37. Portilla-Yandun, J.; Jacome, E. Covariate extreme value analysis using wave spectral partitioning. *J. Ocean. Atmos. Techol.* **2020**, *37*, 873–888. [CrossRef]

38. Goda, Y. Uncertainty in design parameter from the viewpoint of statistical variability. *J. Offshore Mech. Arct. Eng.* **1992**, *114*, 76–82. [CrossRef]

39. Ferreira, J.A.; Soares, C.G. An application of the peaks over threshold method to predict extremes of significant wave height. *J. Offshore Mech. Arct. Eng.* **1998**, *120*, 65–176. [CrossRef]

40. Van Gelder, P.H.A.J.M.; Vrijling, J.K. On the distribution function of the maximum wave height in front of reflecting structures. In Proceedings of the Coastal Structures Conference, Santander, Spain, 7–10 June 1999.

41. Alves, J.H.G.M.; Young, I.R. On estimating extreme wave heights using combined Geosat, Topex/Poseidon and ERS-1 altimeter data. *Appl. Ocean. Res.* **2003**, *25*, 167–186. [CrossRef]

42. Castillo, E. *Extreme Value Theory in Engineering*; Academic Press: Cambridge, MA, USA, 1988.

43. Meucci, A.; Young, I.R.; Breivik, O. Wind and wave extremes from atmosphere and wave model ensembles. *J. Clim.* **2018**, *31*, 8819–8893. [CrossRef]

44. Challenor, P.G.; Wimmer, W.; Ashton, I. Climate change and extreme wave heights in the North Atlantic. In Proceedings of the Envisat and ERS Symposuim, Salzburg, Austria, 6–10 September 2005.

45. Anderson, C.W.; Carter, D.W.T.; Cotton, P.D. Wave climate variability and impact on offshore structure design extremes. *Shell Int. Rep.* **2001**, *1*, 88.

46. Hinkel, J.; Klein, R.J.T. Integrating knowledge to assess coastal vulnerability to sea-level rise: The development of the DIVA tool. *Glob. Environ. Chang.* **2009**, *19*, 384–395. [CrossRef]

47. Kirezci, E.; Young, I.R.; Ranasinghe, R.; Muis, S.; Nicholls, R.J.; Lincke, D.; Hinkel, J. Projections of global scale extreme sea levels and resulting episodic coastal flooding over the 21st Century. *Sci. Rep.* **2020**, *10*, 11629. [CrossRef]

48. Meucci, A.; Young, I.R.; Hemer, M.K.E.; Ranasinghe, R. Projected 21st century changes in extreme wind-wave events. *Sci. Adv.* **2020**, *6*, eaaz7295. [CrossRef]

49. Evans, J.; Ekström, M.; Ji, F. Evaluating the performance of a WRF physics ensemble over South-East Australia. *Clim. Dyn.* **2012**, *39*, 1241–1258. [CrossRef]

50. Dee, D.P.; Uppala, S.M.; Simmons, A.J.; Berrisford, P.; Poli, P.; Kobayashi, S.; Bechtold, P. The ERA-Interim reanalysis:

Configuration and performance of the data assimilation system. *Q. Jnl. Roy. Met. Soc.* **2011**, *137*, 553–597. [CrossRef]

51. Hersbach, H.; Bell, B.; Berrisford, P.; Hirahara, S.; Horányi, A.; Muñoz-Sabater, J.; Simmons, A. The ERA5 global reanalysis. *Q. J. R. Meteorol. Soc.* **2020**, *146*, 1999–2049. [CrossRef]

52. Quilfen, Y.; Chapron, B.; Vandemark, D. The ERS scatterometer wind measurement accuracy: Evidence of seasonal and regional biases. *J. Atmos. Ocean. Techol.* **2001**, *18*, 1684–1697. [CrossRef]

53. Hersbach, H.; Stoffelen, A.; de Haan, S. An improved C-band scatterometer ocean geophysical model function: CMOD5. *J. Geophys. Res.* **2007**, *112*, C03006. [CrossRef]

54. Verhoeh, A.; Portabella, M.; Stoffelen, A. High-resolution ASCAT scatterometer winds near the coast. *IEEE Trans. Geosci. Remote Sens.* **2012**, *50*, 2481–2487. [CrossRef]

55. Chou, K.H.; Wu, C.C.; Lin, S.Z. Assessment of the ASCAT wind error characteristics by global dropwindsonde observations. *J. Geophys. Res. Atmos.* **2013**, *118*, 9011–9021. [CrossRef]

56. Xu, H.; Lin, N.; Huang, M.; Lou, W. Design tropical cyclone wind speed when considering climate change. *J. Struct. Eng.* **2020**, *145*, 04020063. [CrossRef]

57. Young, I.R.; Zieger, S.; Babanin, A.V. Global trends in wind speed and wave height. *Science* **2011**, *332*, 451–455. [CrossRef]

58. Young, I.R.; Ribal, A. Multi-platform evaluation of global trends in wind speed and wave height. *Science* **2019**, *364*, 548–552. [CrossRef]

59. Timmermans, B.W.; Gommenginger, C.P.; Dodet, G.; Bidlot, J.R. Global wave height trends and variability from new multimission satellite altimeter products, reanalyses, and wave buoys. *Geophys. Res. Lett.* **2020**, *47*, e2019GL086880. [CrossRef]

7

A Computational Workflow for Generating a Voxel-Based Design Approach Based on Subtractive Shading Envelopes and Attribute Information of Point Cloud Data

Miktha Farid Alkadri [1,*], **Francesco De Luca** [2], **Michela Turrin** [1] and **Sevil Sariyildiz** [1]

[1] Department of Architecture and Engineering Technology, Faculty of Architecture and the Built Environment, Delft University of Technology, Julianalaan 134, 2628 BL Delft, The Netherlands; M.Turrin@tudelft.nl (M.T.); I.S.Sariyildiz@tudelft.nl (S.S.)

[2] Department of Civil Engineering and Architecture, Tallinn University of Technology, Ehitajate tee 5, 19086 Tallinn, Estonia; francesco.deluca@taltech.ee

* Correspondence: M.F.Alkadri@tudelft.nl

Abstract: This study proposes a voxel-based design approach based on the subtractive mechanism of shading envelopes and attributes information of point cloud data in tropical climates. In particular, the proposed method evaluates a volumetric sample of new buildings based on predefined shading performance criteria. With the support of geometric and radiometric information stored in point cloud, such as position (XYZ), color (RGB), and reflection intensity (I), an integrated computational workflow between passive design strategy and 3D scanning technology is developed. It aims not only to compensate for some pertinent aspects of the current 3D site modeling, such as vegetation and surrounding buildings, but also to investigate surface characteristics of existing contexts, such as visible sun vectors and material properties. These aspects are relevant for conducting a comprehensively environmental simulation, while averting negative microclimatic impacts when locating the new building into the existing context. Ultimately, this study may support architects for taking decision-making in conceptual design stage based on the real contextual conditions.

Keywords: voxel-design approach; shading envelopes; point cloud data; computational design method; passive design strategy

1. Introduction

1.1. General Background

The rapid development of 3D laser scanning technology has reached across multiple-disciplines within design and engineering. However, the practical implementation of this technology is often applied in major fields, such as photogrammetry [1,2], cultural heritage [3–6], and environmental engineering [7,8]. Digital reconstruction as one of the main subjects in these scopes has been used predominantly for building performance assessments [9,10], where the contextual modeling of existing studies is frequently based on a 3D solid modeling context [11,12]. As a consequence, high computational costs and time are required to cover the entire set of complex building forms. On the other hand, the use of point clouds during the early stage of architectural design has not yet been fully explored, especially related to the performance simulation task and design decision support.

As an entity of 3D data scanning, Otepka et al. [13] illustrates the point cloud as a universal denominator for laser scanning and photogrammetric data. Its data structure is principally characterized by position information (XYZ) as a permanent element coupled with auxiliary information attached to it, such as color attributes (RGB), reflection intensity (I), and any abstract information [14]. The prospective applications of these attributes not only represent metadata information of the real environment, but also enable designers and researchers to perform numerous tasks, such as data processing, visualization, and analysis. Moreover, this can help architects further to address environmental design issues, such as solar and shading performance.

The technological advancement of point cloud reveals the relevance of integrating it into the passive design strategy, especially when dealing with generative architecture designs that currently lack several relevant aspects. For example, first, understanding the site characteristics of an existing environment. While 3D site modeling primarily deals with a building-oriented context, surrounding properties, such as vegetation and adjacent buildings, are often neglected [15]. This may not only affect the performance simulation of a proposed design, but also potentially create microclimatic issues when it comes to the real context. Second, the absence of surface properties, such as roughness and material characteristics on a manually-built 3D model, may cause a crucial discrepancy when dealing with environmental simulation between planned and existing buildings [16]. With geometric and radiometric information extracted from point cloud data, this study, therefore, proposes an integrated passive design approach based on shading performances of new and existing contexts.

As a contextual design approach, this study specifically investigates the idea of subtractive shading envelopes that are principally extracted from the concept of solar envelopes initially introduced by Knowles [17]. In this regards, solar envelopes permit architects to design appropriate massing of a new building into the existing environment by guaranteeing desirable sun access for surrounding buildings during the critical period [18], while subtractive shading approach aims to extract potential performances of the existing contexts and integrate it with a 3D volumetric massing of a proposed building based on predefined shading performance criteria.

Since then, various computational methods of solar envelopes, such as descriptive geometry, solar obstruction angle, and constructive solid geometry have been defined [19]. These approaches have successfully demonstrated the concept of solar envelopes into various urban settings (e.g., single building, open space, and urban scale) and multiple functional utilities (e.g., housing, offices, and commercial buildings). It is worth noting that the contextual settings of the existing methods primarily focus on temperate zones of southern and northern hemisphere countries, which have distinct climatic conditions during the four-seasons. This means that design objectives and climatic parameters of most existing methods for solar envelopes become less applicable when it comes to tropical countries, especially for those located on the Equator, such as Indonesia. Since tropical countries present wet and dry seasons all year round, the objective of solar envelopes significantly shifts and aims to minimize the penetration of direct sun access to the buildings, due to high temperatures. For example, housing in Indonesia is typically designed in a way that prohibits direct sunlight from penetrating the dwelling, especially into primary living spaces, so that temperatures are kept low during the day. Consequently, the air conditioner (AC) frequently becomes a short-term solution to mitigate the building's temperature, which unfortunately contributes to the annual increase in energy consumption [20]. Accordingly, shading conditions become considerably relevant for urban forms generation in tropical contexts. This study specifically proposes an environmental design strategy that integrates shading performance aspects and attributes information from point cloud data through a computational workflow of a voxel-based design approach.

Furthermore, the following section will present a theoretical background of the existing studies, starting from solar envelopes, shading envelopes, subtractive solar envelopes, and subtractive solar envelopes based on point cloud data so-called SOLEN approach. This will be followed with a description of a proposed method in Section 2, while case studies in Section 3. Section 4 will comprehensively discuss the findings of the simulation results. Lastly, Section 5 will describe the conclusions, limitations, and future recommendations of the study.

1.2. Related Works

1.2.1. Solar Envelopes

In the remote past, the concept of vernacular architecture has successfully contributed to preserving sustainable building envelopes [21,22]. This can be observed through the development of the Indus Valley, Mohenjo-Daro in India, 2500 BC [23], El-Lahun village in Egypt (1857–1700 BC) [24], and many classical Greek cities, such as Olynthus in North Hill—a city designed to benefit from passive solar energy for the heating of buildings [25]. This strategy was known as solar-oriented homes or so-called "solar architecture". Since then, solar architecture has become an essential guide for designers to develop sustainable urban planning. For example, Andrea Palladio has discussed the proper norms of city planning by considering wide streets for cold climate countries and narrow streets for tropical countries [26]. Additionally, Ildefons Cerdà integrated green areas into the public and private space of Barcelona in his masterplan of the city so as to enhance the comfort of inhabitants [27]. During the industrial revolution, the idea of urban solar policy or refers to the post-war housing was also implemented in France (in 1912), Germany (in 1920), and New York (in 1916). These examples have shown a positive contribution to architectural buildings, not only to reduce the energy consumption of the built environment [28], but more importantly, to support a healthy living environment [29]. Furthermore, the idea of solar accessibility has been elaborated further through the concept of solar envelopes.

By definition, solar envelopes stand for imaginary boundaries that are constructed based on the sun's movement. It is regulated based on specific space-time constraints [18]. According to this principle, solar envelopes can be transformed into geographic and climatic properties within the size of on-site buildings [19]. Geographic properties deal with a group of parameters that define the spatial relationship between the design plot and existing context related to orientation typology, surrounding facades, sidewalks, building height, longitude, latitude, floor area ratio (FAR), setback, shadow fences, and street sizes. On the other hand, climatic properties consist of parameters that determine the geometric transformation of the proposed building based on the time construction, such as cut-off-times, solar angle, sun path, dry bulb temperature, sun access hours, solar altitude, and solar azimuth. These parameters are used not only to generate solar envelopes, but also to identify the character and qualities of the built environment. For example, orientation plays a great role in examining the geometrical shape of solar envelopes, especially when dealing with the street layout in relation to various angular values, colonnades with a variety of direct solar radiations, and solar urban layouts [30]. A seasonal leaf cover from the surrounding vegetation can also affect the geometrical configuration of solar envelopes as it may be considered as a part of geographic elements for violating excessive direct sun access during summer [16]. Besides, the solar angle as a climatic property is used to determine geometric solar envelopes based on the construction planes [31]. It is mostly employed for simple shape plots, with borders aligned with the main cardinal directions in east-west (EW) and north-south (NS) and for the main hours and days, such as the noontime during summer and winter solstice, and spring/autumn equinoxes.

1.2.2. Shading Envelopes

As opposed to solar envelopes, the concept of shading envelopes primarily deals with the solar radiation-reduction to achieve appropriate daylight for urban equatorial climates. This permits architects not only to establish a geometrical configuration of solar shading envelopes, but also to control the direct sun exposure of the building's own façades and surroundings during a critical time. Two different types of shading approaches are identified as follows:

- Building forms

  In this part, the concept of shading envelopes aims to promote a passive design strategy through the form generation within the conceptual design stage. This means that the volumetric shape of proposed buildings is developed based on the consideration of solar shading criteria. An interesting example can be observed through the concept of "shadow umbrella" introduced by Emmanuel [32]. This concept proposed a design approach of shading strategies incorporated with natural elements, such as vegetation and water bodies, aiming to create shading for adjacent buildings and to mitigate the urban heat island for tropical neighborhood areas. Accordingly, a new configuration of urban block shapes with a thermally comfortable can be generated. DeKay [33] addressed a similar strategy with the concept of climatic envelopes. The climatic envelope primarily aims to generate a building mass that guarantees access for diffused daylight and solar energy resources for the surrounding buildings. This concept specifically contains a geometrical intersection between daylight and solar envelopes based on the sky exposure plane and solar protection plane, respectively. In this case, sky exposure plane refers to imaginary sloping planes that allow penetration of natural light and air on the building facades in higher density districts [34]; meanwhile, solar protection plane refers to an inclined plane that is generated from the profile angle, the so-called vertical shadow angle (VSA) [35]. In a similar vein, Capeluto [36] proposed the concept of self-shading envelopes by extending the functional properties of the sky exposure plane through the solar collection envelopes (SCE) model. The geometrical configuration of self-shading envelopes results in cone shapes, due to the required façade inclination and shading orientation. Therefore, the envelope's roof areas should be larger than the bottom part of the envelope geometries. This approach aims to avoid overheating and at the same time, to maximize the self-building protection for a certain period during summer.

- Building components

In addition to building forms, shading approaches are also applied to specific building components, such as windows, cantilevers, and openings on the building façade, based on the determined building shapes. Although this study limits the scope of investigation on a form-finding design solution, some studies on shading mechanisms drive potential efforts to handle more complex projects. For example, Yezioro and Shaviv [37] proposed SHADING as a design and evaluation tool for analyzing mutual shading between buildings and other surrounding properties, such as vegetation. It specifically calculates the insolated fraction on the building surfaces quantitatively and performs a ray-tracing algorithm to identify the shadows visually at a particular time. Similarly, Marsh [38] also used a ray-tracing analysis to identify the external obstruction of solar intensity on the optimized shape of shading geometries. In this case, optimized shading designs have effectively accommodated passive solar control through the building apertures [39,40] Other approaches deal with a graphic solution of shading design tools [41], form-finding of static exterior shading devices called SHADERADE [42], and a cellular method to define optimal shading patterns [43].

Although these approaches may address the aspect of tropical design contexts, they lack some critical aspects during the simulation of shading envelopes. First, the quality of solar radiation (i.e., the quantity of direct sunlight hours) received by surrounding buildings is not taken into account by most methods and tools, due to a fixed period when determining direct sun access. Consequently, all geometrical shapes of existing buildings are treated similarly when receiving the irradiation qualities without considering the obstruction of properties and building orientation of the plot. Second, shadow fences of surrounding buildings' facades are primarily regulated by a Z-axis. This makes the design configuration of the resulting envelopes rely only on the horizontal shading lines. In fact, shading areas of surrounding façades are more complex, especially when dealing with dense areas and multiple urban forms. In order to compensate for these issues, the existing studies present some relevant aspects, such as subtractive mechanism and point cloud data that may be useful for further development. This will be discussed in the following section.

1.2.3. Subtractive Solar Envelopes

As previously mentioned in the general background, this approach specifically subtracts a volumetric matrix of the 3D plot according to solar accessibility criteria. This is done by projecting solar vectors acquired from the number of direct sunlight hours on surrounding building facades. In principle, this mechanism has been addressed by Leide and Schlüter [44,45] via volumetric site analysis (VSA). Their approaches aim to explore urban site information by simulating multiple environmental performances, such as solar radiation, airflow, visibility, thermal comfort, and wind velocity through volumetric insolation analysis (VIA), volumetric visibility analysis (VVA), and computational fluid dynamics (CFD). Such this approach and other related developments [46–48], however, merely focus on the architectural form-finding without any further consideration on the concept and design principles of solar or shading envelopes.

On the other hand, De Luca [49–51] and Darmon [52] have proposed a similar subtractive mechanism based on the performance criteria of solar envelopes or so-called subtractive solar envelopes. This approach specifically involves sun visibility that aims to evaluate sun vectors from a predefined shadow grid of surrounding building windows without any obstruction from other existing buildings. In parallel, ray-tracing analysis is performed from surrounding windows to the voxels within the 3D plot using a Boolean expression (true or false statement). In this operation, the true statement will be executed when sun vectors hit or intersect the 3D polyhedra, and accordingly, the voxels subtraction procedure to the 3D polyhedra can be performed. Meanwhile, the false statement indicates an unsuccessful intersection. This condition means that voxels that are not intersected may contribute to the generation of geometric solar envelopes.

The subtractive solar envelopes ultimately permit architects not only to deal with various geometric configurations based on solar performance analysis, but also to highlight the potential use of voxel-based generative designs for urban environments. However, despite such potential improvements, aspects (such as sun visibility and ray-tracing analysis) pose several critical considerations, especially when addressing the contextual design strategy. For example, the identification of visible sun vectors merely considers the surrounding building contexts while neglecting relevant geometric properties, such as vegetation and other site characteristics (e.g., material properties). This consequently can affect irradiation analysis during the environmental performance simulation between a proposed building and the existing contexts. Besides, the window's grid configuration lacks in representing the insolation values of building facades during the ray-tracing analysis, due to its limited consideration of geometric centroids of each surrounding window.

In order to address these gaps, the existing workflow of subtractive solar envelopes has been improved by incorporating it with the prospective application of 3D laser scanning (point cloud data). By exploiting the practical usability of the point cloud in different fields, the scope of information properties in the real contexts can be improved, and to some extent, it becomes relevant to the specific aforementioned issues. Regarding sun visibility, the 3D point cloud not only captures the most and the least sun-exposed areas through buildings and vegetation, but also investigating material performances of contextual datasets through optical (reflectivity and translucency) and thermal properties (albedo and emissivity). Besides, the ray-tracing analysis between a proposed building and surrounding contexts is performed based on 3D point cloud datasets of the existing context. In other words, it substitutes the surrounding window's grid on the building facades proposed by the existing approach.

1.2.4. Subtractive Solar Envelopes Based on Point Cloud Data (SOLEN)

With the support of geometric and radiometric properties [53,54] stored in a 3D point cloud, the integrated computational workflow between subtractive solar envelopes and attribute information of point cloud data has been established [55,56]. It specifically integrates functional properties of position information (XYZ), color information (RGB), and reflection intensity (I). Each of these attributes caters to different potential tasks. For example, color information (RGB) can be used not only to extract and segment certain areas within the dataset based on its values [57,58], but also to translate them into new information properties [59,60]. This can include converting data attribution of RGB into HSV values to perform the measurement analysis and road maintenance [61] and extracting the semantic information of the indoor environment with automatic room labeling [62,63]. Meanwhile, reflection intensity (I) predominantly deals with surface and spectral properties of the scanned objects [64] as it constitutes the return strength values of laser pulse or backscattered echo for each recorded point [54]. Accordingly, the intensity values can be used not only to map geological layers and pavement lines [54], but also to detect natural phenomena, such as frozen and wet surfaces on roads [65], and the measurement of seasonal snow cover [66]. On the other hand, position information (XYZ) constitutes of geographic coordinate that marks each recorded point's specific location. This attribute plays a great role in synchronizing index between color and intensity values as it can attach to both attributes. Thus, complex areas of the dataset can be precisely extracted based on selected values. In general, these attributes contribute not only to extend particular performances of 3D point cloud data, such as identification of existing material properties [67], but also to extend the applicability of environmental analysis during the conceptual design stage.

Before performing the subtractive solar envelopes, the dataset correction is required to minimize erroneous levels during scanning. In this case, some aspects, such as environmental and meteorological conditions, atmospheric pollution [68,69], unit specification of the scanner, surface properties of the scanned objects, and scanning geometries [70] can principally affect metadata information of the datasets during scanning. While correcting all these variables seems impractical, due to some local constraints (i.e., manufacturers), this study specifically focuses on correcting the acquisition geometry based on the angle of incidence, which is relevant to the proposed subtractive mechanism. Having established the corrected datasets, it can be further used not only to perform the ray-tracing analysis between the 3D polyhedra and selected solar vectors, but also to calculate the material properties of existing contexts that are useful to evaluate the environmental performance of a proposed building. In parallel, insolation analysis is performed to identify the potential solar energy of the resulting solar envelopes.

2. Proposed Methods for Subtractive Shading Envelopes

This study proposes a computational framework that consists of three phases: input, simulation process, and output (see Figure 1). Within a simulation process, five sequential procedures are developed, ranging from A (input parameters) to E (form generation process of self-shading envelopes).

To perform specific tasks in each predefined step, the proposed workflow was supported by several digital tools. For example, Topcon GLS 2000 [71] was employed to collect high-resolution point cloud datasets. It was complemented with Maptek I-Site [72] to perform dataset registrations, coloring, modeling, and most importantly, to facilitate the data transfer from scanner to the workstation in any designated format. Moreover, Cloud Compare (CC) [73] was used, not only for dataset preparation, but also for dataset pre-processing, such as attribute selection, dataset formatting, scalar field features, and the normal surface calculation. In alignment with that, Matlab [74] was specifically used to assist dataset correction (i.e., optimal normal values, intensity correction, and dataset subsampling), while Rhino [75] (coupled with Grasshopper [76] for visual scripting) was employed to develop a 3D geometric model of a proposed building and to perform solar simulation analysis by using a Ladybug [77] component in Grasshopper. Furthermore, a detailed task, dataset formats, and outputs of each step are addressed below.

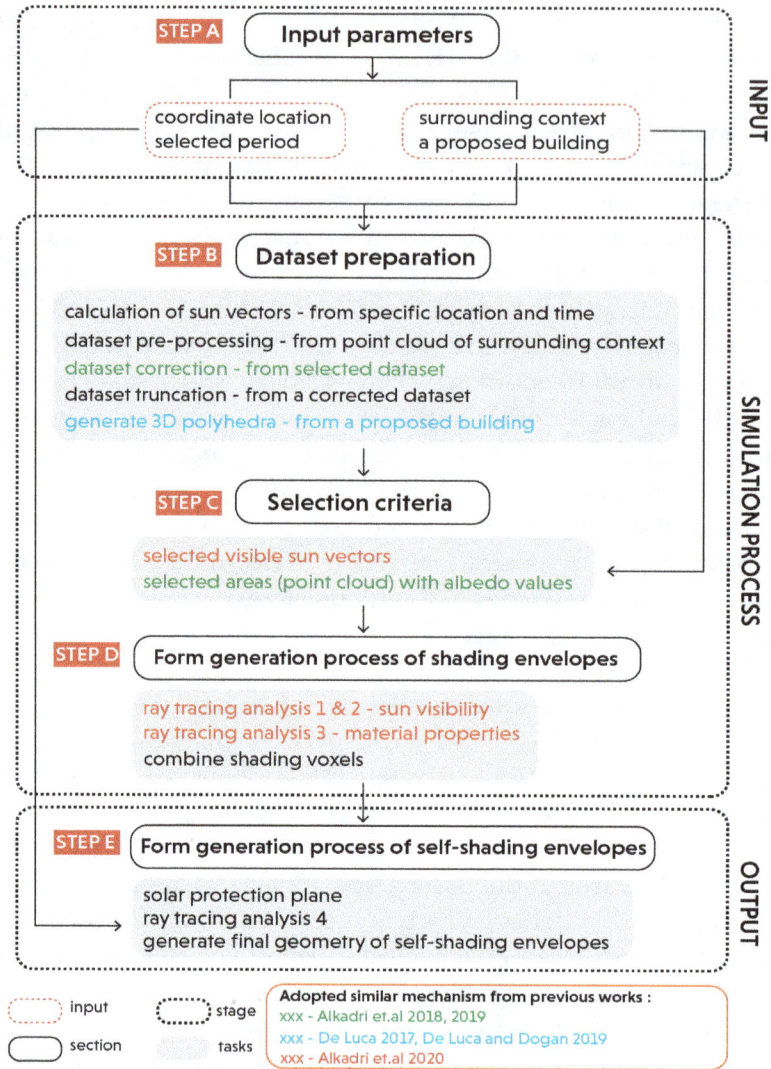

Figure 1. An overview of the proposed computational workflow.

2.1. Stage A Input (Step A—Preparation of Input Parameters)

As a starting point for the computational procedures, the input section refers to step A, which contains a series of parameters that are used to construct contextual settings of the subtractive shading envelopes. Specifically, it consists of climatic properties that correspond to coordinate location (i.e., longitude and latitude position based on World Geodetic System 1984, WGS84) and selected periods (i.e., month, year, day, and hours) (see Figure 2). The longitude and latitude coordinates influence the envelope's geometrical properties on the basis of sun position and solar angle. For example, high-latitude sites are characterized by a small angle of solar altitude and smaller degrees of solar radiation. Moreover, specific periods are required to obtain the number of critical hours of natural illumination that affects a proposed building and surrounding contexts.

On the other hand, geographic properties include surrounding contexts (i.e., existing buildings and vegetation) and a geometric model of the proposed building. In this regard, the surrounding context contains a 3D point cloud data of the existing environment. As a raw dataset, its format properties often rely on the type of 3D scanner, but as long as the required attribute information (i.e., XYZ, RGB, and reflection intensity) is legible, any raw dataset formats are acceptable (i.e., PTX file). Meanwhile, several parameters, such as height, width, floors, setback, and building function, must be established to generate an initial 3D envelope for a proposed building. These inputs are then executed into the following Step B (dataset preparation) based on corresponding parameters and tasks.

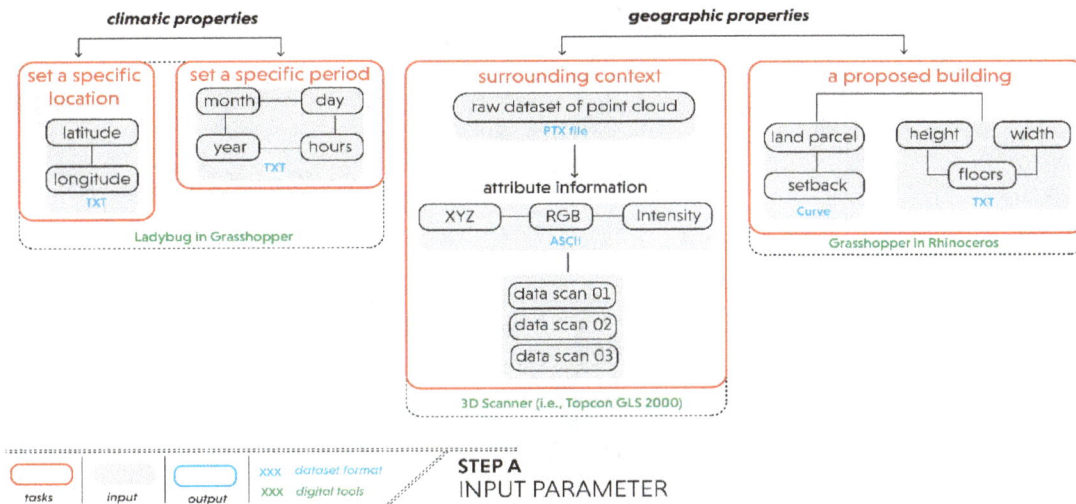

Figure 2. Preparation of input parameters.

2.2. Stage B Simulation Process

This section focuses on translating the raw datasets into a simulation model by examining three main steps (i.e., dataset preparation, selection criteria, and form generation process of shading envelopes). Each step serves specific actions that are performed sequentially based on specific computational tasks. As illustrated in Figure 1, this section adopts several workflows that are partially implemented from previous works. For example, pre-processing point cloud datasets and the calculation of material properties (i.e., albedo values) are indicated with a green text [67]. These works focused on developing a material database of existing contexts and solar radiation analysis based on a small sample of point cloud data. Next, 3D polyhedra and sun visibility analysis are illustrated in the blue text [49,50]. These works primarily investigated a voxel-based generative design of subtractive solar envelopes based on 3D and parametric modeling. Last, the ray-tracing analysis between a proposed building and point cloud data of surrounding contexts is illustrated in the red text [56]. This work refers to a form generation of subtractive solar envelopes that consider surface properties of existing contexts based on geometric and radiometric information of point cloud data. While these previous works address different objectives, some features are still relevant for supporting an integrated design concept and computational workflow for establishing the subtractive shading envelopes.

To illustrate specific tasks in this stage, a detailed discussion of each proposed step is presented below.

2.2.1. Step B—Dataset Preparation

This step aims to prepare all the necessary datasets to be readily used in the simulation model. After establishing the input from climatic and geographic properties, five tasks are required to perform. Two of these tasks (i.e., sun vectors calculation and the initial envelopes generation) can be run in parallel while the other three tasks (i.e., pre-processing datasets, dataset correction, and subsampling dataset) can be executed sequentially (see Figure 3).

Task 1, sun vectors refer to the number of sunlight hours that must be preserved on surrounding facades during the required period. As compared with cut-off times, which refer to a fixed period, sun access duration can be selected from a range of available hours for a specific façade on a specific day or for each day during a specific period. In doing so, sun visibility plays a crucial role when determining the relevant sun vectors.

Task 2, the initial envelope of a proposed building, is generated based on the predefined criteria. In this case, the functional program of a proposed building is projected as a public library, as well as communal space for the local community. To support the main activities, some spaces are established, such as a reading room, meeting room, toilet, and exhibition areas. Accordingly, the building needs to be accessible, and its indoor environment should be thermally comfortable for supporting the daily activities.

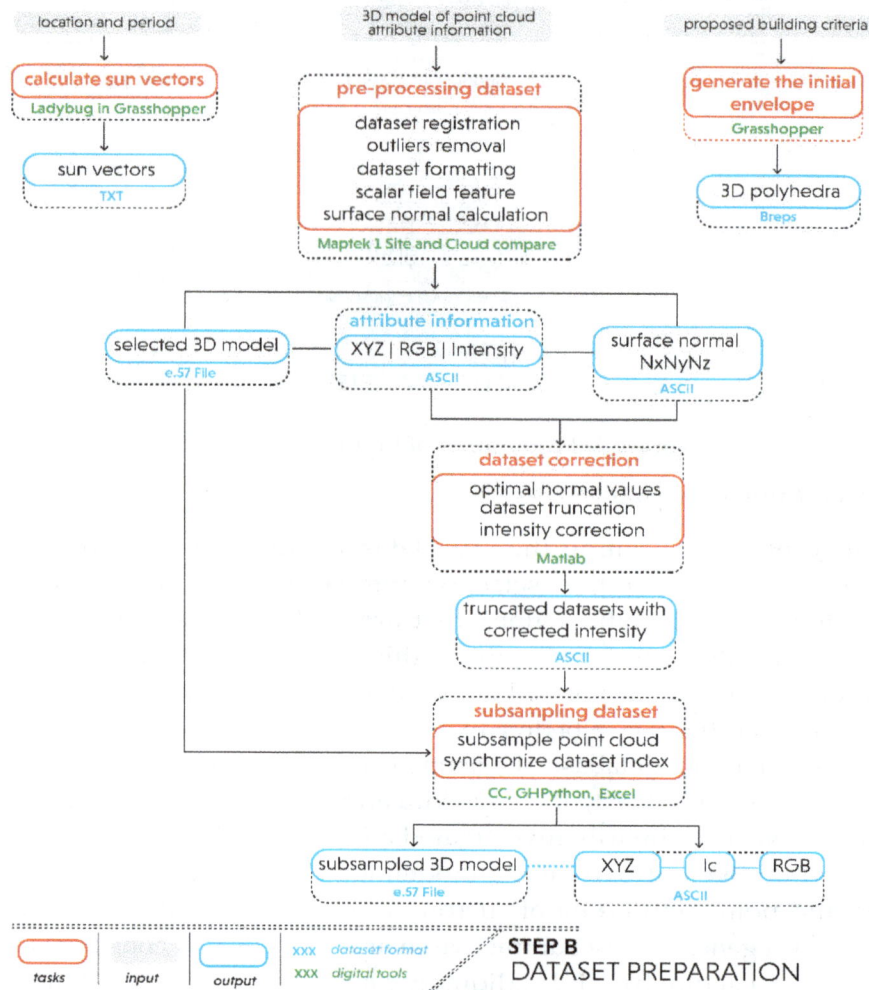

Figure 3. Detailed procedures for the dataset preparation.

Task 3, the pre-processing tasks start with the dataset registration. This process aims to locate recorded datasets into a standard coordinate system based on the reflector position and GPS orientation. Afterwards, the outlier (unnecessary cloud of points) removal is performed not only to clean the boundary of selected datasets, but also to filter noises created during scanning. The dataset formatting also plays an important role in compensating for interoperability issues during the simulation process. In this case, the initial format of 3D raw point clouds (i.e., PTX) and its metadata are converted to E57 and ASCII files, respectively, in order to be accessible for various digital processing tools. In addition to this, the scalar field of the dataset is activated to identify the attached values' scale in each attribute. Last, normal surfaces (NxNyNz) of the dataset are calculated to find the appropriate normal values of each projected point during scanning. As normal values of the raw point cloud are excluded in the typical attribute properties, various angles of incidence, ranging from the sample of 10° to 90°, are firstly computed to each data scan. This is done by using the Hough Normal plugin [78] in CC, due to the original form of unstructured raw point cloud data. In this regard, each point within each data scan has different preliminary normal values. Ultimately, task 3 results in several outputs, such as a 3D model of selected datasets, attribute information of point cloud data with raw intensity values, and the normal surface of each data scan at various incidence angles.

Task 4, the dataset correction, is performed to compensate for the scattering condition of unstructured point clouds during scanning, specifically for amending radiometric properties of the dataset. It is worth noting that this procedure can only be applied on a single scan, due to the potentiality of mixing a reference point of the scanner and intensity values on the merge datasets [54]. The correction step starts with finding the average distribution of optimal projection points from the preliminary normal values. This achieves a reliable normal surface on each applied angle in the data scan. To do so, the following equation [79] is applied with an assumption that the original position of a 3D scanner located at (0,0,0).

$$i = cos^{-1}\left(\frac{\overline{dn}.\overline{dl}}{\left|\overline{dn}\right|\left|\overline{dl}\right|}\right)$$
(1)

where:

i = initial incidence angle
$\overline{dn}$ = direction of the normal surface
$\overline{dl}$ = direction of the laser pulse

After configuring the point distribution from a various range of cosinus products, an evaluation is conducted to the standard deviation of each registered cosine value within the dataset. It aims to identify the pattern of point density to reduce scattered and coarse point clouds through the dataset truncation. Afterwards, the truncated datasets can be used for the intensity correction. In this regard, the angle of incidence becomes a relevant factor for correcting the dataset's acquisition geometry, given that instrumental effects highly affect the raw intensity value of TLS datasets. This procedure is executed based on the following equation [54].

$$I_c = I_{raw} \cdot \frac{1}{cos\ \alpha}$$
(2)

where:

I_c = corrected intensity
I_{raw} = original intensity
α = angle of incidence

Task 5, the subsampling dataset, is performed in CC to reduce the density of points during the simulation. However, this procedure creates interoperability issues when the resulting dataset is matched and visualized to the initial 3D model in Rhino. Specifically, the 3D model cannot directly recognize the input of subsampled datasets, due to different units and scales of the attribute information. Therefore, synchronizing the initial index between the two datasets identifies metadata information in the geometric 3D model. This further permits the extraction of certain areas in the 3D dataset based on selected attributes values.

2.2.2. Step C—Selection Criteria

After preparing the corrected datasets from climatic and geographic properties, this study sets two environmental performance criteria that support the geometric generation: sun visibility and material properties (see Figure 4). Sun visibility plays a crucial role in filtering the sun vectors that have direct access to the dataset of surrounding contexts. To do so, sun vectors generated from the indicated period are multiplied with normal vectors extracted from subsampled datasets. The resulting normal irradiance values are then evaluated on the basis of the projected angle. In this case, irradiance values with equal and larger than 90° are eliminated as they consist of zero and negative cosine values. Accordingly, these values are then excluded within the list of visible sun vectors because the surface of the datasets does not properly absorb their solar energy. Afterward, the resulting values can be used to select the corresponding points within the dataset.

On the other hand, material aspects are used to measure the performance behavior of the existing site's surface properties. This allows architects to identify susceptible areas that may affect the geometrical performance of the proposed design. This study specifically computes albedo values to detect the absorbance percentage of solar energy on surrounding contexts by considering the RGB color and corrected intensity (Ic) of contextual datasets. A detailed procedure regarding the calculation of albedo values can be found in our previous research [67]. Furthermore, the resulting albedo values are filtered based on the threshold below 0.3. Although this setting indicates the low albedo, it can be used not only to identify areas that contain a high level of heat absorbance, but also to analyze and mitigate microclimatic impacts of the surrounding areas especially related to thermal issues and urban heat island (UHI) effect. In order to avoid that, these areas need to be blocked from direct sun exposures by excluding the indicated surfaces with high albedo values. Afterward, the resulting indexes of low albedo values (below 0.3) are synchronized not only with selected normal vectors from the sun visibility to register the corresponding normal surface of the dataset, but also with XYZ attributes to select the matching points within the dataset.

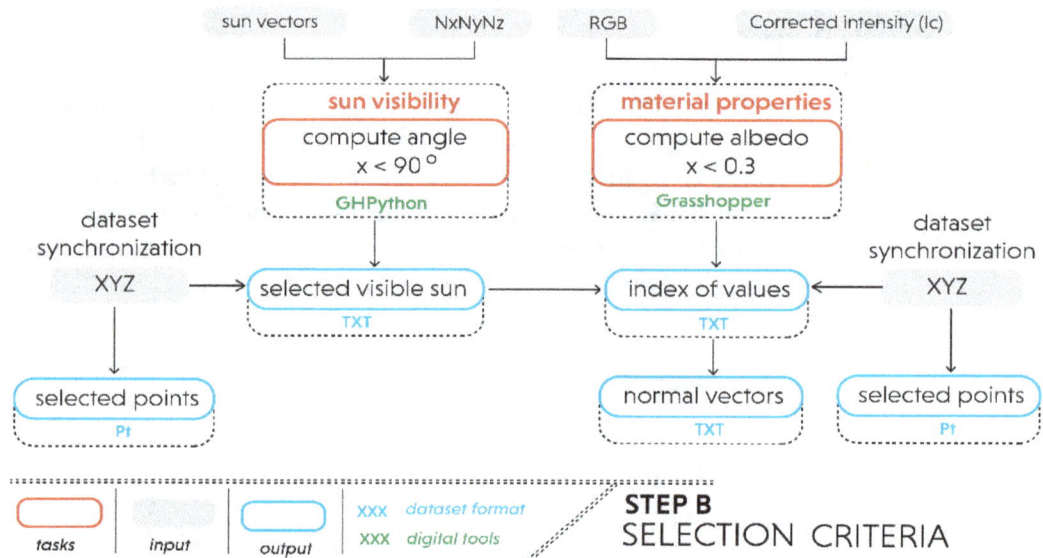

Figure 4. Selection procedures based on the criteria of sun visibility and material properties.

2.2.3. Step D—Form Generation Process of Shading Envelopes

After establishing all the required parameters from the selection criteria, this step focuses on developing the simulation workflow (see Figure 5). It starts with the first ray-tracing analysis that requires input from visible sun vectors, selected corresponding points from sun visibility, and 3D polyhedra of a proposed building. This procedure principally applies a Boolean expression to assign true or false conditions on selected voxels within the 3D polyhedra array. The ray-tracing analysis-01 generates intersecting rays that are then evaluated based on the predefined criteria of direct sun access. In this case, voxels that are not blocking sun access to the proposed buildings or are categorized as an unsuccessful intersection with the 3D polyhedra will be considered part of the shading voxels (refer to voxels-02 in Figure 5). This workflow can also be called as a reverse solar envelope. Meanwhile, voxels that receive sun access will be forwarded to a later step in the second ray-tracing analysis (refer to voxels-01 in Figure 5).

Furthermore, reference points are generated from the voxels-02 to be used in the simulation of ray-tracing analysis-02. As a follow-up to the previous procedure, the ray-tracing analysis-02 aims to maximize the geometric generation of shading voxels. In doing so, by changing the basis projection of initial reference points originated from surrounding buildings to the voxels-02 may compensate for geometric obstruction of polyhedra at a certain projection angle in the ray-tracing analysis-01. Instead of applying this procedure to the original 3D polyhedra, it is used to re-evaluate voxels-01 based on reference points of voxels-02 so as to identify additional voxels (refer to voxels-03) that fulfill the criteria of receiving shading condition. As for the input for material properties, a similar procedure of ray-tracing analysis is also performed by considering the lowest albedo values (ranging from 0 to 0.3) applied to surround contexts. These results in voxels-04 so that in total, three groups of voxels (i.e., voxels-02, voxels-03, and voxels-04) are generated to shade surrounding buildings. These voxels are then combined into one group, voxels-05. To ensure that a shading condition also applies to a proposed building, the self-shading workflow is performed in the following stage.

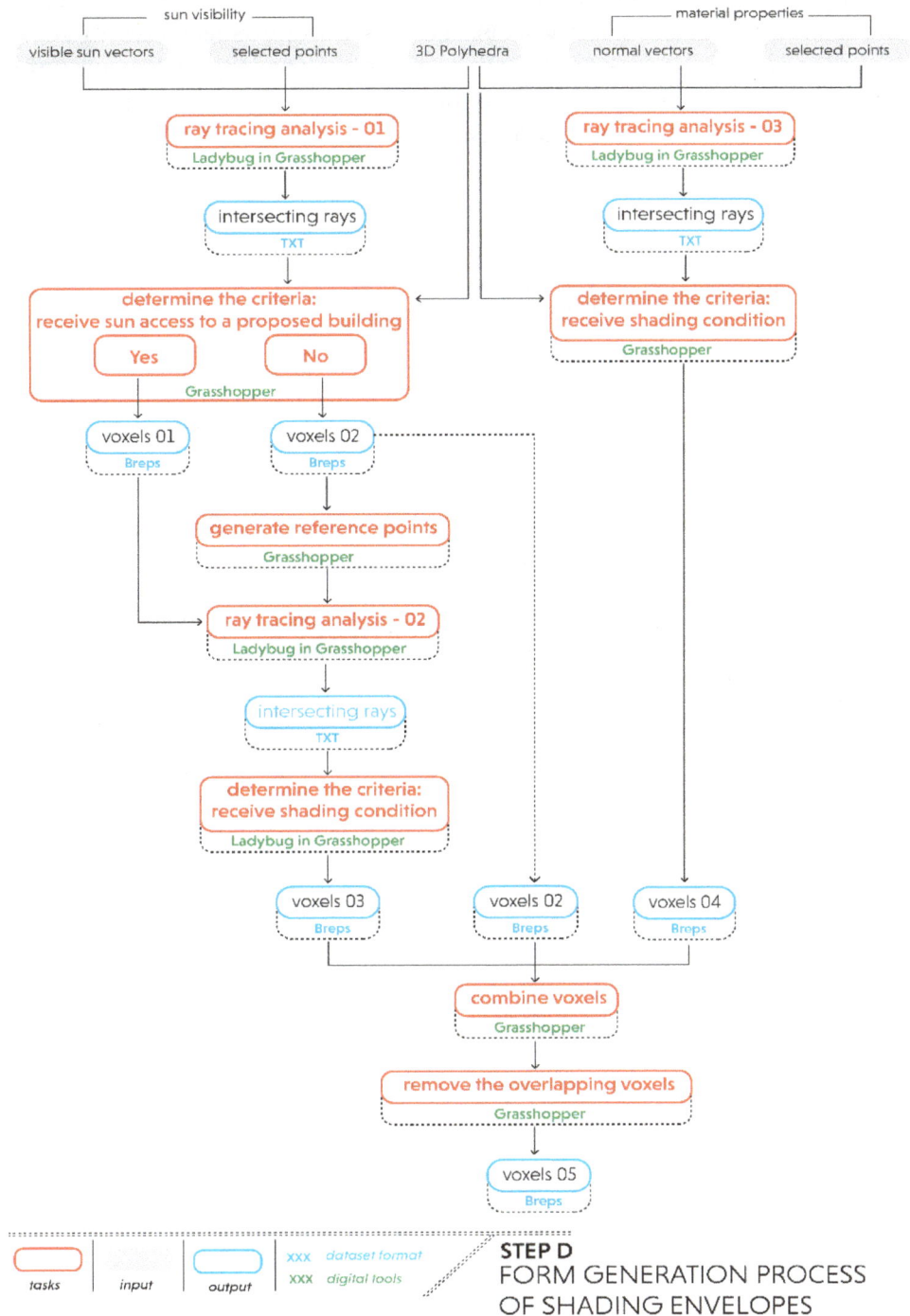

Figure 5. Detailed procedures for the design simulation.

2.3. Stage C Output (Step E—Form Generation Process of Self-Shading Envelopes)

As a last stage of the computational workflow, the output contains final tasks to generate a geometric configuration of self-shading envelopes (see Figure 6). The first task begins with selecting the upper part of the stacking voxels. This upper part acts as a roof or shelter to guarantee a shading condition for all properties under its envelope within a predefined period. Afterward, a solar protection plane can be applied on the bottom surfaces of the upper part of the voxels. It aims to establish reference points that are used as the basis for ray projections. Furthermore, the ray-tracing analysis-04 is performed by considering inputs from reference points of the upper voxels, the remaining stacking voxels (i.e., bottom part), and initial sun vectors calculated for the analysis period under consideration. This simulation will evaluate the remaining voxels by maintaining the one that receives a shading condition while removing the unshaded voxels. The resulting voxels (refer to voxels-06) are then combined with the upper voxels to establish the geometric envelope for each data scan (refer to voxels-07 and voxels-08). After identifying self-shading voxels for each data scan, the final step is to combine all these voxels into a final geometric envelope that represents a final configuration of envelopes (refers to voxels-09).

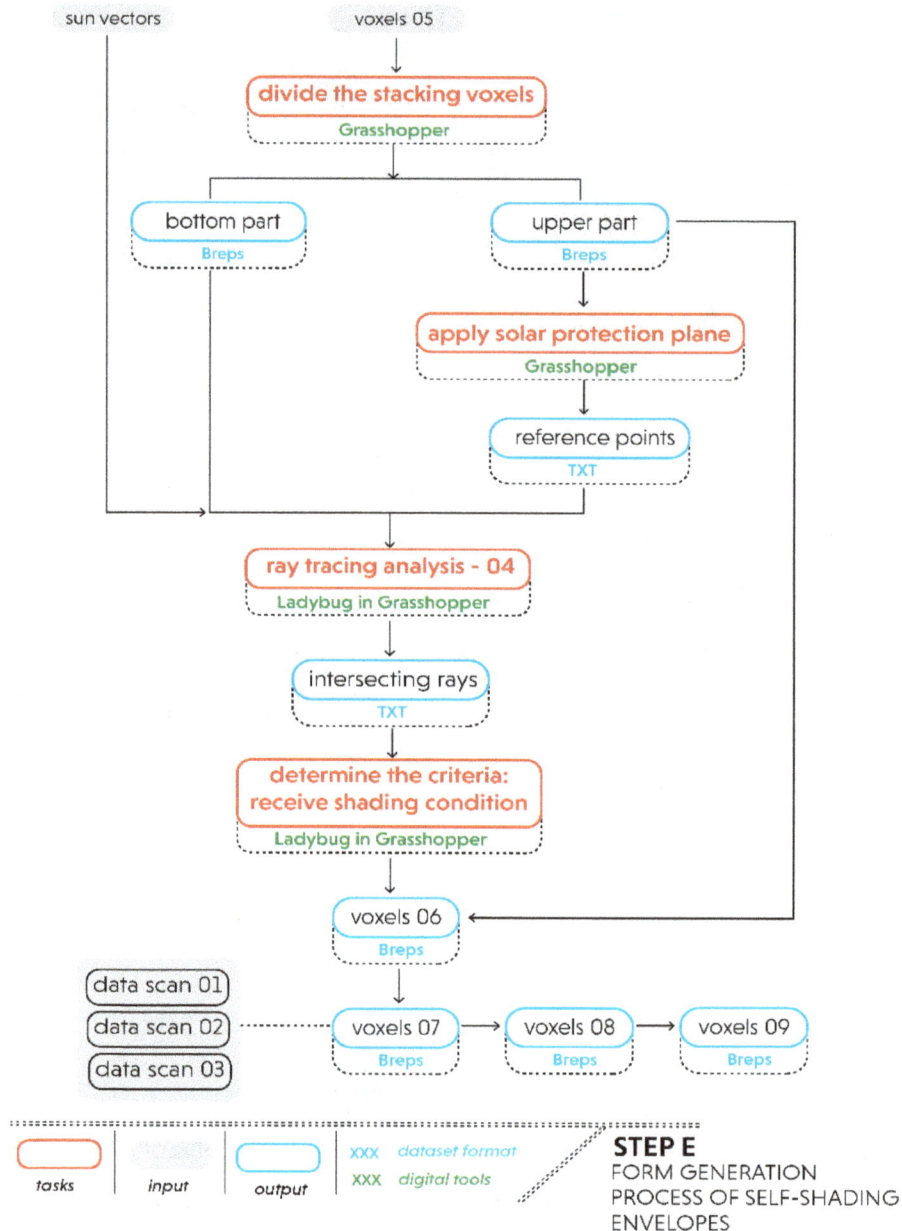

Figure 6. Detailed procedures for generating the final output of self-shading envelopes.

3. Dataset Collection

In regards to the dataset collection of on-site scanning, this study needs to fulfill at least two major aspects, as also mentioned in our previous study [56]. First, the collected point clouds should consist of Terrestrial Laser Scanning (TLS) datasets. It aims not only to accommodate more accurate geometric properties and better reality-based representation than Airborne Laser Scanning (ALS), but also to capture specific areas with various contextual elements, such as vegetation, temporal objects, material, and other elements, that may potentially contribute to the simulation model. Second, metadata information stored in a point cloud should at least contain typical attributes, such as XYZ, RGB, and reflection intensity. Geometric and radiometric properties within these attributes are used to identify material properties of the existing context and to conduct environmental performance analysis.

Furthermore, to demonstrate the proposed workflow, this study collaborates with SHAU, an architectural firm located in Bandung, West Java, Indonesia, to collect 3D point cloud data to design a new public library. This project is located in Citarip, West Java, Indonesia, (6°56'18.4"S 107°35'15.1"E, with ellipsoid reference WGS84) and surrounded by some vegetation and massive walls of neighboring buildings. The collected dataset was gathered using Topcon GLS-2000M incorporated with reflectors and GPS devices to achieve an accurate position during the registration process. As a light-weight and high-speed 3D laser scanner, this tool is also featured with a rugged design instrument (for use in the field) and full-dome field of view (FOV) with a selectable laser class, enabling the user to scan in extreme work environments and eye-safety concerns in dense areas. A detailed specification of the tool can be observed below (see Table 1):

Table 1. Detailed specifications of 3D scanner [71].

Parameters	Performance Specification Unit
System performance	
Maximum range (at 90% reflectivity) GLS-2000M	350 m (standard)
Single point accuracy Distance Angle	3.5 mm (1–150m), 1 sigma 6"
Tilt sensor Type Range	Liquid 2-axis tilt sensor ±6
Target detection accuracy	3" at 50 m
Laser scanning system	
Type	Pulse (time-of-flight) precise scan technology
Laser class	3R (high speed/standard) 1 M (low power)
Field of view (per scan)	Horizontal-360° Vertical-270°
Spot size	4 mm at 20 m (FWHM)
Scan rate	High Speed: Up to 120,000 points/sec
Physical and environmental	
Operation temp Dust/humidity	−5 °C to 45 °C IP54
Scanning control	
Scan time and resolution	Interval 12.5mm: High Speed: 01:46 Standard: 03:31 Low Power: 04:22

This tool's performance specifications allow the dataset collection with only three single scans to capture a sufficient scene for the selected site (see Figure 7). With these three single scans, the dataset's computational performance simulation can also be more manageable and relevant to the currently proposed workflow. Given that this approach is part of the exploratory study, architects need to select not only important information, but also a degree of detailed properties that are relevant to the simulation. In this regard, the setting of each scan approximately consists of four million points per scan within five minutes with a resolution 12.5 × 12.5 mm @10m. The distance between scanners and the designated objects are principally determined based on the approximate coverage of the scanner capacity, which may cover 360° of the horizontal field of view (FOV) with the distance for single-point accuracy 1–150 m. In this case, Scan 2 reaches the closest distance between the scanner and the object with 2.75 m. This is because data scan coverage areas need to capture the surrounding wall and corner spots behind the temporary shelters. Meanwhile, the longest distance is obtained by Scan 01 with 34.9 m, due to the diagonal position of the scanner to the corner of the site. Besides, in this dataset collection, it is worth noting that as long as the entire scenes of the selected site and relevant objects are covered, the scanner can be located at any appropriate distances.

Figure 7. Dataset collection with different views captured in relation to the scanner position.

4. Results and Discussion

After establishing the selected dataset, this section presents the analysis results of the implemented workflow. It follows the five aforementioned steps as follows:

4.1. Step 1—Input Parameters

As previously described in the section about presenting the proposed workflow, input parameters consist of climatic and geographic properties. As for climatic properties, due to a constant temperature over the year in Indonesia [80], this study sets April 21st, 2019, as a sample of the selected period that represents a starting date for a dry season. Although the selected time range is limited, this simulation takes place as an exploratory study that focuses on exploring the feasibility of integrated computational workflow for shading envelopes based on attributes information of point cloud data. This specific duration is furthermore defined for four hours, starting from 11 am to 3 pm, which is averagely representing the highest temperature during the daylight hours. In parallel, a time-step for the simulation is set smaller than one hour to improve the simulation accuracy. This setting ultimately yielded four sun vectors that are simulated on each point of the dataset. Meanwhile, for a proposed building, the 3D polyhedra are extruded from the plot by considering predefined criteria of the buildings and local regulations, such as a 10 m height and 4–6 m setback requirement (see Figure 8). Each polyhedron is set to 2 × 2 × 2 m with the total number 690 units, including the generated asymmetrical shape, due to irregular order of the plot's boundary. In principle, this polyhedron's dimension can be vary depending on the architectural concept and functional program of the proposed building.

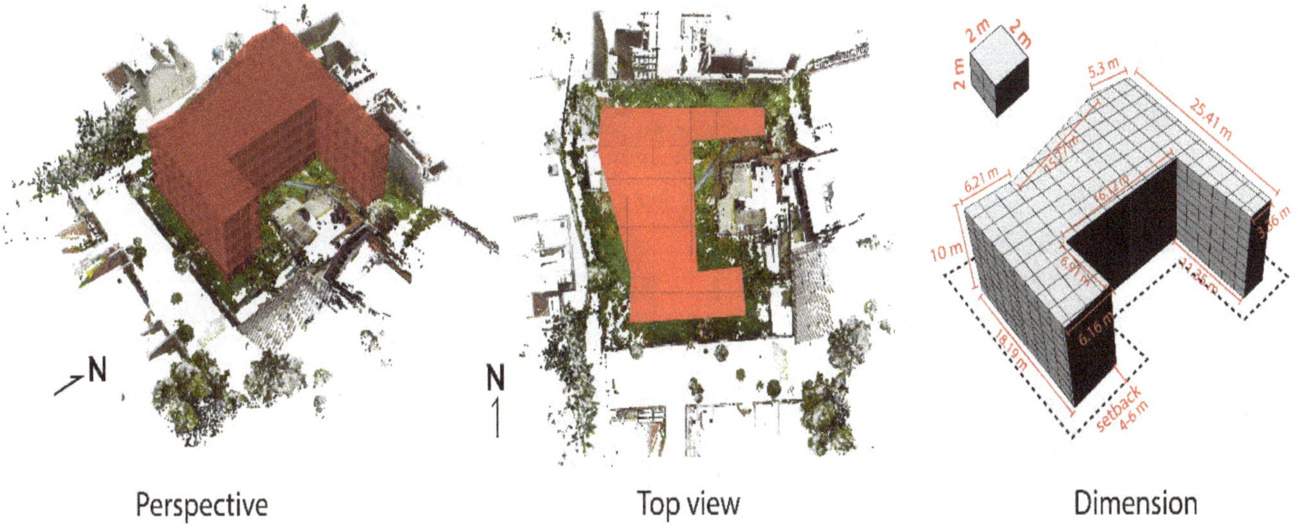

Perspective   Top view   Dimension

Figure 8. Selected dataset and 3D polyhedra for a case study.

4.2. Step 2—Dataset Preparation

While tasks (such as sun vectors calculation and 3D polyhedra generation) have been addressed previously, this section primarily discusses the result of sequential tasks, which consist of dataset correction and dataset subsampling. The dataset correction can be performed after establishing normal values with various incident angles for each selected data scan. According to Figure 9A, all points are plotted on the basis of the cosine values of each incident angle. The pattern shown by all data scans is that points are distributed and assemble at a certain angle. This means that the laser beam projection may correspond very well at certain areas of the dataset during scanning, depending on the position of the scanner and geometric properties of the building surfaces (i.e., roughness). Afterward, to find the optimal normal values within this pattern, a standard deviation of cosine values is calculated based on a sample of points (see Figure 9B). This becomes a basis for truncating the relevant datasets based on their distribution density.

In order to determine a threshold value of the dataset truncation, a standard deviation of the whole population of the dataset is plotted (see Figure 9C). This facilitates us to generate the cosine values pattern resulting in Figure 9B so that the number of points with a high distribution level of density can be identified reliably. As a result, ground areas of the plot (refer to blue-coded pattern in Figure 9B) that mostly contain grass are automatically removed, due to irrelevant properties with the predefined criteria, while datasets with a red pattern are used for the intensity correction (See Figure 10).

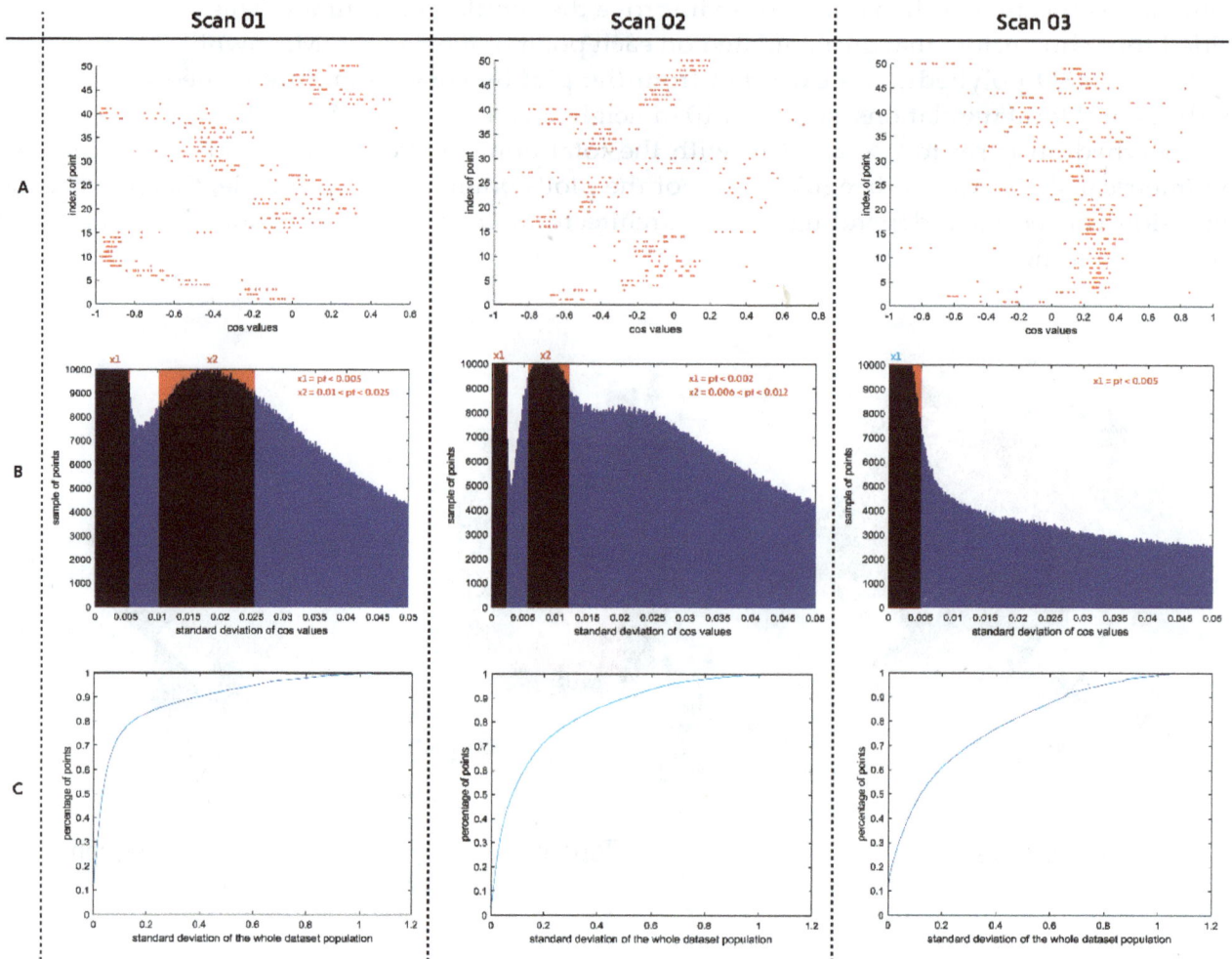

Figure 9. Dataset correction (**A**). Point distributions based on cosine values of each incident angle (**B**). The standard deviation of cosine values based on a group sample of points (**C**). The standard deviation of the whole dataset population.

Figure 10 specifically demonstrates the result of intensity correction from the raw attributes to the corrected one through the scalar field of intensity and intensity distribution values. The dataset transformations of each scan are plotted based on four-color steps, ranging from blue to yellow, which represent the lowest and the highest intensity values. The threshold between these values is represented by light and dark green. The trend clearly shows that the raw intensity of all data scans contains yellow areas (refers to the "before" part), representing high reflective surfaces. However, these areas predominantly assign incorrect surface materials as it should be. For example, some parts of the buildings, such as wood façade and clay roof tiles, are attributed to yellow intensity. In fact, these materials principally contain a high level of emissivity value or low intensity, which means that the return incident energy from these materials to the scanner is highly decreased, due to its spectral reflectance mechanism on rough, dark, and dull surfaces [81]. In addition, the intensity correction also compensates for scanned areas around the scanner, due to the impact of the brightness level of intensity. These areas are massively indicated with yellow-coded values, due to the high level of intensity produced by the scanner at a very short distance. In this case, the scanner's brightness reducer is assumed not entirely and comprehensively applied in these areas, especially within the distance range of 10 m from the scanner position. This is because some local constraints, such as atmospheric variables (i.e., humidity and temperature pressures) and surface roughness of the objects, may influence specular and diffuse reflection of the laser pulse during scanning [54].

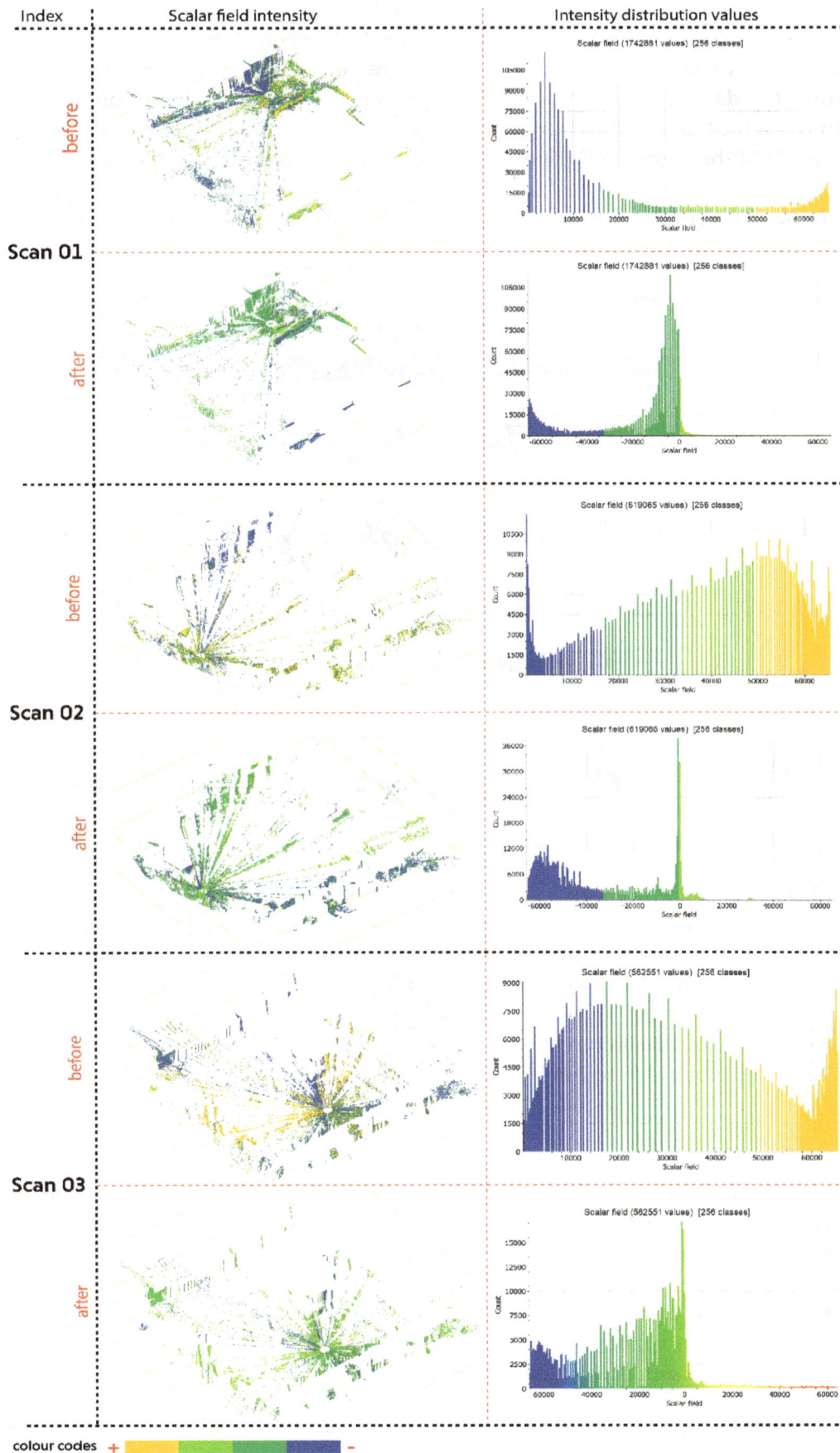

Figure 10. Intensity correction on each data scan.

The corrected datasets are then performed with the subsampling procedure in CC in order to control the unstructured point density when it comes to the simulation. In this case, the distance between points is set to 0.05 or equal to 5 cm, which results, on average, in a decrease point, up to 70% of the truncated datasets. This procedure, however, only works on position information (XYZ) because it can cause interoperability issues when it transforms into the 3D model in Rhino, due to the different nature of the algorithmic operation. To tackle this issue, index information of each point (refers to ID in Figure 11) needs to be extracted beforehand during the truncation process and then used to synchronize the original attributes (i.e., XYZ and RGB) with the corrected ones in the 3D model. The workflow of this dataset transformation is illustrated further in Figure 11.

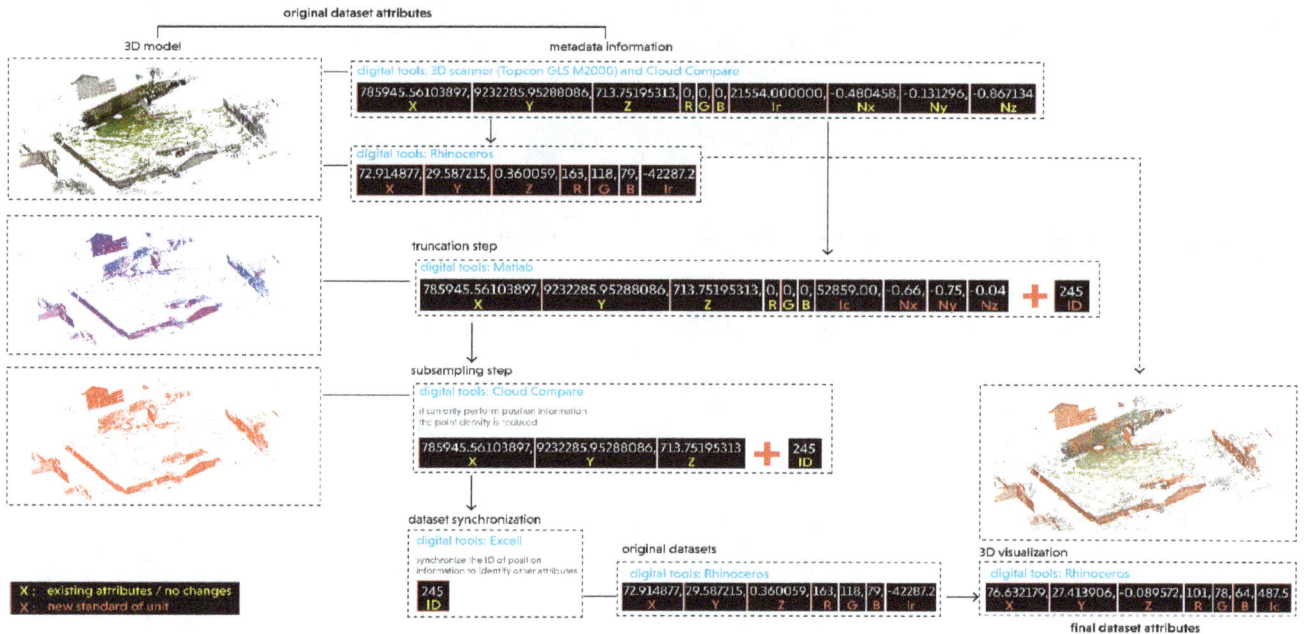

Figure 11. Transformation of the dataset attributes.

In general, Figure 11 demonstrates a unit conversion of the dataset attributes that take place according to the step and digital tool used, starting from the raw data scan to the final stage of the workflow. A major transformation principally occurs when converting the original datasets from the 3D scanner to the 3D model in Rhino, which changes the unit of position (XYZ), color (RGB), and the raw intensity values (Ir). Therefore, the interoperability aspect plays a critical role not only to support the simulation analysis on the metadata information of the dataset (i.e., correction, truncation, and subsampling step), but also to help the visualization of geometric configuration based on the selected attributes.

4.3. Step C—Selection Criteria

Before running the simulation, the subsampled datasets are evaluated based on the criteria of sun visibility and albedo values. These criteria identify relevant points that will be used for the voxels generation of shading envelopes. According to Figure 12, the resulting points for both criteria are significantly decreased as compared to the total points from the subsampled datasets. This is because a majority of subsampling points on the ground areas contain zero and negative sun vector values, which do not fulfill the sun visibility criteria. Accordingly, these points are automatically isolated, while successful ones are used as the basis datasets to perform the criteria of albedo values and the ray-tracing analysis.

In general, the trend in Figure 12 shows that the majority of selected points both for albedo values and the visible sun are similar and overlapping in a certain spot, such as the surrounding areas of the scanner position. On the other hand, each data scan also illustrates different specific areas for point distribution. For example, points for Scan 01 are primarily detected on the edge of the building's roof, while Scan 02 and Scan 03 are partially distributed in the brick wall and clay roofs of the surrounding houses, respectively. Furthermore, the selected datasets for each data scan are then used in the simulation process. This will be presented in the following step.

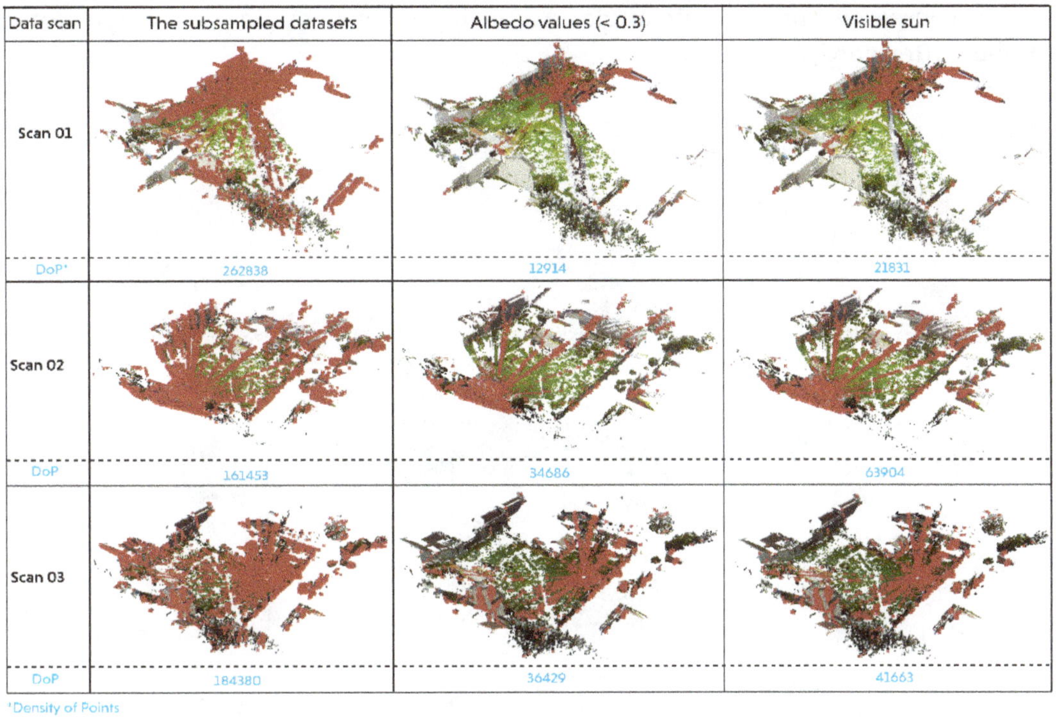

Data scan	The subsampled datasets	Albedo values (< 0.3)	Visible sun
Scan 01			
DoP*	262838	12914	21831
Scan 02			
DoP	161453	34686	63904
Scan 03			
DoP	184380	36429	41663

*Density of Points

Figure 12. The resulting points after performing the selection criteria.

4.4. Step D—Form Generation Process of Shading Envelopes

The resulting datasets that successfully fulfill the predefined criteria in the previous step are used in the simulation by following a series of tasks illustrated in Figure 5. The outputs of these tasks are demonstrated in Figure 13. It consists of five steps to generate volumetric shapes and three steps to simulate the ray-tracing analysis. Beforehand, each data scan is divided into two parts in order to minimize the high computational cost during the simulation. According to Figure 13, some analysis can be further discussed as follows:

- The ray-tracing analysis (see part RTA-01) shows the intersection lines that occur between visible sun vectors, selected points of the existing site, and 3D polyhedra. The result of these intersections is then illustrated in voxels-01, which indicate a group of voxels that fulfill the criteria of receiving direct sun access. In this case, the simulation of Scan 03 yields a lower number of voxels-01 as compared with other scans, due to the density and position of each point to the 3D polyhedra. Specifically, the existing datasets of Scan 03 approximately cover all sides of the 3D polyhedra during the simulation, while Scan 01 only intersects a certain part of the polyhedra because the existing points are predominantly distributed on one side. Conversely, the resulting voxels-02 show a contrasting number of voxels, due to the different sun access criteria.

- Voxels-03 illustrate a simulation result of shading criteria. A major trend is shown by Scan 02, which results in a significantly decreasing number of voxels from the voxels-02. This is because sun vectors originated from the reference points of voxels-02 are massively intersected with the geometric shape of voxels-01.

- Voxels-04 represent a group of polyhedra that block the direct sun access to certain areas of surrounding properties. Each data scan consists of a different voxel configuration depending on the selected areas that have been identified below the albedo values 0.3. On the other hand, voxels-05 constitute as concatenating polyhedra gathered from the resulting voxels-02, voxels-03, and voxels-04. Although the volumetric size of voxels-05 shows a significant improvement, especially for Scan 03, unfortunately, the resulting voxels can only compensate for the shading condition of the surrounding properties and still lack in protecting the proposed building from the direct sun access within a certain angle. Accordingly, a self-shading envelope workflow needs to be applied afterwards.

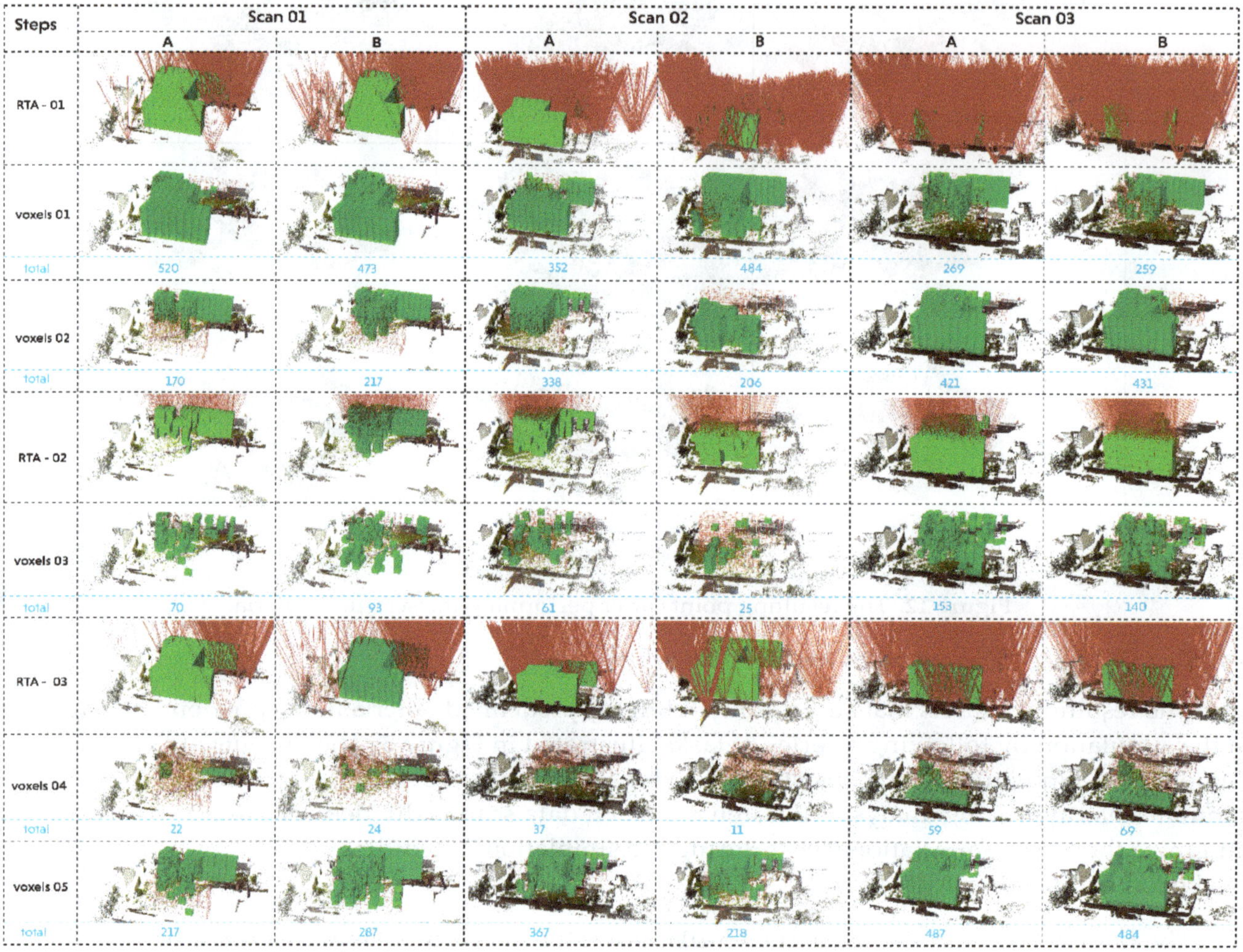

RTA - 01	: ray tracing analysis between visible sun vectors, selected points and 3D polyhedra
RTA - 02	: ray tracing analysis between voxels 01, selected vectors and reference points obtained from voxels 02
RTA - 03	: ray tracing analysis between 3D polyhedra, normal vectors and selcted points gathered from material properties
voxels 01	: the resulting volumetric shapes (voxels) that receive sun access criteria
voxels 02	: the resulting volumetric shapes (voxels) that block direct sun access to surroundings
voxels 03	: the resulting volumetric shapes (voxels) that fulfill shading condition criteria
voxels 04	: the resulting volumetric shapes (voxels) that fulfill shading condition criteria gathered from albedo values
voxels 05	: the resulting volumetric shapes (voxels) after combining voxels 03, 02, 04 and removing the overlapping geometries

Figure 13. The geometric configuration of subtractive shading envelopes based on sequential steps of design simulation.

4.5. Step E—Form Generation Process of Self-Shading Envelopes

In order to perform self-shading envelopes, solar protection planes are applied on the voxels-05 by excluding voxels located on the upper part of the 3D polyhedra. This is because the upper voxels are considered as the roof or shelter for the remaining voxels located on the bottom envelopes. To do so, the upper voxels in each data scan are firstly separated from the set of 3D polyhedra (see Figure 14 step 02), while performing the ray-tracing analysis (see step 03) on the remaining voxels. As a result (see steps 04), voxels-06 illustrate the resulting envelopes that fulfill the criteria of a self-shading mechanism. The trend shows that the resulting voxels-06 for Scan 03 reduce, on average, 50% of the volumetric 3D polyhedra (refers to voxels-05 bottom part), while other data scans also decrease nearly a half of the initial 3D polyhedra. In this case, most of the reducing voxels are located on the edge of the 3D polyhedra or voxels that act as an exterior wall of the 3D polyhedra.

Steps		Scan 01		Scan 02		Scan 03	
		A	B	A	B	A	B
01	Voxels 05 bottom part						
	total	165	227	285	166	296	309
02	Voxels 05 upper part						
	total	52	60	82	52	91	75
03	RT - 04						
04	Voxels 06						
	total	96	158	195	99	119	145
05	Voxels 07						
	total	148	218	277	151	210	220
06	Voxels 08						
	total	259		416		333	
07	Voxels 09						
	total			514			

RTA - 04 : ray tracing analysis between selected sun vectors, bottom part of the voxels 05, and reference points obtained from the upper part of voxels 05
voxels 05 - bottom part : the resulting volumetric shapes (voxels) that are used for the simulation of self-shading envelopes
voxels 05 - upper part : the resulting volumetric shapes (voxels) that are used to perform solar protection plane
voxels 06 : the resulting volumetric shapes (voxels) that receive self-shading condition
voxels 07 : the resulting volumetric shapes (voxels) that are used as a final geometry of the self-shading envelopes
voxels 08 : the resulting volumetric shapes (voxels) after combining dan removing overlapping sub voxel for each data scan
voxels 09 : the final polyhedra after combining dan removing overlapping voxels from each data scan

Figure 14. Geometric configuration of subtractive shading envelopes.

Voxels-06 of each data scan is combined with the upper part voxels that are previously separated in the previous step (refers to voxels-05 step 02). This process results in voxels-07 (refers to step 05), which are then used to perform the merging procedure in step 06—voxels-08. This step also includes elimination procedures for the voxels located in the same location. In so doing, the total number of polyhedra represented by voxels-08 shows the core geometry for each data scan. For example, the initial amount of voxels-08 after combining Scan 1A and 1B is 366 polyhedron, but after eliminating the overlapping polyhedron, it yields 259 polyhedra. Thus, 107 voxels are indicated as overlaps during the merging procedure. Afterward, the voxels-08 of each data scan are merged using the same steps in order to generate the final geometry of self-shading envelopes. Ultimately, the total number of voxels that fulfill the shading criteria for surrounding contexts and the 3D polyhedra is 514 of 690 (see Figure 14).

Furthermore, the resulting geometries of self-shading envelopes can be identified as a core geometry that fulfills the building's main activities based on the daylight condition. For example, architects may plot voxels located on the ground level as a communal space, workshops, and playing areas for children as it requires more open activities while the upper floor can be fulfilled with a reading room or any other activities that match with the required shading. In addition, the roof of this upper floor can be utilized as a solar collector for PV panels that may produce supplementary solar energy for the buildings. This is relevant because Indonesia's geographical condition is located in the Equator can support the electricity production around 1,534 kWh/year for each installed solar panel based on 4.5 kWh/m^2 average daily radiation [82]. As compared to the existing method of self-shading envelopes proposed by Capeluto [36], this approach not only creates potential shading that merely comes from the roof of the envelopes, but also originates from a dynamic form of the envelope facades that correspond to different solar vectors. For these reasons, geometrical configurations of the final envelopes in this study show more variation in different orientations.

In design practices, the proposed approach can be adopted by architects as a further step for designing and analyzing high performing envelopes both for new and existing development areas. The Dutch architecture and urban design firm, MVRDV, has been implemented the basic design principles of this approach into several projects based on solar-oriented designs. As a new development area, for example, the P15 Ravel Plot, which is located in the Zuidas district, 1082 LC, Amsterdam was constructed based on the idea of the optimal line of sight integrated with the three-dimensional landscape and greenery [83]. Meanwhile, in urban scale, MVRDV proposed the idea of solar energy as a part of design intervention for zero energy neighborhoods to the existing historic infrastructures in Bordeaux, France [84]. These examples clearly show a stepping-stone for architectural design practices to explore further the relevance and potential application of a voxel-based design approach in supporting sustainable environmental design. In this regard, point cloud data can be a powerful instrument that fortifies environmental performance simulation of design context during the early design phase.

5. Conclusions

This study investigates the potential application of attribute information stored in point cloud data to support a new computational method for a voxel-based design approach based on shading performance criteria. As a part of the contextual design strategy, the proposed workflow specifically presents form generation based on subtractive shading envelopes and material properties of existing contexts that are used to generate a new method for self-shading envelopes in tropical countries. Ultimately, this integrated approach may support architects in taking a comprehensive design decision during the early stage of the design process based on real contextual datasets. As an exploratory study, this work presents several concluding remarks as follows:

- The attribute information of point cloud data (i.e., XYZ, RGB, and reflection intensity) contributes not only to calculate material properties of existing context, but also to be a part of selection criteria for generating voxel-based subtractive shading envelopes.
- The dataset preparation includes pre-processing and correction steps that help architects minimize the environmental effects of the dataset measurement during scanning and to select reliable and relevant information for the contextual analysis and the simulation process.
- The proposed workflow enables architects to produce more variation of geometrical facades, especially related to the final geometry of self-shading envelopes that are not only depending on the roof perimeters, but also considering reference points that attach to the upper part of each generated voxel.
- The ray-tracing analysis between 3D polyhedra and selected points of surrounding contexts permits one to identify specific areas and voxels that fulfill shading performance criteria during a predefined period.

- As a contextual design approach, self-shading envelopes not only receive environmental performance responses from surrounding buildings to the proposed design, but also deliver feedback from the new building to the existing contexts.

Despite the findings mentioned above, there still some limitations. For example, first, the implementation of this proposed workflow in design practice may require a collaboration with the field of remote sensing, especially related to the dataset collection and preparation. This is because some specific tasks need prior knowledge regarding the dataset pre-processing and particular digital tools. Second, the limitations of simulating highly dense datasets imply the number of solar vectors that need to be reduced, especially when dealing with the ray-tracing analysis procedure. Third, normal surface of the datasets and voxels of the proposed building should be evaluated by each solar vector, but as a consequence, it requires high computational processing that is currently lacking in our 3D modeling tools. Therefore, further research is expected to consider these issues in order to enhance the quality of the simulation results. It is also recommended to implement this study in different urban settings with multiple building functions, such as high-rise or new development areas, so that various urban forms can be further explored.

Author Contributions: Conceptualization, F.D.L. and M.F.A.; methodology, M.T. and M.F.A.; writing—original draft preparation, M.F.A.; writing—review and editing, M.T. and F.D.L.; visualization, M.F.A.; supervision, S.S.; All authors have read and agreed to the published version of the manuscript.

Acknowledgments: We thank Indonesian Endowment Fund for Education (LPDP) for funding a PhD research of the first author under the chair of Design Informatics, Faculty of Architecture and the Built Environment of TU Delft, the European Regional Development Fund grant (the Estonian Centre of Excellence in Zero Energy Resource Efficient Smart Buildings and Districts) ZEBE n. 2014-2020.4.01.15-0016 and the European Commission H2020 grant Finest Twins n. 856602 for supporting the work of the second author. In addition, the authors would like to express their gratitude to SHAU in Bandung for providing access to their projects, Irwan Gumilar from the study program of Geodesy and Geomatics Engineering, ITB and Haidar from PT Asaba for assisting us in collecting the dataset, Yu-Chou Chiang for his technical support during dataset processing, James Nelson in TU Delft urbanism editing team for proofreading the preliminary manuscript, and TU Delft Library for supporting logistics and open access of this journal publication.

References

1.   Shih, N.-J.; Wu, M.-C. A 3D Point-Cloud-Based Verification of as-Built Construction Progress. In Proceedings of the CAAD Futures 2005 Conference, Vienna, Austria, 20–22 June 2005.

2.   Alsadik, B.; Gerke, M.; Vosselman, G. Visibility analysis of point cloud in close range photogrammetry. *ISPRS Ann. Photogr. Remote Sens. Spat. Inf. Sci.* **2014**, 2, 9–16. [CrossRef]

3.   Andriasyan, M.; Moyano, J.J.; Julián, J.E.N.; García, D.A. From Point Cloud Data to Building Information Modelling: An Automatic Parametric Workflow for Heritage. *Remote Sens.* **2020**, 12, 1094. [CrossRef]

4.   Shanoer, M.M.; Abed, F.M. Evaluate 3D laser point clouds registration for cultural heritage documentation. *Egypt. J. Remote Sens. Space Sci.* **2018**, 21, 295–304. [CrossRef]

5.   Remondino, F. Heritage Recording and 3D Modeling with Photogrammetry and 3D Scanning. *Remote Sens.* **2011**, 3, 1104–1138. [CrossRef]

6.   Núñez-Andrés, M.A.; Buill, F.; Costa-Jover, A.; Puche, J.M. Structural assessment of the Roman wall and vaults of the cloister of Tarragona Cathedral. *J. Build. Eng.* **2017**, 13, 77–86. [CrossRef]

7.   Bornaz, L.; Lingua, A.; Rinaudo, F. Engineering and Enviroonmental Applications of Laser Scanner Techniques. In Proceedings of the ISPRS Commission III Symposium Photogrammetric Computer Vision, Graz, Austria, 9–13 September 2002.

8.   Kassner, R.; Koppe, W.; Schüttenberg, T.; Bareth, G. Analysis of the Solar Potential of Roofs By Using Official Lidar Data. In Proceedings of the ISPRS Congress, Beijing, China, 3–11 July 2008.

9.   Carneiro, C.; Morello, E.; Desthieux, G. Assessment of Solar Irradiance on the Urban Fabric for the Production of Renewable Energy using LIDAR Data and Image Processing Techniques. In Proceedings of the Advances in GIScience, Hannover, Germany, 2–5 June 2009.

10.   Desthieux, G.; Carneiro, C.; Camponovo, R.; Ineichen, P.; Morello, E.; Boulmier, A.; Abdennadher, N.; Dervey, S.; Ellert, C. Solar Energy Potential Assessment on Rooftops and Facades in Large Built Environments

Based on LiDAR Data, Image Processing, and Cloud Computing. Methodological Background, Application, and Validation in Geneva (Solar Cadaster). *Front. Built Environ.* **2018**, *4*, 1–22. [CrossRef]

11. Staneva, N.N. Approaches for generating 3D solid models in AutoCAD and solid works. *J. Eng.* **2008**, *3*, 28–31.

12. Shapiro, V. Spatial Automation Laboratoty. In *Solid Modeling*; University of Wisconsin: Madison, WI, USA, 2001.

13. Otepka, J.; Ghuffar, S.; Waldhauser, C.; Hochreiter, R.; Pfeifer, N. Georeferenced Point Clouds: A Survey of Features and Point Cloud Management. *ISPRS Int. J. Geoinf.* **2013**, *2*, 1038–1065. [CrossRef]

14. Richter, R.; Döllner, J. Concepts and techniques for integration, analysis and visualization of massive 3D point clouds. *Comput. Environ. Urban Syst.* **2014**, *45*, 114–124. [CrossRef]

15. Arayici, Y. Towards building information modelling for existing structures. *Struct. Surv.* **2008**, *26*, 210–222. [CrossRef]

16. Alkadri, M.F.; Turrin, M.; Sariyildiz, S. The use and potential applications of point clouds in simulation of solar radiation for solar access in urban contexts. *Adv. Comput. Des.* **2018**, *3*, 319–338.

17. Knowles, R.L. *Energy and Form: An Ecological Approach to Urban Growth, Cambridge*; The MIT Press: Cambridge, MA, USA, 1974.

18. Knowles, R.L. *Sun, Rhytm and Form, Massachusetts*; The MIT Press: Cambridge, MA, USA, 1981.

19. Alkadri, M.F.; De Luca, F.; Turrin, M.; Sariyildiz, S. Understanding Computational Methods for Solar Envelopes Based on Design Parameters, Tools, and Case Studies: A Review. *Energies* **2020**, *13*, 3302. [CrossRef]

20. GlobalABC, IEA and UN. *2019 Global Status Report for Buildings and Construction: Towards a Zero-Emission, Efficient and Resilient Buildings and Construction Sector*; IEA and United Nations Environment Programme: Madrin, Spain, 2019.

21. Chandel, S.; Sharma, V.; Marwah, B.M. Review of energy efficient features in vernacular architecture for improving indoor thermal comfort conditions. *Renew. Sustain. Energy Rev.* **2016**, *65*, 459–477. [CrossRef]

22. Ehzan, M.; Farshchi, M.A.; Ford, A. Vernacular Architecture And Energy Use In Buildings: A Comparative Study. *IJAMCE* **2015**, *2*, 35–42.

23. Lepore, M. The right to the sun in the urban design. *Vitr. Int. J. Arch. Technol. Sustain.* **2017**, *2*, 24–43. [CrossRef]

24. Mazzone, D. The Dark Side of a Model Community: The Ghetto of el-Lahun. *J. Anc. Egypt. Archit.* **2017**, *2*, 19–54.

25. Butti, K.; Perlin, J. *A Golden Threat 2500 Years of Solar Architecture and Technology*; Van Nostrand Reinhold Company: New York, NY, USA, 1980.

26. Giacomo, L. *The Architecture of A. Palladio in Four Books*; John sWatts: London, UK, 1715.

27. Curreli, S.; Coch, H. Solar access in the compact city: A study case in Barcelona. In Proceedings of the 3rd International Conference PALENC 2010 Passive & Low Energy Cooling for the Built Environment, Rhodes Island, Greece, 29 September–1 October 2010.

28. Wang, X.; Wei, Z.-S.; Pang, X. Solar design in the application of the city planning. In Proceedings of the International Conference on Low-Carbon Transportation and Logistics, and Green Buildings, Beijing, China, 12–13 October 2012.

29. Sarkar, A. Low energy urban block: Morphology and urban guidelines. In Proceedings of the 45th ISOCARP Congress, Porto, Portugal, 18–22 October 2009.

30. Littlefair, P.J.; Santamouris, M.; Alvarez, S.; Dupagne, A.; Hall, D. *Environmental Site Layout Planning: Solar Access, Microclimate and Passive Cooling in Urban Areas*; Construction Research Communications Ltd.: London, UK, 2000.

31. Dekay, M.; Brown, G.Z. *Sun, Wind & Light: Architectural Design Strategies*, 3th ed.; John Wiley & Sons: Hoboken, NJ, USA, 2014.

32. Emmanuel, R. A Hypothetical 'Shadow Umbrella' for Thermal Comfort Enhancement in the Equatorial Urban Outdoors. *Arch. Sci. Rev.* **1993**, *36*, 173–184. [CrossRef]

33. DeKay, R.M. Climatic urban design: Configuring the urban fabric to support daylighting, passive cooling, and solar heating. *Sustain. City VII* **2012**, *1*, 619–630.

34. Dinić, M.; Mitković, P. Planning regulations in the USA and their implications on urban design in the central city zone. *Facta Univ. Arch. Civ. Eng.* **2011**, *9*, 289–299. [CrossRef]

35. Szokolay, S.V. *Introduction to Architectural Science: The Basis of Sustainable Design*; Architectural Press: Oxford, UK, 2004.

36. Capeluto, I.G. Energy performance of the self-shading building envelope. *Energy Build.* **2003**, *35*, 327–336. [CrossRef]

37. Yezioro, A.; Shaviv, E. Shading: A design tool for analyzing mutual shading between buildings. *Sol. Energy* **1994**, *52*, 27–37. [CrossRef]

38. Marsh, A. Computer-Optimised Shading Design. In Proceedings of the Eight International IBPSA Conference, Eindhoven, The Netherlands, 11–14 August 2003.

39. Stevanović, S.; Stevanović, D.; Dehmer, M. On optimal and near-optimal shapes of external shading of windows in apartment buildings. *PLoS ONE* **2019**, *14*, e0212710. [CrossRef] [PubMed]

40. Kabre, C. Winshade: A computer design tool for solar control. *Build. Environ.* **1998**, *34*, 263–274. [CrossRef]

41. Etzion, Y. An improved solar shading design tool. *Build. Environ.* **1992**, *27*, 297–303. [CrossRef]

42. Sargent, J.; Niemasz, J.; Reinhart, C.F. In Proceedings of the SHADERADE: Combining Rhinoceros and Energyplus for the Design of Static Exterior Shading Devices, Sydney, Australia, 14–16 November 2011.

43. Kaftan, E.; Marsh, A. Integrating the cellular method for shading design with a thermal simulation. In Proceedings of the International Conference "Passive and Low Energy Cooling for the Built Environment", Santorini, Greek, 19–21 May 2005.

44. Leidi, M.; Schlüter, A. Exploring urban space—Volumetric site analysis for conceptual design in the urban context. *Internatl. J. Archit. Comput.* **2013**, *11*, 157–182.

45. Leidi, M.; Schlüter, A. Volumetric insolation analysis. In Proceedings of the CleanTech for Sustainable Buildings—From Nano to Urban Scale (CISBAT 2011), Lausanne, Switzerland, 14–16 September 2011.

46. Ozel, F. SolarPierce: A Solar Path-Based Generative System. In Proceedings of the Computation and Performance—Proceedings of the 31st eCAADe Conference, Delft, The Netherlands, 18–20 September 2013.

47. Da Veiga, J.; La Roche, P. A Computer Solar Analysis Tool for the Design and Manufacturing of Complex Architectural Envelopes: EvSurf. In Proceedings of the SIGraDi 2002, 6th Iberoamerican Congress of Digital Graphics, Caracas, Venezuela, 27–29 November 2002.

48. Littlefair, P. Passive solar urban design: Ensuring the penetration of solar energy into the city. *Renew. Sustain. Energy Rev.* **1998**, *2*, 303–326. [CrossRef]

49. De Luca, F. Solar form finding. In Proceedings of the 37th Annual Conference of the Association for Computer Aided Design in Architecture: Disciplines and Disruption, ACADIA, Cambridge, UK, 2–4 November 2017.

50. De Luca, F.; Dogan, T. A novel solar envelope method based on solar ordinances for urban planning. *Build. Simul.* **2019**, *12*, 817–834. [CrossRef]

51. De Luca, F.; Voll, H. Computational method for variable objectives and context-aware solar envelopes generation. In Proceedings of the 8th Annual Symposium on Simulation for Architecture and Urban Design, SimAUD, Toronto, ON, Canada, 22–24 May 2017.

52. Darmon, I. Voxel computational morphogenesis in urban context: Proposition and analysis of rules-based generative algorithms considering solar access. In Proceedings of the Advanced Building Skins, Bern, Switzerland, 28–29 October 2019.

53. Kaasalainen, S.; Krooks, A.; Kukko, A.; Kaartinen, H. Radiometric Calibration of Terrestrial Laser Scanners with External Reference Targets. *Remote Sens.* **2009**, *1*, 144–158. [CrossRef]

54. Kashani, A.G.; Olsen, M.J.; Parrish, C.E.; Wilson, N. A Review of LIDAR Radiometric Processing: From Ad Hoc Intensity Correction to Rigorous Radiometric Calibration. *Sensors* **2015**, *15*, 28099–28128. [CrossRef]

55. Alkadri, M.F.; De Luca, F.; Turrin, M.; Sariyildiz, S. Making use of point cloud for generating subtractive solar envelopes. In Proceedings of the eCAADe SIGraDi 2019: Architecture in the Age of the 4th Industrial Revolution, Porto, Portugal, 7 September 2020.

56. Alkadri, M.; De Luca, F.; Turrin, M.; Sariyildiz, I.S. An integrated approach to subtractive solar envelopes based on attribute information from point cloud data. *Renew. Sustain. Energy Rev.* **2020**, *123*, 109742. [CrossRef]

57. Fujita, Y.; Hoshino, Y.; Ogata, S.; Kobayashi, I. Attribute Assignment to Point Cloud Data and Its Usage. *Glob. J. Comp. Sci. Technol.* **2015**, *15*, 2.

58. Zhan, Q.; Yu, L.; Liang, Y. A point cloud segmentation method based on vector estimation and color clustering. In Proceedings of the 2nd International Conference on Information Science and Engineering (ICISE), Hangzhou, China, 4–6 December 2010.

59. Kobayashi, I.; Fujita, Y.; Sugihara, H.; Yamamoto, K. Attribute Analysis of Point Cloud Data with Color Information. *J. Jpn. Soc. Civ. Eng. Ser. F3 Civil Eng. Inform.* **2011**, *67*. [CrossRef]

60. Aijazi, A.K.; Checchin, P.; Trassoudaine, L. Segmentation Based Classification of 3D Urban Point Clouds: A Super-Voxel Based Approach with Evaluation. *Remote Sens.* **2013**, *5*, 1624–1650. [CrossRef]

61. Fujita, Y.; Kobayashi, I.; Chanseawrassamee, W.; Hoshino, Y. Application of Attributed Road Surface Point Cloud Data in Road Maintenance. *J. Jpn. Soc. Civ. Eng. Ser. F3 Civil Eng. Inform.* **2014**, *70*. [CrossRef]

62. Tamke, M.; Blümel, I.; Ochman, S.; Vock, R.; Wessel, R. From point clouds to definitions of architectural space: Potentials of automated extraction of semantic information from point clouds for the building profession. In Proceedings of the eCAADe, Newcastle upon Tyne, UK, 10 September 2014.

63. Turner, E.; Cheng, P.; Zakhor, A. Fast, Automated, Scalable Generation of Textured 3D Models of Indoor Environments. *IEEE J. Sel. Top. Signal Process.* **2014**, *9*, 409–421. [CrossRef]

64. Tan, K.; Cheng, X. Correction of Incidence Angle and Distance Effects on TLS Intensity Data Based on Reference Targets. *Remote Sens.* **2016**, *8*, 251. [CrossRef]

65. Shin, J.-I.; Park, H.; Kim, T. Characteristics of Laser Backscattering Intensity to Detect Frozen and Wet Surfaces on Roads. *J. Sens.* **2019**, *2019*, 1–9. [CrossRef]

66. Kaasalainen, S.; Kaartinen, H.; Kukko, A. Snow cover change detection with laser scanning range and brightness measurements. *EARSeL eProc.* **2008**, *7*, 133–141.

67. Alkadri, M.F.; Turrin, M.; Sariyildiz, S. A computational workflow to analyse material properties and solar radiation of existing contexts from attribute information of point cloud data. *Build. Envir.* **2019**, *155*, 268–282. [CrossRef]

68. Suchocki, C.; Błaszczak-Bąk, W. Down-Sampling of Point Clouds for the Technical Diagnostics of Buildings and Structures. *Geoscience* **2019**, *9*, 70. [CrossRef]

69. Armesto, J.; Riveiro, B.; González-Aguilera, D.; Rivas-Brea, M.T. Terrestrial laser scanning intensity data applied to damage detection for historical buildings. *J. Archaeol. Sci.* **2010**, *37*, 3037–3047. [CrossRef]

70. Soudarissanane, S. The Geometry of Terrestrial Laser Scanning; Identification of Errors, Modeling and Mitigation of Scanning Geometry. Ph.D. Thesis, Faculty of Civil Engineering and Geosciences, TU Delft, Delft, The Netherlands, January 2016.

71. TOPCON. GLS-2000 Series: Multi Functional 3D Laser Scanner. Available online: https://www.topconpositioning.com/sites/default/files/product_files/gls-2000series_broch_7010-2152_reve_team_en_us_lores.pdf (accessed on 6 January 2020).

72. Maptek. *Maptek I-Site Studio*; Maptek: Denver, CO, USA, 2015.

73. Compare, C. Cloud Compare: 3D point cloud and mesh processing software, open source project. 2019. Available online: http://www.cloudcompare.org/ (accessed on 1 April 2020).

74. Matlab. MathWorks. 2020. Available online: https://nl.mathworks.com/products/matlab.html?s_tid=hp_ff_p_matlab (accessed on 2 April 2020).

75. Rhinoceros. Rhinoceros: design, model, present, realize, analyse...2020. Available online: https://www.rhino3d.com/ (accessed on 4 April 2020).

76. Davidson, Scoot. Grasshopper: Algorithmic modelling for rhino. 2020. Available online: https://www.grasshopper3d.com/ (accessed on 4 April 2020).

77. Ladybug. Ladybug tools. 2020. Available online: https://www.ladybug.tools/ (accessed on 5 April 2020).

78. Boulch, A.; Marlet, R. Deep Learning for Robust Normal Estimation in Unstructured Point Clouds. In Proceedings of the Eurographics Symposium on Geometry Processing, Berlin, Germany, 20–24 June 2016.

79. Sasidharan, S. A Normalization scheme for Terrestrial LiDAR Intensity Data by Range and Incidence Angle. *Internatl. J. Emerging Technol. Adv. Eng.* **2016**, *6*, 322–328.

80. Ministry of Foreign Affairs. *Climate change profile: Indonesia*; Ministry of Foreign Affairs of The Netherlands: The Hague, The Netherlands, 2018.

81. Optotherm. Optotherm Thermal Imaging, Optotherm Inc. 2018. Available online: https://www.optotherm. com/emiss-examples.htm (accessed on 9 January 2020).

82. Damayanti, H.; Tumiwa, F.; Citraningrum, M. *Residential Rooftop Solar: Technical and Market Potential in 34 Provinces in Indonesia*; Institute for Essential Services Reform (IESR)—Accelerating Low-Carbon Energy Transition: Jakarta, Indonesia, 2019.

83. Mass, W. What's Next? How Do We Make Vertical Urban Design? In Proceedings of the Council on Tall Buildings and Urban Habitat (CTBUH), Shenzen, China, 16–21 October 2016.

84. MVRDV. Bastide Niel. MVRDV. 2010. Available online: https://www.mvrdv.nl/projects/46/bastide-niel- (accessed on 3 August 2020).

8

Attenuation Factor Estimation of Direct Normal Irradiance Combining Sky Camera Images and Mathematical Models in an Inter-Tropical Area

Román Mondragón [1], Joaquín Alonso-Montesinos [2,3,*], David Riveros-Rosas [1], Mauro Valdés [1], Héctor Estévez [1], Adriana E. González-Cabrera [1] and Wolfgang Stremme [4]

[1] Department of Solar Radiation at the Geophysics Institute of the National Autonomous University of Mexico, Mexico City 07840, Mexico
[2] Department of Chemistry and Physics, University of Almería, 04120 Almería, Spain
[3] CIESOL, Joint Centre of the University of Almería-CIEMAT, 04120 Almería, Spain
[4] Department of Spectroscopy and Remote Perception at the Geophysics Institute of the National Autonomous University of Mexico, Mexico City 07840, Mexico
* Correspondence: joaquin.alonso@ual.es

Abstract: Nowadays, it is of great interest to know and forecast the solar energy resource that will be constantly available in order to optimize its use. The generation of electrical energy using CSP (concentrated solar power) plants is mostly affected by atmospheric changes. Therefore, forecasting solar irradiance is essential for planning a plant's operation. Solar irradiance/atmospheric (clouds) interaction studies using satellite and sky images can help to prepare plant operators for solar surface irradiance fluctuations. In this work, we present three methodologies that allow us to estimate direct normal irradiance (DNI). The study was carried out at the Solar Irradiance Observatory (SIO) at the Geophysics Institute (UNAM) in Mexico City using corresponding images obtained with a sky camera and starting from a clear sky model. The multiple linear regression and polynomial regression models as well as the neural networks model designed in the present study, were structured to work under all sky conditions (cloudy, partly cloudy and cloudless), obtaining estimation results with 82% certainty for all sky types.

Keywords: cloud detection; digitized image processing; artificial neural networks; solar irradiance estimation; solar irradiance forecasting; solar energy; sky camera; remote sensing; CSP plants

1. Introduction

One of the main factors supporting the continued consumption of energy from renewable sources, such as solar energy, is their potential to substitute approximately 4% of the electricity currently generated from burning fossil fuels. Given their high energy consumption, the world's largest economies are aiming to sustainably develop their own power generation processes by implementing new technologies and methodologies. Using rigorous analytical models, research studies in this area have evaluated the future costs, benefits and disadvantages for electricity generation; they have also analyzed how solar energy generation will evolve [1–3]. According to the latest Global Data report, the solar energy capacity is estimated to increase significantly from about 600 gigawatt (GW) in 2019 to about 1600 GW in 2030 following significant additional capacity coming from China, India, Germany, the US and Japan [4].

Over recent years, a wide body of research has been carried out to optimize the solar energy resource using new technologies and taking advantage of regions where there is a high concentration of surface solar irradiance. The beam irradiance has been predicted for cloudless, partially cloudy and overcast skies over the short-term (from 1 to 180 min) using a sky camera, where the average nRMSE values obtained were 24.36%, 20.9% and 19.17%, respectively [5]. G. Reikard calculated the solar irradiance over time horizons of 60, 30, 15, and 5 min, implementing Autoregressive Integrated Moving Average (ARIMA) with errors between 20% and 90% [6]. Solar irradiance forecasting applied to photovoltaic energy production was implemented using the Smart Persistence algorithm in Machine Learning techniques, achieving an nRMSE of 25% on the best panels over short horizons, and 33% over a 6 h horizon [7].

An analysis of energy forecasting in solar-tower plants combining a short-term solar irradiation forecasting scheme with a solar-tower plant model, the System Advisor Model (SAM), was used to simulate the behavior of the Gemasolar and Crescent Dunes plants. The findings showed that the best results appeared for the 90-min horizon, where the annual forecasting energy yield for Gemasolar was 97.34 GWh year while for Crescent Dunes it was 392.57 GWh year [8]. Similarly, cloud abundance forecasting has been studied for timescales of between (1–180 min), resulting in short-term forecasting (of less than one hour) and medium-term forecasting (up to 3 h), which was proven to have an 80% success rate—indeed, it was so successful that an application (portal) tool was developed that helps to increase power plant production [9,10].

Recent studies have presented a method for the probabilistic forecasting of solar irradiance based on the joint Probability Distribution Function (PDF) of irradiance predicted using the Numerical Weather Prediction (NWP) and the irradiance observed; these are based on models of meteorological processes such as atmospheric dynamics, cloud formation and radiative transfer processes [11]. H. Yang and B. Kurtz estimated direct solar irradiance over the short and medium term for different sky conditions using MSG satellite images, obtaining an $nRMSE$ of 21% and an r value above 0.79 for direct irradiance over a period of 0–3 h [12]. Studies performed at U.C. San Diego forecasted solar irradiance based on cloud detection using a sky image system which evaluated cloud performance over thirty-one consecutive winter days, maintaining $nRMSE$ errors of 20% [13].

By developing a clear sky model, it might be possible to know the initial solar energy conditions in the operation area of a power generation plant. Using the parameters included in such a model, the theoretical solar irradiance could be calculated in real time by means of models that methodologically apply numerical algorithms that could be related to clear sky images [14].

Machine-learning-based methodologies, such as genetic algorithms (GA) and neural networks, have been proposed and applied to solar irradiance modeling [15]. Using various indices to compute solar irradiance such as sky camera images, IR data, pyranometer data, pyrheliometers, NOOA/AVHRR data and clarity indexes (amongst others), important research studies have focused their attention on estimating solar irradiance using multiple linear regression models, polynomial regression models and models that use artificial neural networks [16,17]. Another approach utilizes the artificial neural networks method along with the backpropagation algorithm to forecast solar irradiance; the simulation results have shown that mean absolute percentage errors in the four example days of the forecasting are less than 6% [18].

Solar irradiance has a very high degree of variability, owing to many environmental factors, including cloud cover, relative humidity, and air temperature [19]. Solar irradiance variability due to clouds has become one of the main concerns for the electricity grid as the energy market has steadily expanded over recent years [20]. Consequently, the ability to predict the presence of clouds and interpret their relevant characteristics is essential for predicting solar energy variations, and thus mitigate the effects of production fluctuations in concentrated solar energy systems [20]. Solar irradiance attenuation is caused mainly by the presence of gases in the atmosphere, aerosols, clouds and dust particles, amongst others things. In physical terms, it is due to the reflection, absorption and scattering that the radiation suffers along its path, linking surface measurements directly to the volume, shape, thickness, and types of clouds. Most solar thermal power plants now employ a thermal storage medium in order to stabilize the sudden variations between the electricity load and the alternating solar energy, thus improving system operability and stability [21–23]. Therefore, the ability to characterize cloudiness is paramount. In this context, the technical use of sky cameras has been expanding as a tool to forecast solar irradiance. There are records of its use at the University of California Merced's solar observatory station (2012), at CIESOL, University of Almeria, Spain (2013), at the University of California, San Diego (2014), at UNAM in Mexico (2015) and at the University of Singapore (2019), amongst others. Using this technology with models that convert the digitized information into irradiance indices, it is possible to estimate the direct solar irradiance for different sky conditions, as they occur, by modeling a camera calibration system (a process where an image's pixel intensity is related to the amount of solar radiation present at that moment) [5,24,25]. In addition, remote sensing techniques (satellite images and sky cameras) have been combined with radiometric data for sky classification processing [26].

Solar companies recognize that cost remains the main drawback of Concentrated Solar Power (CSP) systems. Consequently, important projects such as CSPIMP (Concentrated Solar Power efficiency IMProvement) have managed to make CSP plants more competitive [27]. R.Chauvin and J. Nou, forecasted the cloud cover for different sky conditions over 5–30 min with a spatial resolution of (1 km^2) using sky images and algorithm development [27,28]. F. Batlles and J. Montesinos established that detecting and classifying clouds as well as determining their trajectory are essential factors in forecasting cloud cover [29]. Likewise, they demonstrated that infrared channels are essential for cloud height assignment and the visible channel is necessary for cloud opacity determination; however, this was solely applied to short-term cloud forecasting using multispectral satellite imagery [29,30]. It is clear that a wide range of different technologies and systems have been used to estimate the solar resource; nonetheless, most publications focus on solar resource assessment and not on estimating the attenuation factor.

In this paper, three models were applied to estimate the direct normal irradiance for Mexico City based on determining the attenuation percentage caused by clouds under different sky conditions. To perform the study, a total sky camera was used (TSI-880 model) from which the digitized image levels were characterized and modelled to determine the attenuation coefficient of this solar irradiance component.

2. Methodology

2.1. Data

The data and images used in this work were obtained from the Solar Irradiance Observatory (SIO) at the Geophysics Institute of the National Autonomous University of Mexico (latitude: 19.32, longitude: −99.17, elevation: 2280 m above sea level). The observatory serves as the Regional Center for the Measurement of Solar Irradiance, part of the World Meteorological Organization (WMO).

Direct Normal Irradiance (DNI) observations were taken over the period from 2016 to 2017 using a CHP1 Kipp&Zonen pyrheliometer. Regarding the pyrheliometer's operation, Kipp&Zonen state that a maximum uncertainty of 2% is expected for hourly totals and 1% for daily totals. These data pass from the instrument's sensor to the Kipp&Zonen CR3000 data logger. The solar height (α) and azimuth angle were recorded (in degrees) every minute.

A total sky camera with a rotational shadow band (a TSI 880 model) providing a hemispheric view of the sky was used to obtain minute-by-minute images from 6:00 a.m. to 6:00 p.m. throughout 2016 and 2017. The device has a 352 × 288-pixel image resolution, represented by 24 bits. The camera's digitized image is composed of RGB channels [red, green, blue], HSV [hue, saturation, brightness] and E [gray scale]. We call these the digitized channels. The RGB values range from [0–255] while the HSVE values range from [0–1].

To work with solar irradiance models, it is necessary to include additional parameters. We included local monthly values of the Linke turbidity index obtained from the SODA web page [31], supported by Meteotest, Switzerland. The data processing and image analysis were carried out using MATLAB® mathematical software.

2.2. Determination of the Attenuation Percentage

In the present study, an approach was carried out that uses sky camera images to estimate solar irradiance for all sky types. This model allows us to determine the amount of lost solar irradiance caused by the cloudiness factor.

To have reference elements for the attenuation percentage of the lost solar irradiance, we start from a clear sky model, where it is possible to calculate 100% of the direct solar irradiance received at each study moment. To estimate the solar irradiance attenuation, we started with the European Solar Radiation Atlas (ESRA) clear sky model, which requires the Linke turbidity index to determine the theoretical local solar irradiance of a clear sky [32,33]. With ESRA, we obtained the maximum DNI clear-sky value corresponding to the date and time for 400 different images.

The sample of 400 images was selected from an image database recorded from February 2016 to January 2017. We chose 400 images taken under different sky types (cloudy, partly cloudy and clear), covering all the seasons of the year. The selected images were taken from sunrise to sunset every hour, guaranteeing different solar zenith elevations and different sky conditions. The camera was correctly maintained to provide good image definition and clarity thus reducing the risk of working with pixels that were not typical of the sky or cloud type when digitally analyzing the images. All the images were selected when the solar elevation was greater than 10 degrees. The images were classified into the following sky types—233 belonging to partially cloudy skies, 68 to cloudy skies and 99 to clear skies.

The attenuation percentage recorded by the pyrheliometer sensor was calculated considering the value of the direct solar irradiance corresponding to the date and the solar altitude at which the images were taken. For any given sky camera image, the maximum DNI value was calculated from the ESRA model and the DNI value obtained from the pyrheliometer was multiplied by the sine of the solar height. Subsequently, we obtained the percentage of attenuation caused by the different cloud types traversing the solar disk at the time the image was taken. According to the pyrheliometer specifications, the attenuation percentage data obtained had an uncertainty of 2% because we considered the recorded DNI data on an hourly basis. Nonetheless, there were clear days where the maximum DNI value obtained by the ESRA model could not be reached. In this way, a small percentage of attenuation is recorded that is important for solar energy capture.

2.3. Digitized Data That Represent the Percentage of Attenuation

A new database was created that included the percentage of DNI attenuation generated by the different sky conditions and the respective cloud-type classification that caused it. Then, we calibrated the DNI measured on the surface with the pixel values of the clouds near the solar disk. These values better represent (at the pixel level) the solar irradiance attenuation caused by the cloud at that moment. With this procedure, we hope to find patterns between the pixel values, the sky type and the estimated DNI value measured [34]. This method offers better image calibration at the pixel level with respect to the DNI attenuation than do other methods. Indeed, other methods study such a large image area around the solar disk that different sky conditions can be present at the same time.

2.4. Extraction of the Image's Digitized Data

In situations where the sun position is not known (due to the presence of clouds), an algorithm was developed that allows us to know the sun position as Cartesian coordinates, which are those that govern the interpretation of the image's spatial position [35]. The solar position algorithm was presented by Reda and Andreas (2013) to obtain the sun's geographical location [36].

Therefore, the sun's position can be ascertained on any day of the year regardless of the sky conditions. In the 400 images we analyzed, different areas were plotted where we observed the same pixel intensity – this was done to identify the pixel value corresponding to each sky type, which would attenuate the DNI 1 to 3 minutes later. The method applied to define these uniform areas was visual using an algorithm elaborated. We checked the uniformity of the areas by extracting the values of all the pixels for each selected area. In this way we could ensure that the pixels had the same values. For conditions with different pixel values, the area was considered non-uniform and a new area was redrawn. We selected areas near the sun's position to avoid those areas where sunlight might provide information that was not specific to the sky type. This is shown in Figure 1.

Figure 1. Selecting areas from the three different images representing each sky condition. The pixels in each area have the same intensity.

During the area selection process, the digitized channel values attributed to each area were extracted. Using an algorithm developed for the purpose, the pixel intensity averages were obtained for the areas, and for each channel [R G B H S V E]. Since the average values of the areas are related to solar height, we added an external geographic variable [A(α)] to our value sets, which describes the selected areas. We will call the set of values for [R G B H S V E A] digitized data (*DD*). With this method, it is possible to obtain the digitized data causing the attenuation recorded on the surface. The temporal space between the selected area and the subsequent position of the sun fluctuates between 1 and 2 min.

2.5. Combination of Digitized Data

The digitized channel values for blue sky during the day are not the same in the morning as they are in the afternoon. The path that the sunlight takes through the atmosphere to the measurement point changes, giving different shades of color in the sky owing to the light dispersion caused by the presence of atmospheric particles. The blue wavelength is mainly scattered in the atmosphere while the raindrops that make up the clouds do not disperse the sunlight wavelengths, hence making them appear white to gray depending on their density (Rayleigh scattering). Accordingly, we used other variables to identify or contrast the cloud or sky pixels, such as dividing the digitized data (e.g., R/B) or multiplying certain digitized data by the sin of the solar height (e.g., V sin (α)), among others. Essentially, we performed these division and multiplication operations on the *DDs* because this is a widely used mechanism in digital image analysis. In our case, each channel is not equally relevant for characterizing the clouds; therefore, it was necessary to manipulate the channels in this way to distinguish the cloud pixels from the sky pixels. These results were then used as the new input variables for the models. In this way, adequate patterns were acquired to determine the DNI attenuation caused by cloud presence.

2.6. Estimation of the DNI Attenuation Factor Using Multiple Linear Regression

An initial model for measuring the surface DNI attenuation factor starts with the digitized image data, developed using the multiple linear regression method. This model is similar to simple linear regression, the difference being that we need to consider more than one explanatory variable $(x_1, x_2, ..., x_n)$. The multiple linear regression model, using the least squares criterion in matrix form, finds the coefficients $(\beta_1, \beta_2, ..., \beta_k)$ that best represent the dependent variable (y) via the independent variables.

Considering that the regression function that relates both variables is linear, the model to be solved in its matrix form is defined by Equation (1):

$$y_{est} = \begin{bmatrix} y_1 \\ y_2 \\ \cdot \\ \cdot \\ \cdot \\ y_n \end{bmatrix} = \begin{bmatrix} 1 & x_{21} & x_{31} & \ldots & x_{k1} \\ 1 & x_{22} & x_{32} & \ldots & x_{k2} \\ \cdot & \cdot & \cdot & \cdot & \cdot \\ \cdot & \cdot & \cdot & \cdot & \cdot \\ \cdot & \cdot & \cdot & \cdot & \cdot \\ 1 & x_{2n} & x_{3n} & \ldots & x_{kn} \end{bmatrix} \begin{bmatrix} \beta_1 \\ \beta_2 \\ \cdot \\ \cdot \\ \cdot \\ \beta_k \end{bmatrix}, \tag{1}$$

where y_{est} is the estimated direct solar irradiance in percentage of attenuation plus an error term between the independent and dependent variables.

To solve the proposed model, an algorithm was developed that uses the tools offered by the software employed to resolve the multiple linear regression. The independent input variables chosen for the model were DD.

Once the 8 independent variables were defined, we will call register to each area selected with their respective [R G B H S V E A] value. In total, 400 registers were selected. To do this, 300 registers were chosen randomly, generating a matrix (M1) with 2,400 data points in total (300 registers multiplied by 8 variables) for training the programmed multiple linear regression model. The algorithm allows us to obtain the regression coefficients (β_i) necessary to integrate the regression function. We used digitized image analysis to combine the DD set, to obtain the maximum correlation coefficient (Equation (2)) for the direct solar radiation measured on the surface in percentage attenuation terms (y_{mea}) and the direct solar radiation estimated by the model (y_{est}).

$$r = \frac{\sigma_{y_{est}y_{mea}}}{\sigma_{y_{est}}\sigma_{y_{mea}}}, \tag{2}$$

where $\sigma_{y_{est}y_{mea}}$ is the covariance between the estimated and measured input databases, $\sigma_{y_{est}}$ is the standard deviation of y_{est} and $\sigma_{y_{mea}}$ is the standard deviation from y_{mea}.

When we observed combination types where the correlation coefficient tended to decrease, we did not continue with those combination structures. Once the best combination was found from the DD providing the best DNI estimation, the multiple linear regression function was validated with the remaining 100 registers (M2, with 800 data points in total—100 registers multiplied by 8 variables), where the same combination of the variables remained in the data validation set.

2.7. Estimation of the DNI Attenuation Factor Using polynomial Regression

Polynomial regression is another model used to estimate the DNI attenuation factor through the sky camera. Similarly, by means of the least squares criterion, it returns the coefficients (p) for a polynomial of degree (n) which is better adapted to the data behavior trend; the length of p is $n + 1$. Unlike the multiple linear model, when applying the polynomial model, one or two independent variables are required to generate the regression function that fits a nonlinear relationship between the value with independent variable exponents (x^n) versus the dependent variable or explanatory variable (y_{est}). The matrix representation is shown in Equation (3).

$$y_{est} = \begin{bmatrix} y_1 \\ y_2 \\ \cdot \\ \cdot \\ \cdot \\ y_m \end{bmatrix} = \begin{bmatrix} x_1^n & x_1^{n-1} & \ldots 1 \\ x_2^n & x_2^{n-1} & \ldots 1 \\ \cdot & \cdot & \cdot \cdot \\ \cdot & \cdot & \cdot \cdot \\ \cdot & \cdot & \cdot \\ x_m^n & x_m^{n-1} & \ldots 1 \end{bmatrix} \begin{bmatrix} p_1 \\ p_2 \\ \cdot \\ \cdot \\ \cdot \\ p_{n+1} \end{bmatrix}. \tag{3}$$

Following graphical observation, the value dispersion between the estimated attenuation factors was calculated using the clear sky model, and the polynomial behavior was observed. By performing mathematical operations combining digitized channels and solar height to obtain a single independent variable, we selected the one that had the highest correlation coefficient in terms of its linear relationship. Once the independent input variable was defined, we created an algorithm that provided the polynomial coefficients that best represented the dependent variable. To train the model, we worked with 300 registers at random, that is, 1200 data points in total according to the combination found. Once the polynomial function was defined, we validated it by making the same channel combination for the 100 registers (400 data points) missing from the database. The polynomial regression using two independent input variables to generate the polynomial function $z = f(x, y)$ (x and y being the independent variables) presented a maximum correlation coefficient of $r = 0.5$ therefore we decided to continue with the polynomial regression of one independent variable.

2.8. Estimation of the DNI Attenuation Factor Using Neural Networks

We also implemented the neural networks method to estimate direct solar irradiance (y_{est}). For this, the Neural Net Fitting function was utilized, a function included in the interactive applications (APPS) contained in the MATLAB tools. An artificial neural network perceptron of interconnected multilayers was used, as shown in Figure 2.

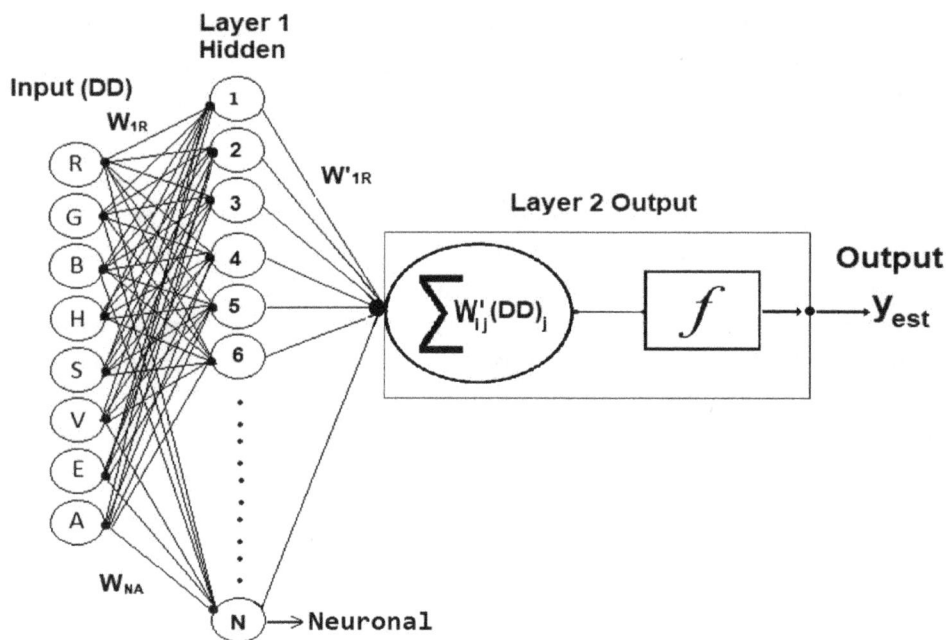

Figure 2. Artificial neural network scheme.

The input data used for executing the neural networks were the same independent variables (x_j) as in the previous models. Each data entry in the neuron is multiplied by validating its corresponding weight or significance (w_{ij}). All the weighted inputs are summed, and a neuron activation function is activated, such that it generates neuronal learning; when passing to the last neuronal layer, the output data are obtained - in our case, the (y_{est}) results. A representation of the artificial neural networks model is described, as in Equation (4), where N is the total number of neurons [37,38].

$$y_{est} = f_i \left(\sum_{j=0}^{N} w_{ij} x_j \right). \tag{4}$$

The Levenberg-Marquardt algorithm was used to train the neurons since it has a faster MSE convergence speed, facilitating the computation time [37,38]. To maintain the same proportion as in the previous models, we used 300 registers to train the neural network; 50 registers were occupied for validation and 50 for testing (400 registers in total). Multiple combinations were performed between the dependent variables to find the digitized data combination that returned the r and $nRMSE$ with the best results.

Once the best input values were found to explain the dependent variable, the neural network was trained by modifying the number of hidden layers such that the r and $RMSE$ results had the best relationship between the estimated and the measured solar irradiance.

3. Results

The results from the proposed models are presented to estimate the attenuation factor of the direct solar irradiance at the pixel level by means of images. To develop the models, we worked with 400 images chosen from a file containing a year's worth of images captured every minute from February 2016 to January 2017 with their respective DNI surface measurements.

To quantify the models' validity, the statistical estimation analysis was obtained for the three cases: $RMSE$—the root mean square error (5), in which the results are shown as a value between 0 and 1 (0 being 0 attenuation and 1 being 100% attenuation), and where N is the total number of estimations; $nRMSE$—the normalized mean square error (6), measured in percent (%); MBE—the mean deviation (7), with the same unit as the $RMSE$; and $nMBE$—given by (8), expressed in (%) and the dimensionless correlation coefficient Equation (2).

$$RMSE = \sqrt{\frac{1}{N}\sum_{i=1}^{N}(y_{est} - y_{mea})^2}, \tag{5}$$

$$nRMSE = \left[\frac{RMSE}{y_{max} - y_{min}}\right]100, \tag{6}$$

$$MBE = \frac{1}{N}\sum_{i=1}^{N}(y_{est} - y_{mea}), \tag{7}$$

$$nMBE = \left[\frac{MBE}{(y_{max} - y_{min})}\right]100. \tag{8}$$

3.1. Multiple Linear Regression Model

Table 1 displays the digitized data (DD) combinations of M1, carried out randomly, that had the highest r value from which the best combination was chosen; this was based on the behavior of the digitized channels in performing the linear regression training with 2400 data points, and then validating the model for M2 with the remaining 800 data points.

In Table 1, we observe how the DD combination that allows a high correlation between the measured and estimated data dominates the image's red component. Once the best digitized data (DD) combination was selected and used as the model's independent input values, we proceeded to validate Equation (9) on the M2 matrix.

$$y_{est} = \beta_1 + \beta_2(DD)_1 + \beta_3(DD)_2 + ...\beta_n(DD)_{n-1}. \tag{9}$$

Table 2 shows the coefficients obtained when developing the multiple linear regression during the model training and when applied to the same DD combinations performed in M1.

Table 1. Combinations of digitized data for training the multiple linear model.

Num	M1. Combination of Digitized Data (DD)	RMSE	nRMSE	MBE	nMBE	r
1	R/E R/B R/V G/R R/G E/B G/B E/R R	0.207	33.761	3.6×10^{-15}	5.9×10^{-13}	0.782
2	R/E R/B R/V G/R R/G E/B G/B R	0.207	33.779	-4.0×10^{-15}	-7.0×10^{-13}	0.781
3	R/E G/B E/B S/H(sin(α)) R/V R/A(sin(α))	0.209	34.085	-5.0×10^{-17}	-9.0×10^{-15}	0.777
4	R/E R/B R/V G/R R/G E/B G/B	0.209	34.188	-8.0×10^{-15}	-1.0×10^{-12}	0.776
5	R/E G/B E/B S/H(sin(α)) R/V A/H(sin(α))	0.209	34.155	9.0×10^{-16}	1.0×10^{-13}	0.776
6	E/R B/V H	0.214	34.908	-2.0×10^{-15}	-3.0×10^{-13}	0.764
7	R/E R/B R/V	0.214	34.955	7.0×10^{-16}	1.0×10^{-13}	0.763
8	R/E G/B V(sin(α))	0.215	35.059	7.0×10^{-16}	1.0×10^{-13}	0.762
9	R/E G/H(sin(α)) B/R S/H(sin(α)) V(sin(α))	0.215	35.111	-8.0×10^{-16}	-1.0×10^{-13}	0.761
10	R/E	0.217	35.443	2.9×10^{-15}	4.8×10^{-13}	0.756

Table 2. Validation of the model and the multiple linear regression coefficients.

Num	Coefficients ($\beta_1, \beta_2, ..., \beta_n$)	RMSE	nRMSE	MBE	nMBE	r
1	−11.98, 3.95, 2.53, 0.0008, −11.79, −0.18, −4.58, 5.82, −0.002, 17.04	0.243	42.36	−0.054	−9.46	0.677
2	20.57, −21.04, 14.45, 0.0006, −5.07, 2.29, −11.48, 1.18, −0.002	0.243	42.37	−0.055	−9.61	0.678
3	−4.97, 2.91, −6.27, 11.81, −0.08, −0.01, −0.05	0.243	42.34	−0.059	−10.29	0.681
4	25.47, −34.22, 17.60, −0.002, −5.84, 11.54, −25.35, 11.89	0.244	42.47	−0.060	−10.57	0.681
5	−5.10, 3.66, −6.12, 11.98, −0.05, -0.015, 0.0002	0.244	42.58	−0.053	−9.22	0.673
6	6.50, −3.68, −0.01, 0.44	0.251	43.72	−0.057	−9.95	0.650
7	−4.13, 5.91, −0.007	0.254	44.26	−0.060	−10.58	0.645
8	−4.51, 4.79, 0.38, −0.12	0.253	44.17	−0.059	−10.35	0.643
9	-2.35, 3.77, −0.0001, −0.743, 0.0028, −0.087	0.251	43.67	−0.053	−9.23	0.647
10	−4.31, 4.92	0.249	43.47	−0.064	−11.24	0.662

Within the model validation, we noticed that the highest correlation between y_{mea} and y_{est} was not necessarily presented in the same combination as that which gave the highest training value ($r = 0.782$). Considering the same combination as in the test data, there is a value of $r = 0.677$ with an $nRMSE = 42.36\%$, while for combination No. 3, a value of $r = 0.681$ was obtained with an $nRMSE = 42.34\%$ and an overestimated value of $nMBE = 10.29\%$ (negative), indicating it was a slightly greater correlation than that selected in the training.

The training (M1) and test (M2) graphs (Figure 3) display the behavior of the multiple linear regression model on a linear trend; both graphs have a high concentration of points in larger attenuations, but far from the linear trend.

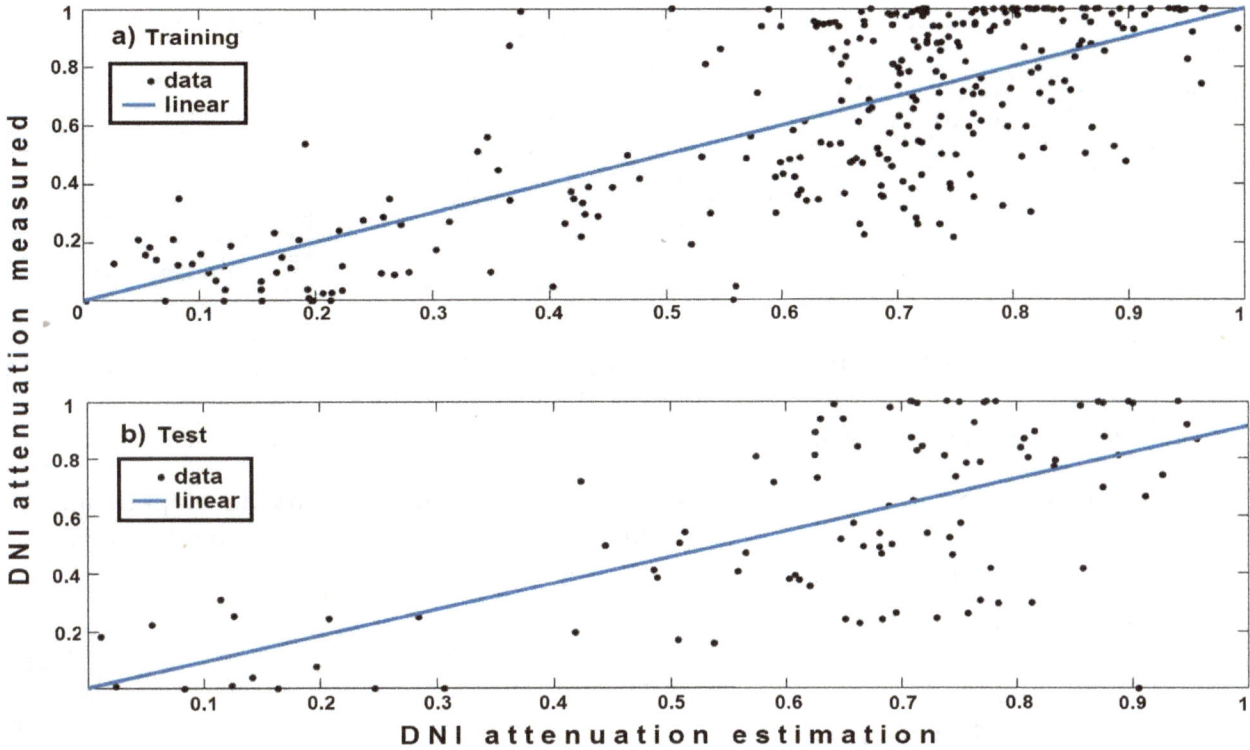

Figure 3. Dispersion between the measured data and the estimated direct normal irradiance (DNI) (%) on a linear trend. (**a**) Training of the multiple linear regression model (M1). (**b**) Test of the multiple linear regression model (M2).

3.2. Polynomial Regression Model

As discussed previously, it is essential to introduce an input variable to develop the polynomial regression. The input variable was selected after creating combinations between the *DDs*. After 50 tests between the *DD* combinations, we noticed that the *r* value tendencies did not present better results. For this reason, in Table 3, there are only 4 combinations shown - these were chosen because they had the highest *r* value.

Table 3. Combination of the digitized data as the input variable for the polynomial regression.

Num	M1. Combination of Digitized Data (*DD*)	*RMSE*	*nRMSE*	*MBE*	*nMBE*	*r*
1	(R/G/B) × E	0.2133	34.77	-6.54×10^{-16}	-1.07×10^{-13}	0.766
2	R/E	0.2174	35.44	2.99×10^{-15}	4.80×10^{-13}	0.756
3	R/B	0.2191	35.71	-6.00×10^{-13}	-1.00×10^{-15}	0.602
4	R/G	0.2466	40.21	-1.00×10^{-15}	-2.00×10^{-13}	0.669

Once the input variable was defined as a *DD* combination, by means of an algorithm developed, the characteristic polynomial function (Equation (10)) was obtained that best defined the percentage of attenuation suffered by the solar irradiance, as measured on the surface.

$$y_{est} = p_1(DD)^n + p_2(DD)^{n-1} + ... + p_n + p_{n+1}. \tag{10}$$

Considering the *x*-axis as the percentage of measured DNI attenuation and the *y*-axis as the percentage of estimated DNI attenuation, we used a polynomial adjustment created in MATLAB to compute the polynomial which best described our scatter plot for its highest value at *r* = 0.766; the grade was *n* = 9, which is plotted in the same graph (Figure 4).

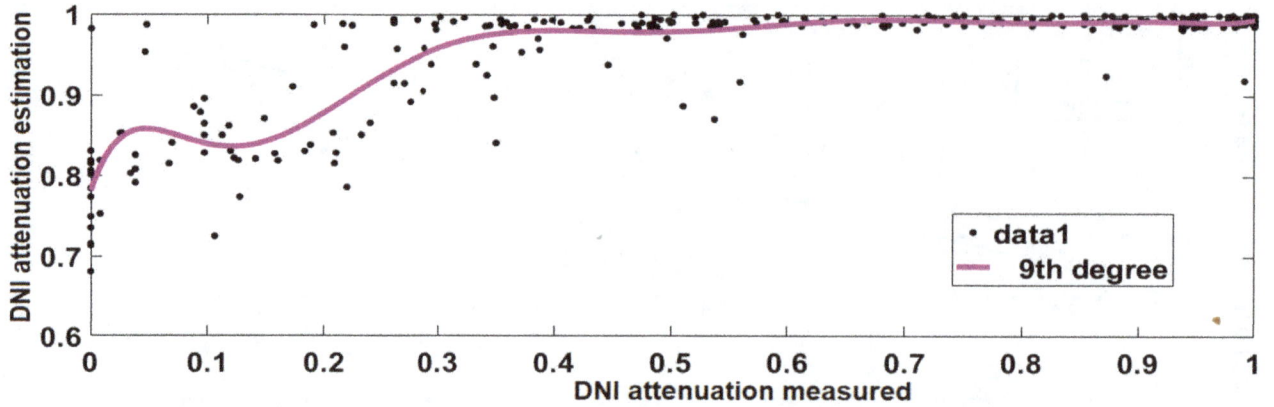

Figure 4. Dispersion between the measured and the estimated DNI data (%) with polynomial adjustment.

For this reason, the polynomial regression for the M1 matrix was developed, applying a polynomial of degree 9 (Figure 5) and, using a developed algorithm, the polynomial coefficients (p) were obtained. Figure 6 shows the model validation applied to the M2 matrix.

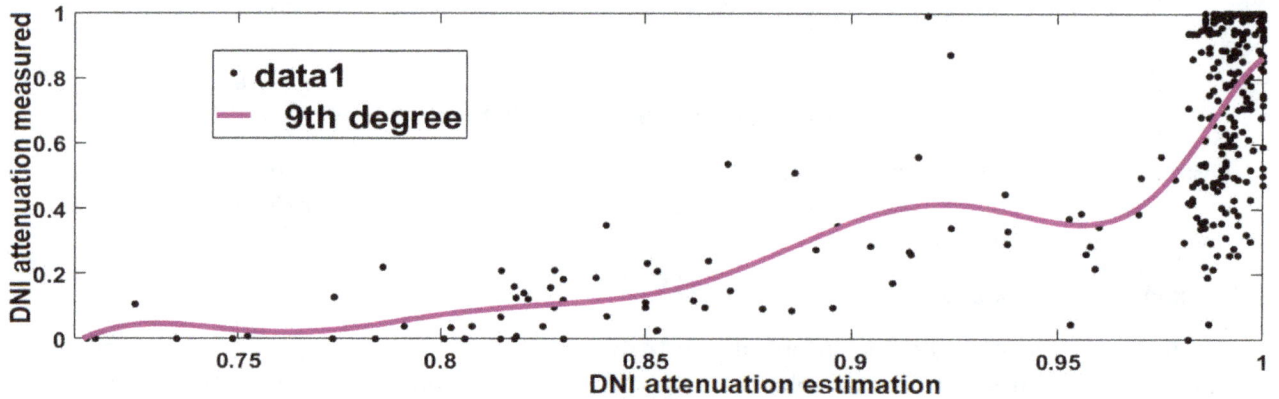

Figure 5. Adjustment of polynomial grade 9 for the M1 training data.

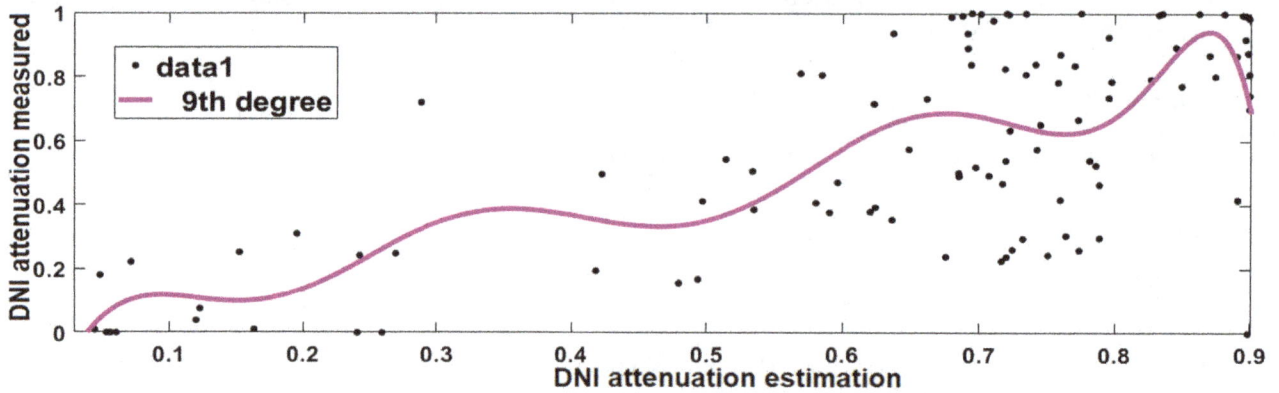

Figure 6. Validation of the polynomial regression model for M2.

Predictably, both polynomial adjustment plots present a similar trend but with a slight precision with respect to the scattering of the points. Table 4 shows the polynomial coefficients obtained from the M1 training and the statistical results when validating the model with the M2 matrix.

Table 4. Validation of the model and the polynomial regression coefficients

Num	Polynomial Coefficients ($p_1, p_2, ..., p_{n+1} * 1.0e + 08$)	RMSE	nRMSE	MBE	nMBE	r
1	$-0.0487, 0.3984, -1.4433, 3.0394, -4.00993, 3.6721, -2.1846, 0.8323, -0.1842, 0.0181$	0.2457	42.77	-0.0611	-10.633	0.6756

Regarding the input variables in the polynomial model, the ratios between the RGB channels multiplied by the gray scale data (E) predominated; this did not depend on the solar height recorded in each image. Analyzing Table 4, we note that, with the polynomial model, a relationship coefficient between the y_{mea} and the y_{est} of $r = 0.6756$ was obtained with an $nRMSE = 42.77\%$ and an overestimated value of the $nMBE = 10.633\%$ (in negative), showing a very similar fit to that of the multivariable linear regression model.

3.3. Neural Networks Model

Starting from the complete DD set, the number of variables was reduced, obtaining different combinations among the DDs to identify which contributed less to the DNI estimation. Using Matlab APPS to develop the model by means of artificial neural networks, we acquired the following statistical results (Table 5), where the term AD refers to all the data, TRA refers to the training data, VAL refers to the validation data and TST refers to the test data.

Table 5. Combination of the digitized data as input parameters for the neural networks model.

Combination of Digitized Data (DD)	R(AD)	R(TRA)	R(VAL)	R(TST)	RMSE(TRA)	RMSE(VAL)	RMSE(TST)
R G B H S V E A	0.771	0.776	0.816	0.636	0.207	0.194	0.254
R G H S V E A	0.779	0.782	0.793	0.743	0.205	0.198	0.228
R G B S V E A	0.774	0.797	0.715	0.787	0.194	0.245	0.211
R G B H V E A	0.771	0.783	0.735	0.750	0.207	0.222	0.197
R G B H S E A	0.782	0.808	0.714	0.713	0.197	0.231	0.207
R G B H S V A	0.767	0.768	0.746	0.803	0.207	0.232	0.190
R G B H S V E	0.774	0.776	0.815	0.694	0.207	0.199	0.236
R B H S E A	0.773	0.809	0.692	0.725	0.190	0.246	0.251
R G H S E A	0.779	0.823	0.579	0.793	0.191	0.262	0.186
R G B S E A	0.756	0.785	0.731	0.616	0.204	0.230	0.268
R G B H E A	0.780	0.795	0.701	0.811	0.199	0.235	0.195
R G B H S A	0.770	0.814	0.624	0.706	0.193	0.246	0.249
R G B H S E	0.779	0.775	0.801	0.769	0.209	0.198	0.203
R H S V E A	0.779	0.782	0.789	0.736	0.209	0.191	0.211
R B S V E A	0.764	0.776	0.727	0.781	0.210	0.220	0.218
R B H V E A	0.714	0.714	0.747	0.651	0.233	0.222	0.255
R B H S E A	0.761	0.766	0.715	0.818	0.209	0.236	0.204
R B H S V A	0.768	0.761	0.778	0.810	0.215	0.205	0.191
R B H S V E	0.765	0.785	0.736	0.681	0.204	0.226	0.236
G B H S V E A	0.774	0.774	0.758	0.811	0.206	0.221	0.196
G H S V E A	0.751	0.760	0.755	0.667	0.220	0.207	0.229
G B S V E A	0.758	0.763	0.757	0.721	0.211	0.223	0.232
G B H V E A	0.770	0.768	0.806	0.677	0.207	0.211	0.227
G B H S E A	0.715	0.786	0.559	0.624	0.205	0.280	0.285
B H S V E A	0.765	0.760	0.796	0.720	0.209	0.218	0.224
B S V E A	0.743	0.730	0.759	0.852	0.224	0.233	0.176
B H V E A	0.754	0.790	0.657	0.687	0.205	0.235	0.253
B H S E A	0.732	0.776	0.616	0.491	0.218	0.242	0.233
B H S V A	0.769	0.751	0.808	0.790	0.210	0.213	0.207
B H S V E	0.765	0.770	0.750	0.765	0.204	0.237	0.212

Considering Table 5, we see that different DD combinations are presented, which are input information parameters required by the neural network for the training. It operates by finding patterns of interest between the DD and the percentage of attenuation values for the measured DNI (y_{mea}) allowing the neural network to learn.

The model results were obtained by training the network, which occupied 10 neurons in a perceptron of two layers, (1. The hidden learning layer and 2. The output layer). After one hundred input parameter combinations were introduced into the network, the combination [R G B H S E A] was selected with the highest r value to make computational runs with a greater number of neurons – this was done to achieve better network learning and thus, generate the computational model that obtained the estimated DNI attenuation percentage with the best approximation to the surface measurement. Table 6 shows that the final computational model for estimating the DNI attenuation percentage using artificial neural networks was defined for a network model containing 60 neurons in the hidden learning layer.

Table 6. Results of the neuronal network training varying the number of neurons.

Combination (DD)	Neurons	R(AD)	R(TRA)	R(VAL)	R(TST)	RMSE(TRA)	RMSE(VAL)	RMSE(TST)
R G B H S E A	10	0.782	0.808	0.714	0.713	0.197	0.231	0.207
R G B H S E A	20	0.773	0.798	0.705	0.719	0.201	0.228	0.233
R G B H S E A	30	0.775	0.786	0.701	0.713	0.877	0.198	0.248
R G B H S E A	40	0.780	0.781	0.777	0.768	0.206	0.203	0.219
R G B H S E A	50	0.772	0.805	0.721	0.597	0.198	0.220	0.271
R G B H S E A	60	0.818	0.875	0.657	0.804	0.159	0.267	0.204
R G B H S E A	70	0.795	0.830	0.696	0.722	0.185	0.231	0.227
R G B H S E A	80	0.733	0.746	0.648	0.796	0.226	0.246	0.229
R G B H S E A	90	0.782	0.820	0.620	0.864	0.190	0.264	0.173
R G B H S E A	100	0.793	0.836	0.580	0.829	0.184	0.249	0.212

Functionally, the model was chosen because, as seen in Figure 7, its training evaluation represented the highest training and test data values, with a validation of $r = 0.657$, a $RMSE = 0.267$ and a test value of $r = 0.804$, with a $RMSE = 0.204$ together with the training regression $r = 0.875$, with an $RMSE = 0.159$ and the analysis of r for all data $r = 0.818$ (of approx. 0.82). The points dispersion in the neural networks model shows a better fit to the linear trend proposed in the graphs. Consequently, we determined that the neural networks model offered better results than the multivariable linear regression model and the polynomial model.

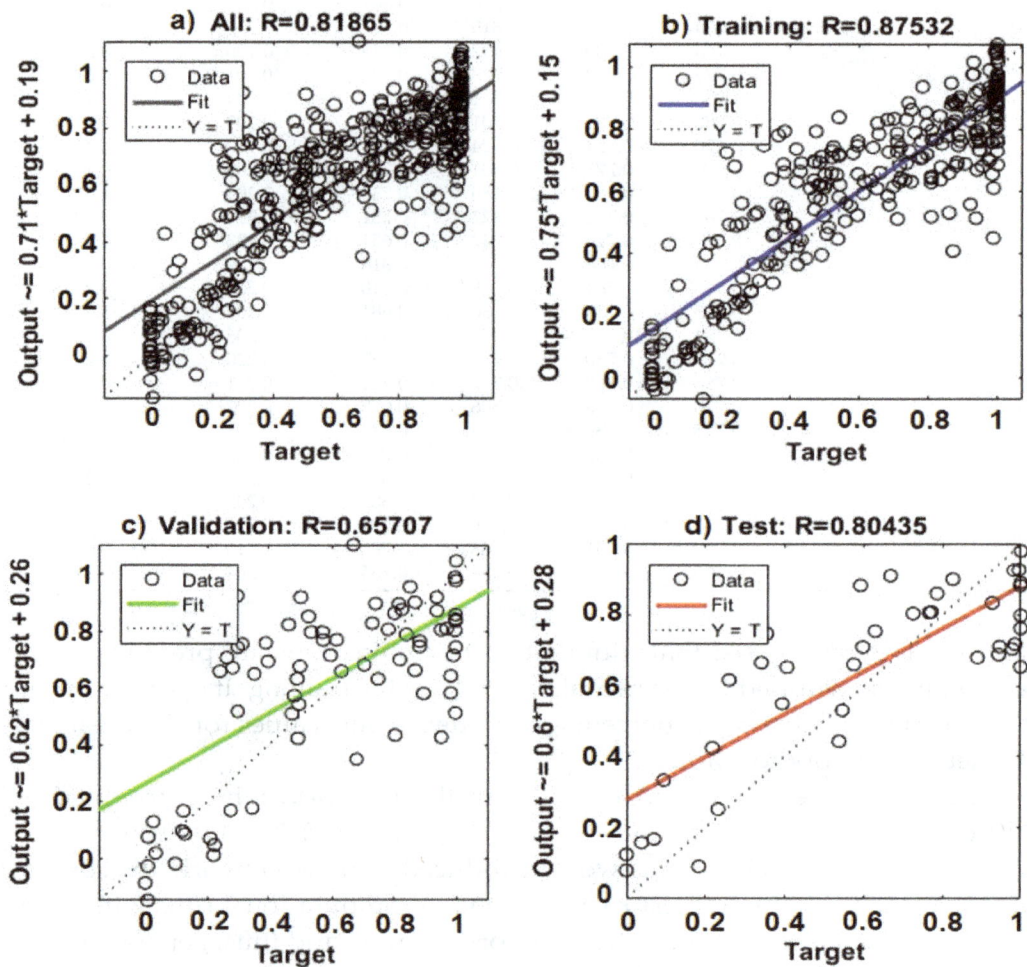

Figure 7. Validation of the neural networks model. (**a**) AD (all data) (**b**) TRA (training data) (**c**) VAL (validation data) and (**d**) TST (test data).

3.4. Functioning of the Models

As we can see in Figures 3, 5 and 6 the models better estimate DNI attenuation for partly cloudy and totally cloudy skies. With respect to the linear and exponential trend lines, we see that the correlation between the attenuation of the measured DNI and the estimated DNI is greater in the range of values between 0.6 to 1 (considering 1 as 100% attenuation). Obtaining attenuations above 0.6, we performed digitized analysis of the images where the percentage of cloudiness was very dense or we analyzed images where low clouds were present, usually nimbostratus or cumulonimbus-types in different seasons and at different times of the year. Values below 0.6 indicated that we were assessing images where high clouds, clear skies or low-density clouds were present, analogous to different seasons and times of the year. These types of cloud, usually altostratus or cumulus clouds which were not very dense, attenuate the DNI lesser degrees.

Regarding Figure 7, which corresponds to the neural networks model, we see that the training graph exhibits a more uniform correlation between the attenuation of the measured and estimated DNI for all sky conditions in all the range of values, from clear skies (0% attenuation) to cloudy skies (1 ; i.e., 100% attenuation). In terms of validation and testing, we understand that the model functions better (as with the 2 previous models) for partly cloudy and cloudy skies.

4. Discussion

We have determined that, of the three methods considered and applied to estimate the percentage of DNI attenuation under all sky conditions, the neural networks model provided the best results in terms of the correlation coefficient obtained, achieving a calibration system for the digitized data and the DNI attenuation. To improve the model's efficiency, more parameters need to be integrated to help explain the dependent variable, such as adding cloud classification. Similarly, it is important to experiment by adding more layers within the neural network so that learning by patterns of interest becomes more and more accurate, thus obtaining output data comparable to what is measured or known.

For all three methods, the estimates might be better if thousands of images were processed, thus minimizing and compensating for errors. Another important point is that we should consider the presence of different cloud layers. Viewing the cloud cover from a terrestrial viewpoint, each cloud has a particular value in the digitized channels. The cloud is seen with particular digitized data and, therefore, it has an attenuation coefficient. However, it is possible to have the same (or similar) digitized data for a cloud but with a different attenuation coefficient because there are further cloud layers above the low cloud, hidden from the sky camera. Such situations can occur and, consequently, they increase the error in the models.

5. Conclusions

In this paper, we presented three models for estimating the DNI attenuation percentage under all sky conditions. Images were taken using a sky camera with a rotational shadow band (model TSI 880). This is the first study carried out for Mexico City that estimates direct solar irradiance based on determining the attenuation resulting from the percentage of cloudiness that is present.

Estimation by means of the multiple linear regression model offers an r reliability of approximately 0.70 while it can compute an error of 42%; however, it also achieves an approximation of $r = 0.78$, decreasing the normalized error to 33%. The best combination was modeled to obtain the final function of the linear model. The red component, used as the input parameter, turned out to be the most predominant for the model, showing that the cloudiness contrast can be detected more easily on this channel than on the other channels. The final model function consisted of 10 summed terms, of which 9 multiply the respective combination between the (DD) by the coefficients extracted from the multiple linear regression.

With regard to the polynomial model, the best model representation obtained was a polynomial function of degree $n = 9$, achieving a certainty of $r = 0.76$ with a normalized error of 35%. The most important combined digitized data for generating a single independent input variable are the RGB channels and the pixel values in grayscale. Similarly, for the multiple linear regression, the final function consisted of 10 summed terms, where 9 of them respectively multiply the coefficients of the polynomial by the value of (DD) raised to a certain exponential degree.

The model built using neural networks achieved a better approximation of the data for the direct solar irradiance attenuation percentage. Through neural networks, it is possible to have data reliability of $r = 0.82$ in a layer of 60 neurons. The best DD combination forms part of the information data for the neuronal model learning. Unlike the two previous cases, the DD combinations are integrated directly. The input data are entered separately and mostly occupy the defined digitized data, excluding the V channel, which represents the brightness in the image. The final model turns out to be an execution code which determines the weights W_{ij} and assigns them to the digitized data with respect to the already known y_{mea} data.

It was possible to develop a first DNI estimation system using sky images applicable in real time for Mexico City. From these sky cam images, we developed a tool that can be applied to any measurement system, complementing the solar irradiance information that reaches the Earth's surface. Our work serves as a preliminary study that helps anticipate drops in solar irradiance. Consequently, if this system is installed in a solar plant, the operator can predict how much irradiance might be lost as a result of passing clouds. This system may also contribute to the classification of clouds depending on the level of attenuated irradiance and in determining the percentage of local cloudiness in the sky on a particular day. The study complements a previous work that predicted solar irradiance over the short or medium term in the hope of determining the vector movement of clouds. Accordingly, the present article aims to be a reference for predicting DNI in the CSP environment.

Author Contributions: Formal analysis, M.V. and H.E.; Investigation, J.A.-M., D.R.-R. and M.V.; Methodology, R.M., J.A.-M., A.E.G.-C. and W.S.; Resources, D.R.-R., H.E. and A.E.G.-C.; Software, R.M., A.E.G.-C. and W.S.; Supervision, J.A.-M. and D.R.-R.; Validation, J.A.-M.; Visualization, W.S.; Writing—original draft, R.M.; Writing—review & editing, R.M. and W.S. All authors have read and agreed to the published version of the manuscript.

Acknowledgments: The authors would like to express their gratitude to CIESOL at the University of Almería, Spain, to the Solar Irradiance Observatory at the Geophysics Institute of the National Autonomous University of Mexico (UNAM) and to CONACyT.

References

1.   Mellit, A.; Massi Pavan, A. A 24-h forecast of solar irradiance using artificial neural network: Application for performance prediction of a grid-connected PV plant at Trieste, Italy. *Solar Energy* **2010**, *84*, 807–821. [CrossRef]

2.   Hernández-Moro, J.; Martínez-Duart, J.M. Analytical model for solar PV and CSP electricity costs: Present LCOE values and their future evolution. *Renew. Sustain. Energy Rev.* **2013**, *20*, 119–132. [CrossRef]

3.   Kabir, E.; Kumar, P.; Kumar, S.; Adelodun, A.A.; Kim, K.H. Solar energy: Potential and future prospects. *Renew. Sustain. Energy Rev.* **2018**, *82*, 894–900. [CrossRef]

4.   Globaldata. Available online: http://www.globaldata.com/global-solar-photovoltaic-capacity-expected-to-exceed-1500gw-by-2030-says-globaldata (accessed on 27 October 2019)

5.   Alonso-Montesinos, J.; Batlles, F.; Portillo, C. Solar irradiance forecasting at one-minute intervals for different sky conditions using sky camera images. *Energy Convers. Manag.* **2015**, *105*, 1166–1177. [CrossRef]

6.   Reikard, G. Predicing solar radiation at high resolutions: A comparison of time series forecasts. *Solar Energy* **2009**, *83*, 342–349. [CrossRef]

7.   Huertas-Tato, J.; Centeno, M. Using Smart Persistence and Random Forest to Predict Photovoltaic Energy Production. *Energies* **2019**, *12*, 100. [CrossRef]

8.   Alonso-Montesinos, J.; Polo, J.; Ballestrín, J.; Batlles, F.J.; Portillo, C. Impact of DNI forecasting on CSP tower plant power production. *Renew. Energy* **2019**, *138*, 368–377. [CrossRef]

9. Alonso-Montesinos, J.; Batlles, F. Short and medium-term cloudiness forecasting using remote sensing techniques and sky camera imagery. *Energy* **2014**, *73*, 890–897. [CrossRef]

10. Alonso, J.; Ternero, A.; Batlles, F.J.; López, G.; Rodríguez, J.; Burgaleta, J.I. Prediction of cloudiness in short time periods using techniques of remote sensing and image processing. *Energy Procedia* **2014**, *49*, 2280–2289. [CrossRef]

11. Kakimoto, M.Y.; Shin, H.; Ikeda, R.; Kusaka, H. Probabilistic Solar Irradiance Forecasting by Conditioning Joint Probability Method and Its Application to Electric Power Trading. *IEEE Trans. Sustain. Energy* **2019**, *10*, 983–993. [CrossRef]

12. Alonso-Montesinos, J.; Batlles, F. Solar radiation in the short-and medium—Term under all sky conditions. *Energy* **2015**, *83*, 387–393. [CrossRef]

13. Yang, H.; Kurtz, B.; Nguyen, D.; Urquhart, B.; Chow, C.W.; Ghonima, M.; Kleissl, J. Solar irradiance forecasting using a ground-based sky imager developed at UC San Diego. *Solar Energy* **2014**, *103*, 502–524. [CrossRef]

14. Nou, J.; Chauvin, R.; Thil, S.; Eynard, J.; Grieu, S. Clear-sky irradiance model for real-time sky imager application. *Energy Procedia* **2015**, *69*, 1999–2008. [CrossRef]

15. Li, J.; Ward, J.; Tong, J.; Collins, L.; Platt, G. Machine learning for solar irradiance forecasting of photovoltaic. *Renew. Energy* **2016**, *90*, 542–553. [CrossRef]

16. Sahin, M.; Kaya, Y.; Uyar, M. Comparison of ANN and MLR models for estimating solar radiation in Turkey using NOAA/AVHRR data. *Adv. Space Res.* **2013**, *51*, 891–904. [CrossRef]

17. Fu, C.; Cheng, H. Predicting solar irradiance with all-sky image features via regression. *Solar Energy* **2013**, *97*, 537–550. [CrossRef]

18. Watetakarn, S.; Premrudeepreechacharn, S. Forecasting of Solar Irradiance for Solar Power Plants by Artificial Neural Network. In Proceedings of the 2015 IEEE Innovative Smart Grid Technologies-Asia (ISGT ASIA), Bangkok, Thailand, 6 November 2015.

19. Wojtkiewicz, J.; Katragadda, S.; Gottumukkala, R. A Concept-Drift Based Predictive-Analytics Framework: Application for Real-Time Solar Irradiance Forecasting. In Proceedings of the IEEE Big Data 2018, Seattle, WA, USA, 10 Decemebr 2018; pp. 5462–5464.

20. Peng, Z.; Yu, D.; Huang, D.; Heiser, J.; Yoo, S.; Kalb, P. 3D Cloud detection and tracking system for solar forecast using multiple sky imagers. *Solar Energy* **2015**, *118*, 496–519. [CrossRef]

21. Alonso-Montesinos, J.; Batles, F. The use of sky camera for solar radiation estimation based on digital image processing. *Energy* **2015**, *90*, 377–386. [CrossRef]

22. Tapakis, R.; Charalambides, A. Equipment and methodologies for cloud detection and classification: A review. *Solar Energy* **2013**, *95*, 392–430. [CrossRef]

23. Martínez-Chico, M.; Batlles, F.; Bosch, J. Cloud classification in a mediterranean location using radiation data and sky images. *Energy* **2011**, *36*, 4055–4062. [CrossRef]

24. Marquez, R.; Coimbra, C. Intra-hour DNI forecasting based on cloud tracking image analysis. *Solar Energy* **2013**, *91*, 327–336. [CrossRef]

25. Gohari, M.I.; Urquhart, B.; Yang, H.; Kurtz, B.; Nguyen, D.; Chow, C.W.; Kleissl, J. Comparison of solar power output forecasting performance of the total sky imager and the university of California, San Diego sky imager. *Energy Procedia* **2014**, *49*, 2340–2350. [CrossRef]

26. Alonso-Montesinos, J.; Batles, F.; López, G.; Ternero, A. Sky camera imagery processing based on a sky classification using radiometric data. *Energy* **2014**, *68*, 599–608. [CrossRef]

27. Chauvin, R.; Nou, J.; Thil, S.; Traoré, A.; Grieu, S. Cloud detection methodology based on a sky-imaging system. *Energy Procedia* **2015**, *69*, 1970–1980. [CrossRef]

28. Chauvin, R.; Nou, J.; Thil, S.; Grieu, S. Cloud motion estimation using a sky imager. In Proceedings of the SolarPACES 2015, Cape Town, South Africa, 16 October 2015. [CrossRef]

29. Batles, F.; Alonso, J.; López, G. Cloud cover forecasting from METEOSAT data. *Energy Procedia* **2014**, *57*, 1317–1326. [CrossRef]

30. Escrig, H.; Batlles, F.J.; Alonso, J.; Baena, F.M.; Bosch, J.L.; Salbidegoitia, I.B.; Burgaleta, J.I. Cloud detection, classification and motion estimation using geostationary satellite imagery for cloud cover forecast. *Energy* **2013**, *55*, 853–859. [CrossRef]

31. Soda-pro. Available online: http://www.soda-pro.com/web-services/atmosphere/linke-turbidity-factor-ozone-water-vapor-and-angstroembeta (accessed on 22 April 2019)

32. Rigollier, C.; Bauer, O.; Wald, L. On the clear sky model of the ESRA-European Solar Radiation Atlas—With respect to the heliosat method. *Solar Energy* **1999**, *68*, 33–48. [CrossRef]

33. Riveros, D.; Gonzalez, A.; Valdes, M.; Zarzalejo, L.; Ramírez, L. *Analysis of Linke Turbidity Index from Solar Measurements in Mexico*; American Institute of Physics: College Park, MD, USA, 2018. [CrossRef]

34. Luiz, S.; Wangenheim, A.; Bueno, E.; Comunello, E. The use of euclidean geometric distance on RGB color space for the classification of sky and cloud patterns. *Atmos. Ocean. Technol.* **2009**, *27*, 1504–1517.

35. Marquez, R.; Gueorguiev, V.; Coimbra, C. Forecasting of global horizontal irradiance using sky cover indices. *Sol. Energy Eng.* **2013**, *135*, 011017–011022. [CrossRef]

36. Reda, I.; Andreas, A. Solar Position Algorithm for Solar Radiation Application. National Renewable Energy Laboratory (NREL). 2003. Available online: https://rredc.nrel.gov/solar/codesandalgorithms/spa/ (accessed on 25 March 2019)

37. López, G.; Gueymard, C.A.; Bosch, J.L.; Rapp-Arrarás, I.; Alonso-Montesinos, J.; Pulido-Calvo, I.; Barbero, J. Modeling water vapor impacts on the solar irradiance reaching the receiver of a solar tower plant by means of artificial neural networks. *Solar Energy* **2018**, *169*, 34–39. [CrossRef]

38. Rosiek, S.; Alonso-Montesinos, J.; Batlles, F. Online 3-h forecasting of the power output from a BIPV system using satellite observations and ANN. *Electr. Power Energy Syst.* **2018**, *99*, 261–272. [CrossRef]

9

Industry Experience of Developing Day-Ahead Photovoltaic Plant Forecasting System Based on Machine Learning

Alexandra I. Khalyasmaa [1,2], **Stanislav A. Eroshenko** [1,2], **Valeriy A. Tashchilin** [1],
Hariprakash Ramachandran [3], **Teja Piepur Chakravarthi** [4] and **Denis N. Butusov** [5,*]

[1] Ural Power Engineering Institute, Ural Federal University named after the first President of Russia B.N. Yeltsin, 620002 Ekaterinburg, Russia; a.i.khaliasmaa@urfu.ru (A.I.K.); s.a.eroshenko@urfu.ru (S.A.E.); v.a.tashchilin@urfu.ru (V.A.T.)

[2] Power Plants Department, Novosibirsk State Technical University, 630073 Novosibirsk, Russia

[3] Department of Electrical and Electronics Engineering, Bharath Institute of Higher Education and Research, Chennai 600073, India; ad.registrar@bharathuniv.ac.in

[4] Department of Computer Science and Engineering, Bharath Institute of Higher Education and Research, Chennai 600073, India; tejapiepur@bharathuniv.ac.in

[5] Youth Research Institute, Saint Petersburg Electrotechnical University "LETI", 197376 Saint Petersburg, Russia

[*] Correspondence: dnbutusov@etu.ru

Abstract: This article highlights the industry experience of the development and practical implementation of a short-term photovoltaic forecasting system based on machine learning methods for a real industry-scale photovoltaic power plant implemented in a Russian power system using remote data acquisition. One of the goals of the study is to improve photovoltaic power plants generation forecasting accuracy based on open-source meteorological data, which is provided in regular weather forecasts. In order to improve the robustness of the system in terms of the forecasting accuracy, we apply newly derived feature introduction, a factor obtained as a result of feature engineering procedure, characterizing the relationship between photovoltaic power plant energy production and solar irradiation on a horizontal surface, thus taking into account the impacts of atmospheric and electrical nature. The article scrutinizes the application of different machine learning algorithms, including Random Forest regressor, Gradient Boosting Regressor, Linear Regression and Decision Trees regression, to the remotely obtained data. As a result of the application of the aforementioned approaches together with hyperparameters, tuning and pipelining of the algorithms, the optimal structure, parameters and the application sphere of different regressors were identified for various testing samples. The mathematical model developed within the framework of the study gave us the opportunity to provide robust photovoltaic energy forecasting results with mean accuracy over 92% for mostly-sunny sample days and over 83% for mostly cloudy days with different types of precipitation.

Keywords: feature engineering; forecasting; graphical user interface software; machine learning; photovoltaic power plant

1. Introduction

Modern regional electric power systems (EPS) are characterized by an increasing share of renewable energy sources (RES). In most of the developed countries, state-supporting mechanisms are implemented for RES development, including fixed tariffs that determine the price per kilowatt/hour, mark-ups, green certificates and other mechanisms. In Russia, the competitive tendering mechanism for the supply contract for the wholesale market has become most widespread, in which the owners of power generation facilities operating on the basis of RES receive a monthly guaranteed payment for capacity. By an order of the Government of the Russian Federation, target indicators of the installed capacity of such generation in the total structure of generating capacities were determined to be 5,871 MW until 2024. At the beginning of 2018, its installed capacity excluding hydroelectric power plants in the UES of Russia amounted to 1.59 GW and in the world, 941.0 GW, and the assessment of the technically affordable energy potential of RES in Russia from various sources is estimated to be from 5–25 billion tons of oil equivalent per year, that is, an estimated 55% of the annual energy consumption.

The task of RES power generation implementation is directly related to the task of electric energy generation forecasting, since the lack of renewable energy sources' reliable forecasts entails the need to constantly maintain a full reserve of active power in the power system [1] (in the amount of available capacity of RES), which actually means the need for an extra regulation response from thermal generation and its operation in uneconomical modes and/or regulation of the power grid congestion, which in turn causes the problem of switched on power generation excess capacities not only at the regional level, but also on a national scale. The problems of energy production forecasting at power generation facilities using various types of RES are associated with the problem of the stochastic nature of their operation modes. Such a task is multifactorial with a large number of poorly formalized and linguistic data, since it is based on meteorological and climatological data, the generalized nature of which also has a strong influence on the result of energy production forecasting [2].

The need to predict the RES generation is fixed at the state level, according to order No. 91 dated 11 February, 2019 "On approval of requirements for energy consumption forecasting and the formation of electric energy and active power balances for a calendar year and particular periods within a year", " ... The volume of electric energy production in the forecasted energy balance of the power system should be determined for wind and solar power plants - on the basis of monthly data on the average long-term value of electrical energy production by these power plants for the last three years, and in the absence of these data (including the power plants under construction), in accordance with the proposals of the owners on the formation of a consolidated forecasted balance ... ". At the same time, in the dispatch centers in Russia, the task of photovoltaic power plant (PVPP) generation forecasting has not been fully addressed yet. Currently, in the short-term planning of power system operation modes in order to compensate for the stochastic decrease in power output by RES-based generation facilities [3], the volume of EPS active power reserves is increased by the total capacity declared by the owners of RES-based power generation facilities.

In order to increase the efficiency of power system operation modes' short-term planning, in terms of power system constraints monitoring and allocating active power reserves, it is necessary to create tools for PVPP generation forecasting for short-term (one day ahead) forecasting. PVPP owners are also interested in developing forecasting tools. Under existing conditions, this will allow not only solving the problems of selecting the composition of the switched-on power generation equipment, but also ensuring effective planning of the main power generation equipment maintenance.

The above emphasizes the relevance of the study and the need to harmonize the process of introducing PVPPs into the power systems, and also reveals a number of fundamentally new problems and tasks requiring the development of new approaches to their solution from the point of view of information-analytical and mathematical principles of raw data processing and analysis [4], especially in the case of using open-source weather data, extracted from weather prediction models of the local hydrological and meteorological data providers.

Except for the poor formalization and linguistic representation of open-source weather data, the problem of weather forecasting is greatly associated with the total coverage of the area by measurements of meteorological stations and posts [5]. Evidently, sparsely populated areas have an insufficient number of available weather data acquisition points, which makes the open-source weather forecasts less reliable, making the problem of RES-based power generation forecasting more challenging.

In [6], a review of various approaches to electrical energy generation forecasting as well as an analysis of the influence of the forecasting accuracy on the power system control efficiency are described. In [7], a detailed review of existing approaches to solar power plants' electrical energy output forecasting is provided.

On the one hand, due to the chaotic nature of weather variations, traditional forecasting methods may not provide the required level of forecasting accuracy. Moreover, the initial dataset may be subjected to various distortions caused by the peculiar features of such power plants' operation modes. For example, in [8], the influence of dust on solar panels' efficiency is analyzed, and in [9], the effect of snow deposits.

In addition, uneven distortions in the collected data may be caused by partial shadowing of solar panels, as shown in [10]. On the other hand, today a large number of different sensors are available, including satellite data. An example of the application of open satellite data to predict the available power of a solar power plant is given in [11].

The use of new types of data allows us to improve traditional forecasting approaches. For example, in [12], the application of the analog ensemble method for the prediction of the solar power plant energy output was described, and in [13], its modification was analyzed for open-source meteorological data. The application of numerical weather prediction (NWP) algorithms for the evaluation of the magnitude of solar irradiation is described in [14]. The implementation of the network of weather monitoring systems allows one to increase the accuracy of such forecasting, an example of which is presented in [15].

The collection of retrospective data and the development of machine learning methods allow us to identify new hidden relationships between parameters and increase the accuracy of electrical energy generation forecasting. M. Abuella and B. Chowdhury [16] describe the use of multiple linear regression for predicting the solar power plant electrical energy output based on advanced meteorological data. The use of linear regression for solving a similar problem is also described in [17]. Along with linear regression, traditional methods of working with time sequences can be used [18].

The rapid development of machine-learning technologies opens up new possibilities for the improvement of forecasting technologies. A new extreme machine learning algorithm proposed in [19] was successfully applied to solve the problem described in [20].

Along with machine-learning technologies, various algorithms for identifying model parameters are used. With the help of such models, the forecast of generated electrical energy is further carried out. In [21], a comparison of various sky models from the point of view of solar irradiation forecasting is provided. In [22,23], various models of solar panels were investigated from the point of electrical energy production.

Despite the great relevance and interest in solar energy forecasting, proved by a large number of regular publications, today, there are a few software packages that provide this functionality. One of the most popular tools for modeling and analyzing the operation of solar panels is the HOMER software package, a system for modeling combined PV systems that allows one to determine the optimal power system configuration.

In scientific literature, you can find many examples of the application of this software package for solving specific applied problems, for example, to optimize the joint operation of a PV plant with a biofuel installation [24]. You can also find examples of HOMER application to analyze the operation of solar power plants located in different geographical positions, for example, in Georgia [25], the island of Saint Martin [26], Indonesia [27], and India [28]. A detailed analysis of existing software systems and their capabilities is given in [29].

Unfortunately, most of these software systems are not applicable to Russian conditions mostly due to the lack of available meters throughout the territory of the country. More importantly, nowadays Russia is actively in the process of implementing new solar power plants, and the main problem is the availability of initial and retrospective data for developing a forecasting model.

In this regard, there is a need to develop a specialized software package adapted to Russian realities and allowing forecasting of solar irradiation at the installation site of solar panels with subsequent day-ahead forecasting of electrical energy production.

In the presented study, the authors provided a possible solution to the problem of solar power plants generation forecasting, based on the generalized open-source weather data, lacking the necessary features, characterizing specific meteorological events and conditions. A forecast is obtained by implementing a multi-stage procedure of machine learning algorithms applied to get the forecast, which is sufficiently reliable for power system control and short-term operational planning.

The rest of the article is organized as follows. The second part considers solar power generation specific features in terms of the technological and exogenous factors, which influence the solar power generation forecast. The third part addresses the detailed problem formulation and initial multi-source dataset characteristics, containing solar geometry calculated values, power plant measurements and open-source weather data.

The authors compared multiple machine-learning algorithms and provided the algorithms' hyperparameters optimization to find the best composition of the algorithms and their parameters for sunny and cloudy days. Finally, a step-by-step procedure was introduced for better cloudy days forecasting, and the practical implementation results were discussed.

2. Solar Power Forecasting Peculiar Features

PVPP is a complicated technical system, containing electrical equipment of direct (DC) and alternating current (AC) with its own automated control systems, relay protection systems, switchgear equipment, etc. Powerful PV plants with an installed capacity above 1 MW typically work in conjunction with interconnected bulk power systems, providing electrical energy in-feed in peak and half-peak hours.

Being a part of the bulk power system incurs technical and operational rules and constraints, which are imposed by the adjacent power system and are to be strictly followed. From a technical point of view, power network topology, power system frequency and voltage level play a crucial role in PV plant electrical energy output. This means that the operation mode of the PV power plant is influenced not just by external meteorological factors, but by external and internal technological conditions, driven by the power system operation mode and the PV power plant itself.

2.1. PV Power Plant Internal Technological Factors

2.1.1. Photovoltaic Panel: Specific Features

The main PVPP element is a photovoltaic (PV) panel. The generated output of the PV panel is determined by various factors, including the power plant configuration, solar irradiation and ambient temperature.

2.1.2. Electrical Circuits of PV Power Plant

There are various topologies for connecting solar panels, and the specific power plant configuration is typically determined at the design stage. Generally, the string configuration is most often used, where several panels are sequentially connected into a string with a voltage of 12–240 V DC. Each string has a DC/DC with MPPT trackers. Several strings are connected in parallel to a DC/AC inverter providing pulse width modulation (PWM) with power output to the AC side [30].

Among the factors that influence PV generation, there are hardly-formalized heterogeneous parameters, which are given in Table 1.

Table 1. Sources of uncertainty at the level of PV power plant.

Parameter	Range
Rated voltage of PV panels (2–48 V)	0.80–1.05
Converter and HV power transformer losses	0.88–0.98
Different characteristics (different producers) of PV panels	0.98–0.99
PV Panel mismatch with declared passport specifications	0.97–0.995
Diode leakage currents	0.99–0.997
Losses in DC/AC cable lines	0.96–0.98
Degradation of PV panels (1%/year)	0.70–1.00

2.2. PV Power Plant External Factors

2.2.1. Solar Irradiation

The key stage in PV plant energy output forecasting is to determine the main energy characteristic, namely, solar irradiance, which depends on many stochastic factors. The total energy flux density of solar irradiation at the surface of the earth incident on the tilted surface of the solar panel is the sum of direct, diffused and reflected irradiation. Each of these components is a difficult-to-predict parameter, depending on both atmospheric and climatic phenomena [31].

2.2.2. External Factors: Meteorological Data

The initial dataset for PV plant energy output forecasting is composed of different data sources:

1. PV plant technical data, including power output history
2. Meteorological actual data retrospective
3. Meteorological forecasting data retrospective
4. Irradiance retrospective data acquired from PV plant

As long as the data is collected from multiple sources and some features are typically not available for weather forecasts, data uncertainty may occur. For example, cloudiness in weather forecasts is typically provided in percentage [%]. Figure 1 provides a typical case of 2 days (16.10.2017 and 17.10.2017), illustrating a possible variation of the solar irradiation based on practically similar cloudiness data. In Figure 1, the red line corresponds to the cloudiness, while the blue bar chart illustrates solar irradiation for 2 sequential days, measured by the pyranometer.

Figure 1. Actual solar irradiance variation in similar cloudiness conditions.

Another important point is the quality of meteorological data. Up-to-date NWP models are based on actual meteorological data, provided by weather stations, spread all over the territory that is being considered. That means that the greater the redundancy of the meteorological measurements, the greater the accuracy of the weather forecast. The formulated principle imposes a computational challenge for under-populated territories with poorly developed weather stations [32].

2.3. Forecasting Problem Specification and Goals of the Study

As it was discussed, the problems with solar power plant energy output forecasting deals are:

(1)   PV plant is an integrated technological system, composed of non-linear electric circuit components, industrial automation and control systems, operating the functional state of AC and DC electrical installations

(2)   The availability of the primary energy source is highly stochastic. Different prediction time horizons correspond to different prediction models as well as different initial data that can be used to improve the prediction accuracy

(3)   The PV power plant forecasting problem deals with multi-source heterogeneous data. Power output measurements are typically considered together with local weather station measurements, which are extracted from data storage facilities of the automated control system of the PV power plant

So, while pursuing the goal of PV energy forecasting accuracy improvement, the following tasks have been solved:

•   Investigation and justification of various mathematical approaches for day-ahead energy forecasting problems;

•   Development of the PV energy forecasting software tool, dealing with heterogeneous multi-source data, acquired from local measurement systems and open-source weather data;

•   Commitment to PV output forecasting accuracy of not less than 80%, which corresponds to the standard 20% admissible deviation from power system operation plan [33]

3. Problem Statement and Available Data

The development of RES in the world's energy systems is one of the main factors that raises requirements for the collection and analysis of their data, in particular, introducing special additional requirements for sensors and collection and data read-out systems [34,35].

Earth-observing systems have progressed over the past decades in terms of image quality and image frequency [36]. Every satellite and drone system has its own limitations, namely, the number of satellites, weather and daylight for optical systems; vegetation for SAR systems; etc., but despite the limitations, this progress has led the remote sensing industry to this data volume, and the stated repetitive images frequency could provide a full daily scope of the earth surfaces using high-resolution images [37]. Nowadays, data from optical, infrared, radio, and microwave remote-sensing devices have revolutionized the meteorology and climatology, as they provide potentially global coverage and therefore improve access to areas that have a limited number of weather stations (areas with rare data) or not covered by routine observations at all. The remote sensing data supports traditional observations and is widely used in NWP, enhancing and improving weather forecasting, etc. [38], and the remote sensing science has become an essential and versatile tool for natural resource managers and researchers in government agencies, environmental institutions and industry [39].

Despite the great potential of modern methods and tools for remote sensing, unfortunately, the costs of their application are not justified in all production industries. Today, RES generation facilities are in most cases private facilities, which are financed from the owners' funds. Not every owner of RES generation is financially able to use satellite earth observation systems to make forecasts.

In this case, generation owners carry out generation forecasting based on open meteorological data, which often, due to data quality, leads to errors and, as a consequence, problems with the generating facility participation in the energy market. Such data have the following disadvantages:

•   open meteorological data delivered by the meteorological provider for the current day are averaged actual data received from a meteorological station and/or a meteorological desk away from the solar power plant, which leads to an error in determining the solar insulation flux density forecast.

- the use of current measurements obtained from meteorological sensors installed on the PVPP to reduce errors in the forecasting task is impossible without the complex statistical algorithms and the numerical models for forecasting weather conditions, which in turn represent a "substitution" of functions and services delivered by the meteorological provider;
- the data composition delivered by the meteorological provider is limited by the parameters of air temperature, wind speed and direction, and cloudiness quantitative and/or qualitative characteristics; even in the case of a numerical model for forecasting weather conditions, the data from the local meteorological station will not be enough for correction, since the cloud characteristics auto-monitoring function at local meteorological desks is usually not implemented.

All of the above problems form the goal of this study: increasing the PVPP generation forecasting accuracy based on open meteorological data.

In the current study, the PV forecasting problem refers to day-ahead active power forecasting (electrical energy) generated by a particular real grid-scale PV power plant based on the retrospective data [40].

3.1. Problem Formulation

Assuming the following initial dataset:

$$S = \begin{bmatrix} \overbrace{x_{1,1}\ x_{1,2}}^{\text{source 1}} & \dots & \overbrace{x_{1,a}\ x_{1,a+1}}^{\text{source 2}} & \dots & \overbrace{x_{1,b}\ x_{1,b+1}}^{\text{source 3}} & \dots & y_1 \\ x_{2,1}\ x_{22} & \dots & x_{2,a}\ x_{2,a+1} & \dots & x_{2,b}\ x_{2,b+1} & \dots & y_2 \\ \dots\ \dots & \dots & \dots\ \dots & \dots & \dots\ \dots & \dots & \dots \\ x_{l,1}\ x_{l,2} & \dots & x_{l,a}\ x_{l,a+1} & \dots & x_{l,b}\ x_{l,b+1} & \dots & y_l \end{bmatrix} \tag{1}$$

where y_j is the predicted parameter; x_{ij} is a feature, corresponding to the parameter; l is the number of observations in the sample; and b is the number of features. All the data is aligned in time.

The goal is to build a mathematical model that will determine the value of the new parameters y_j according to the corresponding features x_{ij} with a given threshold accuracy. In other words, the task is to build a model f, which, having received the input x, would predict the answer y.

3.2. Initial Data Sample Description

In the given problem formulation, the initial dataset includes 16 features, stored in a single database for the period from September 26, 2017 to February 5, 2019. The data was acquired from a real operating PV power plant, located in the south of the Russian Federation. Among the features, we used calculated parameters, measured data, as well as the open-source weather data, acquired from weather providers:

Time, date: 29.09.2017–05.02.2019
Coordinates: Latitude 46.398642, Longitude 48.515582
Calculated parameters:

- solar declination angle, [deg.], range [−23.45, 23.45];
- sunrise time, [hour], range [4.97, 8.58];
- sunset time, [hour], range [16.87, 20.61];
- solar zenith angle cosine, range [0, 0.92];
- solar altitude angle, [deg.], range [0, 66.03];
- solar constant, 1367 [Wh/m^2];
- solar irradiation at the top of the atmosphere, [Wh/m^2], range [0, 1213.47];

Measured data:

- PV power plant hourly actual generation, [kWh], range [0, 12 919.2];

- solar irradiation, [Wh/m^2], range [0, 982.70]

External source data (NWP data from open-source weather provider):

- cloudiness, [p.u.], range [0, 1], step 0.125;

- ambient temperature, [°C], range [−17, 42];

- humidity, [%], range [7, 100];

- wind speed, [m/s], range [0, 15]

The complete dataset contained 11 892 pcs. of samples. The pre-processing stage of the forecasting algorithm presupposed removal of the night-hours samples in order to make the PV power generation dataset more stationary. After night-hour removal, the total amount of the samples was obtained to be equal to 6038 pcs. As far as the data was not sufficient for a 2-year period, it was finally decided to take into account the complete year data from 26 September 2017 to 21 October as a training set and a period from 22 October 2018 to 5 February 2019 as a testing set. Initial consideration of a complete year helped the model to understand the variations of the weather conditions of separate months. In further calculations, this trained model was used for hyperparameters tuning of machine learning algorithms, addressed in the present article.

Solar radiation at the PVPPs is typically measured by the horizontally mounted pyranometers. For the certification of pyranometers, the ISO 9060 standard is used. High-precision instruments were used at the PV plant under consideration, corresponding to the ISO spectrally flat class A. The technical specifications are shown in Table 2.

Table 2. Remote-sensing device technical specifications.

Technical Specifications	Pyranometer
ISO 9060:1990 class	Spectrally flat class A
Response time (95%)	<5 s
Zero offsets	<7 W/m^2
thermal radiation (200 W/m^2)	<2 W/m^2
Non-stability (change/year)	<0.5%
Non-linearity (100 to 1000 W/m^2)	<0.2%
Directional response (up to 80° with 1000 W/m^2 beam)	<10 W/m^2
Temperature response	< 1% (−20 °C to + 50 °C)
Tilt response (0° to 90° at 1000 W/m^2)	<1%
Sensitivity	7 to 14 V^{-6}/W/m^2
Accuracy of bubble level	<0.1°
Spectral range (50%)	285 to 2800 m^{-9}
Maximum operational irradiance	4000 W/m^2

All the data were stored in a database with 1 h time resolution, conditioned by the external weather data time resolution constraints. The influencing parameters of the PV output forecasting problem are obtained using the correlation heat map, which is provided in Figure 2.

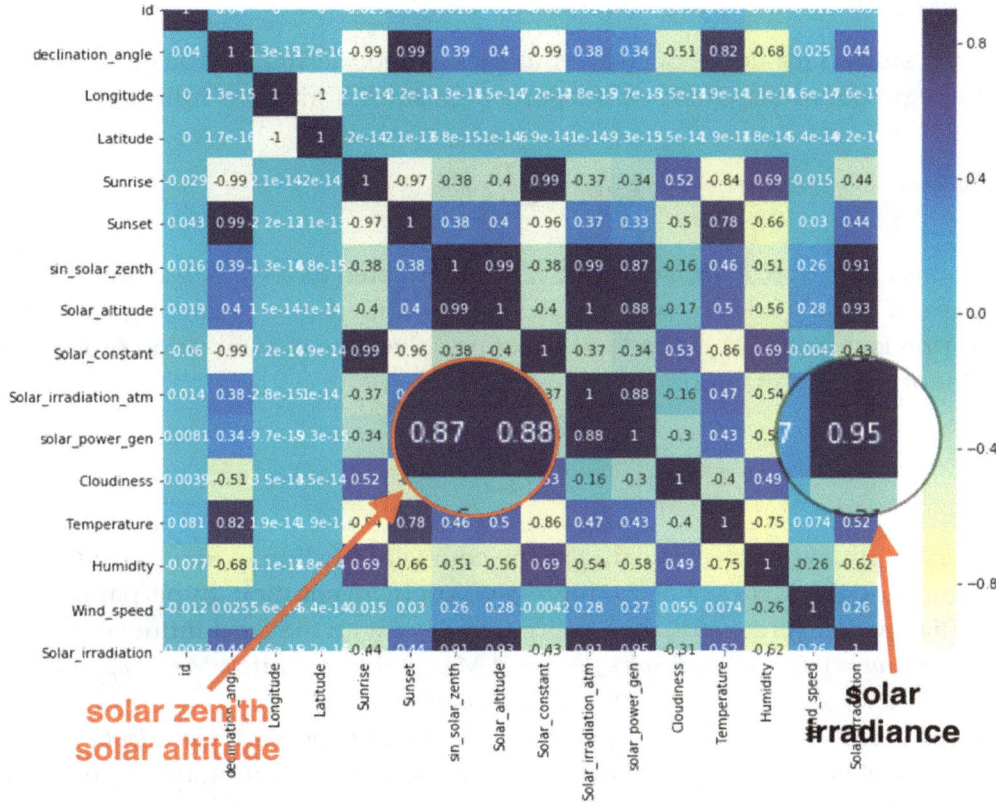

Figure 2. The correlation matrix of the parameters/features.

As one can see from Figure 2, solar zenith angle and solar altitude angle are the major parameters after solar irradiance in the prediction of the PV energy output. It is known from practice that cloudiness is also one of the important parameters.

4. Mathematical Models Description

For the given problem formulation, the following mathematical models were used and tested: random forest regressor; gradient boosting regressor; decision trees regressor; and linear regression.

4.1. Random Forest

Random forest is an algorithm that provides fittings of many decision trees for different sub-samples of the initial dataset at the stage of training and can be generally described by the following procedure [41]:

For each $n = 1, \ldots , N$ (N-the number of tree in the forest):

- generate a sub-sample X_n using bootstrap procedure
- build a decision tree b_n for X_n subsample

The resulting regressor $F(x)$ is given as follows:

$$F(x) = \frac{1}{N} \sum_{i=1}^{N} b_i(x) \tag{2}$$

where N is a number of decision trees; $b_i(x)$ is a decision tree.

4.2. Gradient Boosting

For the given study, Gradient boosting is implemented via the Adaptive Boosting Algorithm (AdaBoost). The regressor of the Gradient Boosting algorithm is given as follows [41]:

$$F(x) = \sum_{i=1}^{m} \gamma_m h_m(x) \tag{3}$$

where $h_m(x)$ is a basic function, a decision tree, typically treated as a weak learner of the algorithm.

In the course of the algorithm, each added tree is aimed at minimizing the loss function L, generated at the previous step, F_{m-1}. Gradient boosting solves the minimization problem by using the negative gradient of the loss function:

$$F_m = F_{m-1}(x) - \gamma_m \sum_{i=1}^{n} \nabla_F L(y_i, F_{m-1}(x_i)) \tag{4}$$

where γ_m is a step length, which is calculated in the course of the line search procedure.

In order to increase the accuracy of the regression problem solution, hyperparameter tuning was applied to the initial model. As a result, the hyperparameters with the most influence were estimated to be equal to: *Learning rat* = 0.01; *Min_samples_leaf* = 2; *Max_feature* = 'auto'; *Max_depth* = 35; *Alpha* = 0.9; *Min_samples_split* = 25; *n_estimators* = 2000; and *subsample* = 0.7.

The optimal value of *max_depth* was experimentally found to be 35. If *max_depth* value is increased, overfitting of the model takes place; when data noise is taken into account, this results in degradation of the performance of the model. The optimal value of the *learning* rate was stated to be 0.01. A value below 0.01 also causes an overfitting effect and leads to dramatic degradation of forecasting accuracy.

4.3. Decision Trees

The Decision Tree approach is implemented via an optimized version of the CART algorithm, which is implemented by the following procedure [41]:

- partitioning of the sample space according to the training and label vectors $x_i \in R^n$ ($i = 1, \ldots, I$) and $y \in R^l$, respectively;

Let the data in node m of the decision tree be referred to as Q. For each potential data split $\theta = (j, t_m)$ consisting of a feature j and the marginal value t_m, partition the data into $Q_{left}(\theta)$ and $Q_{right}(\theta)$ subsets:

$$Q_{left}(\theta) = (x, y)|x_j \leq t_m, \quad Q_{right}(\theta) = Q \backslash Q_{left}(\theta) \tag{5}$$

The impurity at node m of the Decision Tree is estimated based on the impurity function, and the decision tree parameters are selected in accordance with impurity minimization criteria.

Within the scope of the regression problem, determination of locations for future splits is carried out by estimating minimal Mean Squared Error and Mean Absolute Error:

$$H(X_m) = \frac{1}{N_m} \sum_{i \in N_m} (y_i - \bar{y}_m)^2, \quad H(X_m) = \frac{1}{N_m} \sum_{i \in N_m} |y_i - \bar{y}_m| \tag{6}$$

where X_m is the training data in node m of the Decision Tree.

Decision Tree model hyperparameters optimization lead to the following results: *Max_depth* = 16; *Min_samples_split* = 16; *Min_samples_leaf* = 15; *Max_features* = 'auto'; *Random_state* = '16'.

Model parameters were experimentally verified for the given training sample. *Max_depth* was optimized to increase model fitting, but not to overfit the data sample.

4.4. Linear Regression

The Linear regression model is considered as a basic simple regressor in order to correspond to the algorithm complexity with its computational efficiency. The linear model under consideration is described by the following equation [41]:

$$Y = \beta_0 + \beta_1 X_1 + \ldots + \beta_k X_k + \varepsilon \tag{7}$$

where $\beta_{1\ldots k}$ are regression coefficients, and ε is regression error.

The linear regression model is based on the ordinary least squares model ('OLS'). Linear regression models trained along with Polynomial Featuring demonstrated better performance, so this model is also taken into consideration.

The obtained results are moderately fitted when the power is "2". When the power is "3", the data set is overfitted.

4.5. Quality Metrics of the Models

The algorithm we used to test the accuracy of the prediction model is *r2_score*; it is also known as the coefficient of determination. The *r2_score* (i.e. coefficient of determination) is the subtraction of the residual sum of squares of the predicted and actual values divided with the total sum of squares.

$$R^2(y, \widetilde{y}) = 1 - \sum_{i=1}^{n} (y_i - \widetilde{y}_i)^2 \bigg/ \sum_{i=1}^{n} (y_i - \overline{y})^2 \tag{8}$$

where y_i is the actual value of the PV power plant output, kWh; and $\widetilde{y}_i$ is the predicted value of PV power plant output, kWh.

Summary and results of the application of the proposed algorithms to a particular sample day forecasting with and without hyperparameter tuning and pipelining are provided in Figures 3–6 and Tables 3–6.

A particular sample day, depicted in Figures 3–6, corresponds to early October, representing the median between summer and winter solstice in terms of the sunrise and sunset time. The forecasting procedure for stable weather days scores above 90% for all the tested algorithms, which corresponds to the state-of-the-art practice.

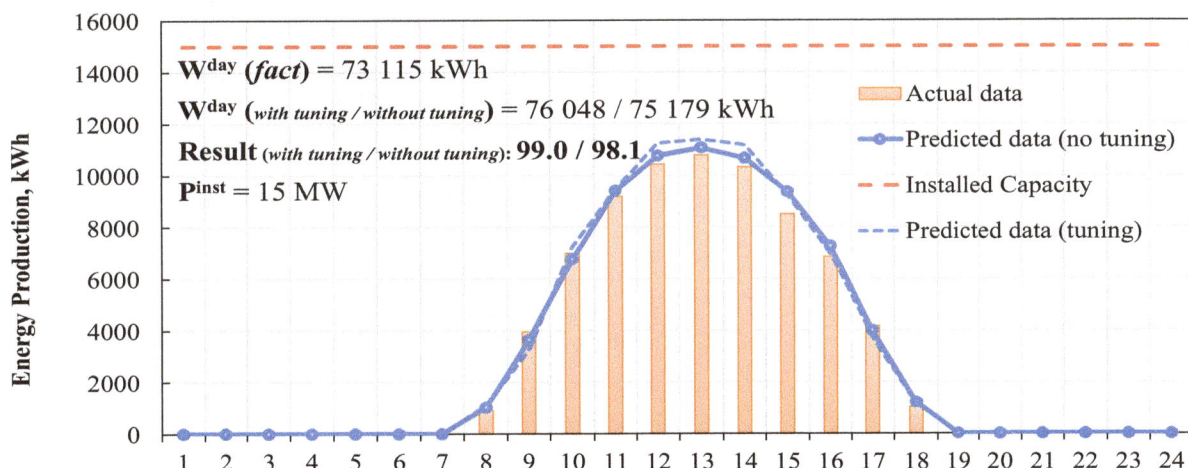

Figure 3. One-day forecasting example with Random Forest regressor.

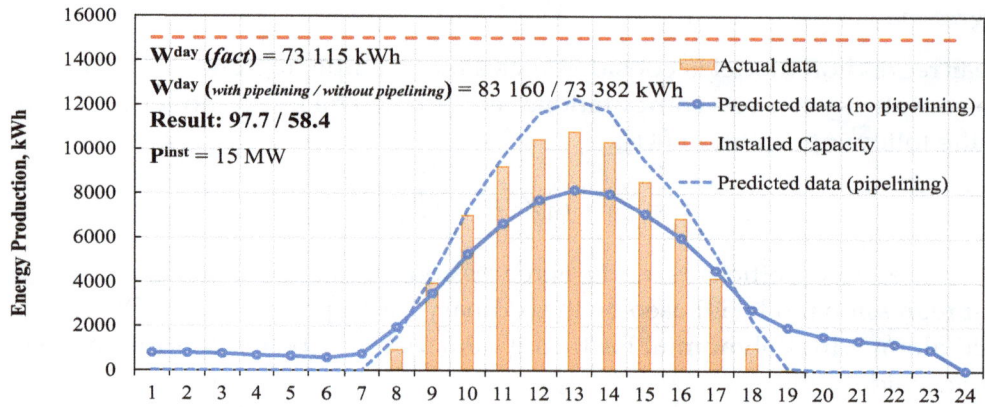

Figure 4. One-day forecasting example with Linear Regression.

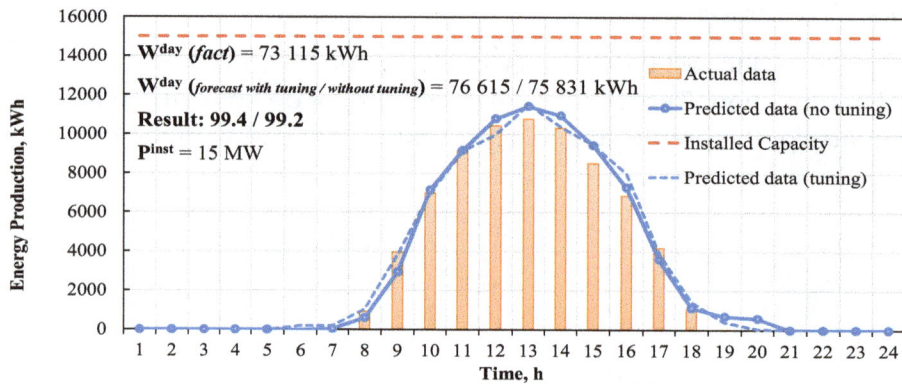

Figure 5. One-day forecasting example with Gradient Boosting regressor.

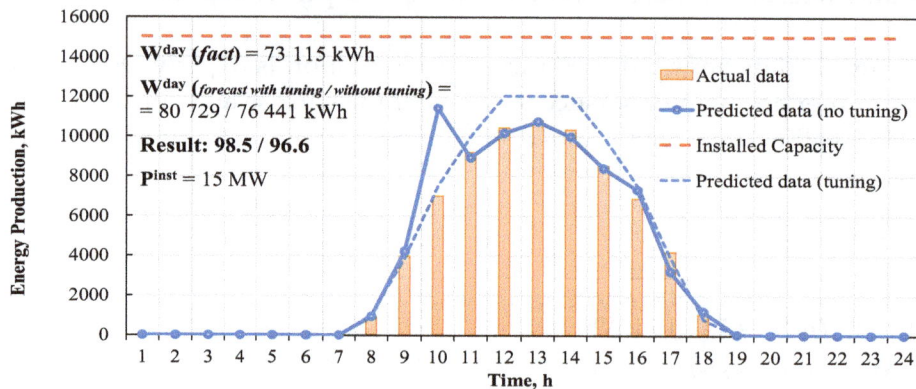

Figure 6. One-day forecasting example with Decision Tree regressor.

Table 3. Sample day-1 analysis: Random Forest.

Parameter	Default Parameters	Tuned Parameters
Score, %	88.60–98.00	82.00–99.00
CPU Time, ms	529	450
Wall time, ms	540	451.8
Max.time consumed, ms	80	555
One-day score, %	98.10	99.00

Table 4. Sample day-1 analysis: linear regression.

Parameter	Without Pipelining	With Pipelining
Score, %	55.00–58.00	94.20–94.50
CPU Time, ms	11	162
Wall time, ms	10.2	107
Max.time consumed, ms	13.8	180
One-day score, %	58.40	97.70

Table 5. Sample day-1 analysis: gradient boosting.

Parameter	Default parameters	Tuned parameters
Score, %	93.25–93.37	99.20–99.50
CPU Time, ms	561	50 600
Wall time, ms	576	51 800
Max.time consumed, ms	600	55 000
One-day score, %	99.20	99.40

Table 6. Sample day-1 analysis: decision trees.

Parameter	Default parameters	Tuned parameters
Score, %	88.60–90.00	91.45
CPU Time, ms	69.2	50.0
Wall time, ms	71.0	51.8
Max.time consumed, ms	80.0	55.0
One-day score, %	96.60	98.50

5. Prediction for Bad Weather Conditions

It is known that the weakest points of PV energy output forecasting are bad weather days predictions. The bad weather data is caused as a result of uneven cloud cover, moisture or also the snow and rain that degrade the solar panels' efficiency. For a given location, all these issues take place from September to December. The box plot diagrams of the prediction accuracy are provided in Figure 7. The accuracy of the forecasting for the given months typically equals to 60–70%, which does not meet the requirements and needs to be addressed.

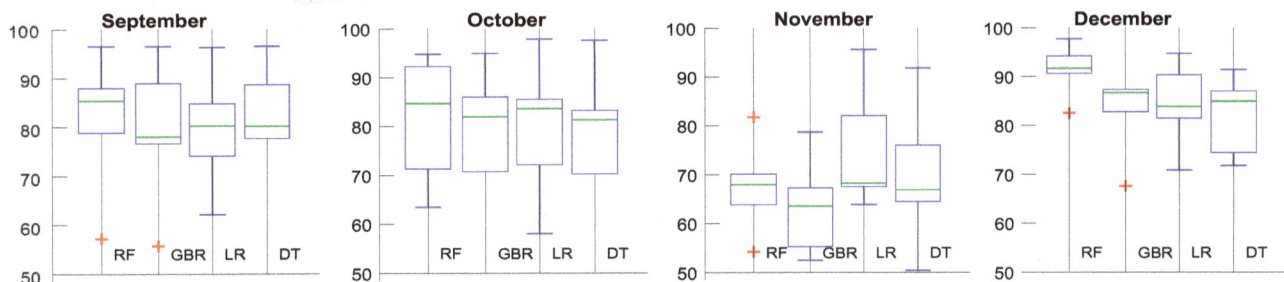

Figure 7. Accuracy box plots for proposed machine-learning models (1-year period).

The problem of extremely uncertain weather conditions is considered on the basis *of a winter day with sporadic clouds*.

For the scenario, provided in Figure 8, the cloudiness and, correspondingly, PV power plant energy output along with solar irradiation are completely uncertain. The clouds are scattered all over the region of PV power plant geographical location. The sudden and unique movements of the clouds are conditioned by the high wind speeds (more than 17 m/s), which produce transient variations of PV power plant electrical energy production and result in noisy data occurrence.

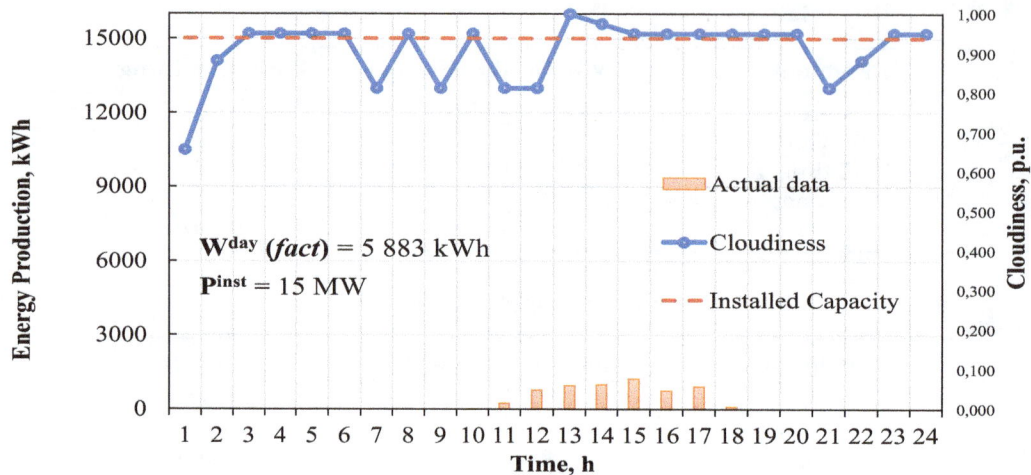

Figure 8. PV power plant energy output plotted versus weather conditions.

In order to make the machine able to predict "bad weather" days, the following points are to be taken into account:

1. In order to predict the PV energy output in sudden cloud motion conditions, the machine learning algorithm is required to be trained along with the noisy data.
2. The noisy data is generally considered when the machine is trained with overfitted data, which leads to the consideration of the smallest variations in the cloudiness.

For the first time, the proposed models were tested without hyper parameters tuning in order to check whether the models work with the same efficiency even when subjected to the different situations and uncertain conditions.

The prediction gives a clear perspective of how uncertain a data set could be and how many calculation efforts the machine has to involve to predict the PV power plant energy output values. As a result, the *Linear Regressor* along with *Decision Trees regressor* did not produce an adequate solution of the PV energy output prediction problem due to high uncertainty and noise in the dataset.

Gradient Boosting Regressor along with *Random Forest regressor* without hyper parameters tuning resulted in the average score of 20%, which cannot be considered as a viable result for power system operation modes planning. After hyper parameters tuning, the machine is taking a lot of time (i.e. 50 seconds) to fit the noisy data with a learning rate of "0.0089" and a decision tree depth of "35".

After running a series of calculation experiments with "bad weather" days, one can conclude that in order to eliminate data uncertainty and model overfitting, the model requires a different feature (or structure) except hyper parameters tuning for "bad" weather conditions.

6. Bad Weather Days Predictor

After scrutinizing the prediction results, we have concluded that uncertainty mostly comes from data values, which have very low PV energy output compared to other peak data points. Coming back to feature correlation analysis, we assumed that uncertain data values can be predicted by firstly predicting the solar irradiance, which is also proportional to the PV plant power output.

By predicting the horizontal solar irradiance, the following sources of uncertainty are eliminated:

*   solar irradiance diffusion and reflection;
*   electrical circuits of the PV power plant; and
*   the state of solar panels (shadow, degradation, etc.).

So, the bad weather days prediction methodology takes the following steps:

1. Predict the factor using a regressor model (*K*).

2.   Predict the solar irradiation using a regressor model (I).
3.   Obtain the cloudiness variance for the period (V):

   a.   If $(V > 1)$, take $(V \times K)$
   b.   If $(3 \times 10^{-3} < V < 5 \times 10^{-3})$, take $(0.5 \times K)$
   c.   If $(5 \times 10^{-3} < V < 5 \times 10^{-2})$, take $(0.01 \times K)$
   d.   If $(5 \times 10^{-2} < V < 0.1)$, take $((V \times 100 + 0.3) \times K)$
   e.   If $(0.1 < V < 0.5)$ or $(V > 1.5)$ or $(3 \times 10^{-3} < V < 0)$, (K).

4.   After checking and obtaining the factor, multiply the factor with the predicted solar irradiation PSI. The multiplied value is the solar power generation predicted value:

$$PSG = [PSI] \times [\text{Resuling Factor on Variance}] \tag{9}$$

The flowchart of the presented algorithms is given in Figure 9.

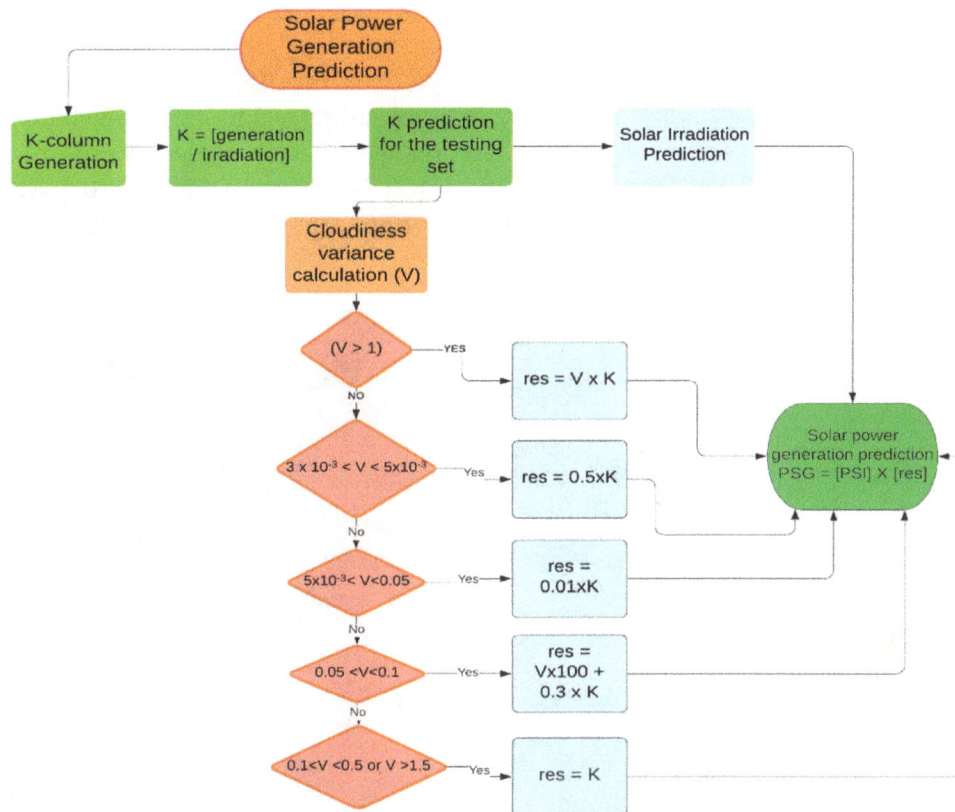

Figure 9. Flow-chart of the K-factor algorithm.

The next important feature of the algorithm is using separate training sets based on month separation. Pre-processing the training set with different month selection is carried out separately for "Jan to Sept" dataset and "Oct to Dec" dataset.

From January to September, heavy snowfall is not likely to occur for a given geographical location, which gives the opportunity to assume the reduction of noisy values in the data set. From October to December, snowfall and foggy conditions are present in the given region of the given data, resulting in noisy data occurrence. Thus, the model is trained separately with noisy conditions and non-noisy ones, resulting in improvement of the confusion matrix. The total $r2\_score$ of the proposed algorithm is estimated to be around 80%.

From October to December, snowfall and foggy conditions are present in the given region of the given data, resulting in the occurrence of noisy data. Therefore, the model is trained separately with noisy conditions and non-noisy ones, resulting in an improvement of the confusion matrix. The total *r2_score* of the proposed algorithm is estimated to be around 80%. Normal weather days can be predicted with higher accuracy and without requiring the factor-based algorithm. The authors used Linear Regression with Polynomial Featuring for good weather days forecasting. After making a large number of observations of different results, we analyzed that the Gradient Boosting Regressor without hyperparameter tuning outperforms all other models. The algorithms used in the K-factor model, depending on the weather conditions, are listed in Table 7.

Table 7. PV energy output prediction algorithms.

Period	Weather	Factor Usage	Model
Jan–Sept	Good	Not required	LR + Polynomial Featuring
Jan–Sept	Bad	Required	GBR + Hyper parameter tuning
Oct–Dec	Good	Required	GBR + Hyper parameter tuning
Oct–Dec	Bad	Required	GBR + Hyper parameter tuning

The short-term PVPP forecasting system developed within the framework of the study was implemented by LLC "Prosoft systems", an industrial automation and metering systems producer, as a program unit of "Energosphera" software package, providing smart metering systems management [42]. The satellite snapshot of the PVPP under consideration is given in Figure 10. At the moment, the forecasting system is being piloted at a real PV power generation facility, located in Astrahan city in the Russian Federation.

Figure 10. PV power plant satellite snapshot (Google Maps®).

Meteorological data is acquired in a 1-h time resolution from the external weather provider and includes cloud coverage, ambient air temperature, humidity, wind direction, and wind speed. Examples of day-ahead forecasts, generated by "Short-term Forecast of Solar Power Station Generation" program unit, which uses the developed approach, are presented in Figure 11 for the following types of weather conditions: clear, cloudy, and overcast, respectively.

*the model provided negative r2_score

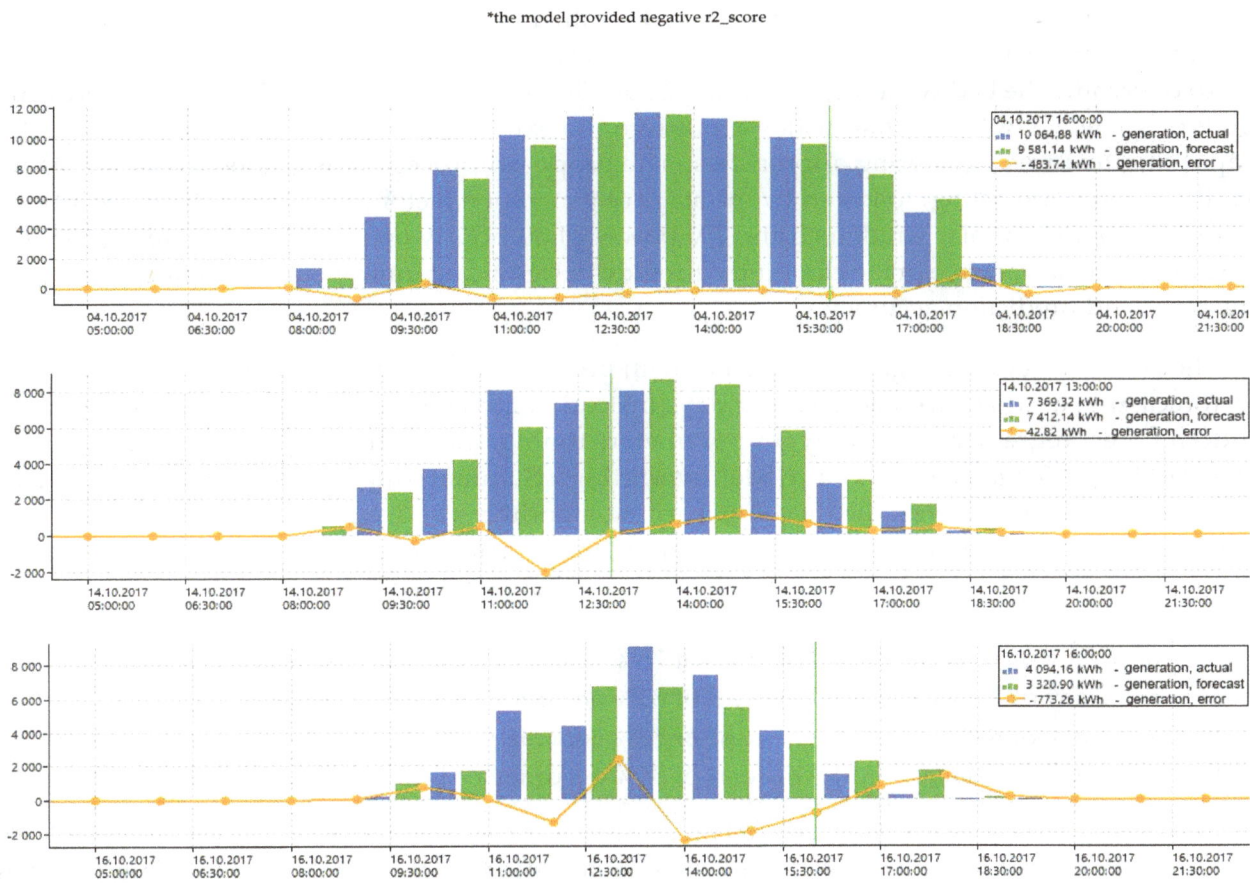

Figure 11. Energosphera: Photovoltaic power plant output forecasting.

The mean forecasting error reduced to the installed capacity of the PVPP for the time period starting from 1 October, 2017 to 31 December, 2017 was estimated to be 4.6%, which is comparable with the forecasts of global practice.

7. Conclusions

The PV power plant forecasting problem deals with multi-source heterogeneous data as far as the initial dataset is composed of the measurements, which are acquired from PV power plant metering systems, and external source weather forecasting data.

The problem was addressed by applying four different mathematical models: Random Forest regressor, Gradient Boosting Regressor, Linear Regression, and Decision Trees regression. Based on computational experiments with hyper parameters optimization and pipelining of the algorithms, the optimal structure and settings of the PV plant energy output forecasting system were identified together with the application restrictions for each of the algorithms.

During computational experiments, it was found that parameters tuning allows improvement of the algorithm performance for all non-ensemble algorithms: for linear regression from 55% to 94%, and for decision trees from 88 to 91%, while the accuracy of ensemble algorithms, such as gradient boosting on decision trees and random forest, did not change significantly.

Within the scope of the study, it was proven that the application of the universal model, applied either for good or bad weather days, may result in significant degradation of the short-term forecasting accuracy, hence, in order to improve the predictive properties of the system, several models are to be developed for various weather conditions. Moreover, it was found that good weather days when the meteorological data is assumed to be noise-free are accurately predicted by using any of the presented mathematical models with an accuracy rate of 90% and higher.

Due to the lack of features in the dataset, bad weather days are characterized by high uncertainty, which may decrease the predicting properties of the system.

To overcome the bad weather forecasting issue, the structure of the algorithm was improved by introducing a novel two-stage forecasting procedure and extracting a new feature from the raw dataset by applying feature engineering approaches. The proposed procedure is composed of the stage of solar irradiation forecasting, followed by the stage of generation factor prediction, which describes the relationship between solar irradiance and PV power plant hourly energy output. A resulting factor scaled down to the variance of the cloudiness provides a significant improvement of forecasting system robustness and prediction accuracy.

The newly introduced algorithm together with proper training sets formulation, resulted in mean 83% forecasting accuracy for bad weather days instead of 20% for Gradient Boosting Regressor and Random Forest regressor without hyper parameters tuning, demonstrating dramatic improvement of the model performance without model overfitting. Summarizing the performance of K-factor algorithm in comparison with the machine learning algorithms addressed in this paper, after taking the mean of five cross-validations with 6038 samples, the K-factor algorithm improves the performance of the addressed machine learning approaches in the following way:

- 92% accuracy of K-factor model instead of 78% accuracy of Random Forest regressor:
- 85% accuracy of K-factor model instead of 83% accuracy of Linear regressor:
- 89% accuracy of K-factor model instead of 73% accuracy of Gradient Boosting regressor;
- 81% accuracy of K-factor model instead of 56% accuracy of Decision Trees regressor.

The results obtained for K-factor model meet the requirements of the transmission and distribution power system operators in terms of 20% admissible deviations of the power system operation plan.

Based on the exhaustive calculations, it was decided to use Linear regression for good weather days forecasting and a factor-based prediction model using Gradient Boosting Regressor for bad weather days in order to sustain robustness and eliminate overfitting.

The presented system of short-term PV energy output forecasting is universal and can be used at any existing PV generation facilities as a part of the Energosfera 8.0 software package (LLC, Prosoft-Systems LLC). Currently, Prosoft-Systems together with the research team of Ural Federal University is developing a system, providing online correction of the short-term forecasts, based on the current measurements of solar irradiation and cloud motion. It is expected that the system will allow the owners of solar power plants to participate in intra-day trading procedures at the wholesale electricity and capacity market.

With the development of generating capacities based on RES, the uncertainty degree in planning the power system operating modes increases significantly. Today, reliable tools are required to predict the generation of power plants using, in particular, solar energy obtained by remote sensing [43]. For short time periods from 1 to 6 h, the generation forecast can be significantly improved by using the current data obtained by direct (proximate) observation (remote sensing) methods. When combining numerical weather forecasting systems with real-time data, forecast deviations caused by inaccuracies in numerical weather forecasting models can be corrected several hours ahead.

Author Contributions: Conceptualization, A.I.K. and S.A.E.; data curation, S.A.E., V.A.T., T.P.C. and D.N.B.; formal analysis, V.A.T. and T.P.C.; funding acquisition, A.I.K.; investigation, S.A.E., H.R., T.P.C. and D.N.B.; methodology, A.I.K., S.A.E. and H.R.; project administration, A.I.K. and H.R.; resources, V.A.T. and D.N.B.; software, A.I.K., S.A.E. and V.A.T.; supervision, A.I.K. and H.R.; validation, V.A.T., T.P.C. and D.N.B.; visualization, T.P.C.; writing–original draft, A.I.K. and H.R.; writing–review & editing, S.A.E. and D.N.B. All authors have read and agreed to the published version of the manuscript.

Acknowledgments: The authors are thankful to the anonymous Referees for their insightful suggestions.

References

1. Gigoni, L.; Betti, A.; Crisostomi, E.; Franco, A.; Tucci, M.; Bizzarri, F.; Mucci, D. Day-Ahead Hourly Forecasting of Power Generation from Photovoltaic Plants. *IEEE Trans. Sustain. Energy* **2018**, *9*, 831–842. [CrossRef]
2. Sangrody, H.; Sarailoo, M.; Zhou, N.; Tran, N.; Motalleb, M.; Foruzan, E. Weather forecasting error in solar energy forecasting. *IET Renew. Power Gener.* **2017**, *11*, 1274–1280. [CrossRef]
3. Conte, F.; Massucco, S.; Schiapparelli, G.; Silvestro, F. Day-Ahead and Intra-Day Planning of Integrated BESS-PV Systems Providing Frequency Regulation. *IEEE Trans. Sustain. Energy* **2020**, *11*, 1797–1806. [CrossRef]
4. Huang, C.; Wang, L.; Lai, L.L. Data-Driven Short-Term Solar Irradiance Forecasting Based on Information of Neighboring Sites. *IEEE Trans. Ind. Electron.* **2019**, *66*, 9918–9927. [CrossRef]
5. Vincent, E. Larson. Chapter 12—Forecasting Solar Irradiance with Numerical Weather Prediction Models. In *Solar Energy Forecasting and Resource Assessment*; Academic Press: Cambridge, MA, USA, 2013; pp. 299–318.
6. Orwig, K.D.; Ahlstrom, M.L.; Banunarayanan, V.; Sharp, J.; Wilczak, J.M.; Freedman, J.; Haupt, S.E.; Cline, J.; Bartholomy, O.; Hamann, H.F.; et al. Recent Trends in Variable Generation Forecasting and Its Value to the Power System. *IEEE Trans. Sustain. Energy* **2015**, *6*, 924–933. [CrossRef]
7. Glassley, W.; Jan, K.; Van Dam, C.C.; Shiu, H.; Huang, J.; Braun, G.; Holland, R. *California Renewable Energy Forecasting, Resource Data and Mapping*; Publication Number: CEC-500-2014-026; California Energy Commission: Sacramento, CA, USA, 2012; pp. 1–135.
8. Maghami, M.R.; Hizam, H.; Gomes, C.; Radzi, M.A.; Rezadad, M.I.; Hajighorbani, S. Power loss due to soiling on solar panel: A review. *Renew. Sustain. Energy Rev.* **2016**, *59*, 1307–1316. [CrossRef]
9. Andrews, R.W.; Pollard, A.; Pearce, J.M. The effects of snowfall on solar photovoltaic performance. *Sol. Energy* **2013**, *92*, 84–97. [CrossRef]
10. Woyte, A.; Nijs, J.; Belmans, R. Partial shadowing of photovoltaic arrays with different system configurations: Literature review and field test results. *Sol. Energy* **2003**, *74*, 217–233. [CrossRef]
11. Jang, H.S.; Bae, K.Y.; Park, H.; Sung, D.K. Solar Power Prediction Based on Satellite Images and Support Vector Machine. *IEEE Trans. Sustain. Energy* **2016**, *7*, 1255–1263. [CrossRef]
12. Alessandrini, S.; Monache, L.D.; Sperati, S.; Cervone, G. Analog ensemble for short-term probabilistic solar power forecast. *Appl. Energy* **2015**, *157*, 95–110. [CrossRef]
13. Zhang, X.; Li, Y.; Lu, S.; Hamann, H.F.; Hodge, B.-M.; Lehman, B. A Solar Time Based Analog Ensemble Method for Regional Solar Power Forecasting. *IEEE Trans. Sustain. Energy* **2019**, *10*, 268–279. [CrossRef]
14. Kakimoto, M.; Endoh, Y.; Shin, H.; Ikeda, R.; Kusaka, H. Probabilistic Solar Irradiance Forecasting by Conditioning Joint Probability Method and Its Application to Electric Power Trading. *IEEE Trans. Sustain. Energy* **2019**, *10*, 983–993. [CrossRef]
15. Andrade, J.R.; Bessa, R.J. Improving Renewable Energy Forecasting With a Grid of Numerical Weather Predictions. *IEEE Trans. Sustain. Energy* **2017**, *8*, 1571–1580. [CrossRef]
16. Abuella, M.; Chowdhury, B. Solar power probabilistic forecasting by using multiple linear regression analysis. *SoutheastCon* **2015**, 1–5. [CrossRef]
17. Hong, T.; Wang, P.; Willis, H.L. A Naïve multiple linear regression benchmark for short term load forecasting. In Proceedings of the 2011 IEEE Power and Energy Society General Meeting, Detroit, MI, USA, 24–28 July 2011; pp. 1–6. [CrossRef]
18. Prema, V.; Rao, K.U. Development of statistical time series models for solar power prediction. *Renew. Energy* **2015**, *83*, 100–109. [CrossRef]
19. Liang, N.; Huang, G.; Saratchandran, P.; Sundararajan, N. A Fast and Accurate Online Sequential Learning Algorithm for Feedforward Networks. *IEEE Trans. Neural Netw.* **2006**, *17*, 1411–1423. [CrossRef]
20. Golestaneh, F.; Pinson, P.; Gooi, H.B. Very Short-Term Nonparametric Probabilistic Forecasting of Renewable Energy Generation—With Application to Solar Energy. *IEEE Trans. Power Syst.* **2016**, *31*, 3850–3863. [CrossRef]
21. Shukla, K.N.; Rangnekar, S.; Sudhakar, K. Comparative study of isotropic and anisotropic sky models to estimate solar radiation incident on tilted surface: A case study for Bhopal, India. *Energy Rep.* **2015**, *1*, 96–103. [CrossRef]

22. Kittisontirak, S.; Dawan, P.; Atiwongsangthong, N.; Titiroongruang, W.; Chinnavornrungsee, P.; Hongsingthong, A.; Sriprapha, K.; Manosukritkul, P. A novel power output model for photovoltaic system. *iEECON* **2017**, 1–3. [CrossRef]

23. Huang, C.-J.; Huang, M.-T.; Chen, C.-C. A Novel Output Model for Photovoltaic Systems. *Int. J. Smart Grid Clean Energy* **2013**, *2*, 139–147. [CrossRef]

24. Gautam, J.; Ahmed, M.I.; Kumar, P. Optimization and Comparative Analysis of Solar-Biomass Hybrid Power Generation System Using Homer. In Proceedings of the 2018 International Conference on Intelligent Circuits and Systems (ICICS), Phagwara, India, 19–20 April 2018; pp. 397–400.

25. Ghose, S.; Shahat, A.E.; Haddad, R.J. Wind-solar hybrid power system cost analysis using HOMER for Statesboro, Georgia. *SoutheastCon* **2017**, 1–3. [CrossRef]

26. Mahmud, N.; Hassan, A.; Rahman, M.S. Modelling and cost analysis of hybrid energy system for St. Martin Island using HOMER. In Proceedings of the 2013 International Conference on Informatics, Electronics and Vision (ICIEV), Dhaka, Bangladesh, 17–18 May 2013; pp. 1–6.

27. Rajani, A.; Darussalam, R.; Pramana, R.I.; Santosa, A. Simulation of PV-Biogas Integration on Hybrid Power Plant using HOMER: Study Case of Superior Livestock Breeding Center and Forage of Animal Feed (BBPTU-HPT) Baturraden. In Proceedings of the 2018 International Conference on Sustainable Energy Engineering and Application (ICSEEA), Tangerang, Indonesia, 1–2 November 2018; pp. 69–74.

28. Vendoti, S.; Muralidhar, M.; Kiranmayi, R. HOMER Based Optimization of Solar-Wind-Diesel Hybrid System for Electrification in a Rural Village. In Proceedings of the 2018 International Conference on Computer Communication and Informatics (ICCCI), Coimbatore, India, 4–6 January 2018; pp. 1–6.

29. Wijeratne, P.; Yang, R.J.; Too, E.; Wakefield, R. Design and development of distributed solar PV systems: Do the current tools work? *Sustain. Cities Soc.* **2019**, *45*, 553–578. [CrossRef]

30. Pannase, V.R.; Nanavala, H.B. A review of PV technology power generation, PV material, performance and its applications. In Proceedings of the 2017 International Conference on Inventive Systems and Control (ICISC), Coimbatore, India, 19–20 January 2017; pp. 1–5. [CrossRef]

31. Javed, A.; Shabir, H.; Ali, H.; Darwade, R.; Gite, B. Predicting Solar Irradiance Using Machine Learning Techniques. In Proceedings of the 2019 International Wireless Communications & Mobile Computing Conference (IWCMC), Tangier, Morocco, 24–28 June 2019; pp. 1458–1462.

32. Schönhuber, M.; Cuervo, F. About the Impact of NWP Models' Temporal Resolution on Rain Attenuation Forecasts. In Proceedings of the 2019 URSI Asia-Pacific Radio Science Conference (AP-RASC), New Delhi, India, 9–15 March 2019; pp. 1–3.

33. Ministry of Power and Energy of Russian Federation. On implementation of the requirements for power systems and electrical installation reliability security. In *Guidelines on Power Systems Stability*, 3rd ed.; Ministry of Energy of Russian Federatio: Moscow, Russia, 20 August 2018. (In Russian)

34. Lukaitis, V.Y. *Autonomous Power Generation Facilities, Hybrid Structures Comprising Renewable Energy Sources*; Lukaitis, V.Y., Glushkov, S.Y., Eds.; Interindustry Scientific and Production Company Energospectechnic (ISPC Energospectechnic): Moscow, Russia, 2019; Volume 2, Issue 2. [CrossRef]

35. Chen, W.; Liu, Y.; Wang, N. A Novel Grouping Aggregation Algorithm for Online Analytical Processing. In Proceedings of the 2012 National Conference on Information Technology and Computer Science, China, 16–18 November 2012. [CrossRef]

36. Chu, Y.; Cao, G.; Hayat, H. Change Detection of Remote Sensing Image Based on Deep Neural Networks. In Proceedings of the 2016 2nd International Conference on Artificial Intelligence and Industrial Engineering (AIIE 2016), Nanjing, China, 20–21 November 2016. [CrossRef]

37. Kussul, N.; Skakun, S.V.; Lavreniuk, M.; Shelestov, A.Y. Deep Learning Classification of Land Cover and Crop Types Using Remote Sensing Data. *IEEE Geosci. Remote Sens. Lett.* **2017**. [CrossRef]

38. Kuleshov, Y. Use of Remote Sensing Data for Climate Monitoring in WMO Regions II and V (Asia and the South-West Pacific). Australian Bureau of Meteorology, 1 June 2017. Available online: https://www.wmo.int/pages/prog/wcp/ccl/opace/opace2/documents/TT-URSDCM_Use_ Remote_Sensing_DataClimateMonitoringRAII-V.pdf (accessed on 30 August 2020).

39. K, R.E.; Townsend, P.A.; Gross, J.E.; Cohen, W.B.; Bolstad, P.; Wang, Y.Q.; Adams, P. Remote sensing change detection tools for natural resource managers: Understanding concepts and tradeoffs in the design of landscape monitoring projects. *Remote Sens. Environ.* **2009**, *113*, 1382–1396. [CrossRef]

40.   Eroshenko, S.; Khalyasmaa, A.; Snegirev, D. Machine learning techniques for short-term solar power stations operational mode planning. *E3S Web Conf.* **2018**, *51*, 5. [CrossRef]
41.   Machine learning in Python: Web-portal. Available online: https://scikit-learn.org/ (accessed on 30 August 2020).
42.   Prosoft-System, Engineering Company. Available online: https://www.prosoftsystems.ru/en/news/energosfera-8_0-software-package-expands-the-scope-of-capabilities (accessed on 30 August 2020).
43.   Edenhofer, O.; Pichs-Madruga, R.; Sokona, Y. *Special Report on Renewable Energy Sources and Climate Change Mitigation*; IPCC: Geneva, Switzerland, 2011; p. 1075. ISBN 978-92-9169-131-9.

10

On the Land-Sea Contrast in the Surface Solar Radiation (SSR) in the Baltic Region

Anders V. Lindfors [1],*, **Axel Hertsberg** [1],†, **Aku Riihelä** [1], **Thomas Carlund** [2], **Jörg Trentmann** [3] **and Richard Müller** [3]

[1] Finnish Meteorological Institute, P.O. Box 503, 00101 Helsinki, Finland; axel.hertsberg@gmail.com (A.H.); aku.riihela@fmi.fi (A.R.)

[2] Swedish Meteorological and Hydrological Institute, 60176 Norrköping, Sweden; thomas.carlund@smhi.se

[3] Deutsche Wetterdienst, 63067 Offenbach, Germany; joerg.trentmann@dwd.de (J.T.); richard.mueller@dwd.de (R.M.)

* Correspondence: anders.lindfors@fmi.fi

† Current address: Eniram Oy, 00101 Helsinki, Finland.

Abstract: The climatological surface solar radiation (SSR; also called global radiation), which is largely dependent on cloud conditions, is an important indicator of the solar energy production potential. In the Baltic area, previous studies have indicated lower cloud amounts over seas than over land, in particular during the summer. However, the existing literature on the SSR climate or how it translates into solar energy potential has not paid much attention to how the SSR behaves quantitatively in relation to the coastline. In this paper, we have studied the climatological land–sea contrast of the SSR over the Baltic area. For this, we used two satellite climate data records, CLARA-A2 and SARAH-2, together with a coastline data base and ground-based pyranometer measurements of the SSR. We analyzed the behaviour of the climatological mean SSR over the period 2003–2013 as a function of the distance to the coastline. The results show that off-shore locations on average receive higher SSR than inland areas and that the land–sea contrast in the SSR is strongest during the summer. Furthermore, the land–sea contrast in the summer time SSR exhibits similar behavior in various parts of the Baltic. For CLARA-A2, which shows better agreement with the ground-based measurements than SARAH-2, the annual SSR is 8% higher 20 km off the coastline than 20 km inland. For summer, i.e., June–August, this difference is 10%. The observed land–sea contrast in the SSR is further shown to correspond closely to the behavior of clouds. Here, convective clouds play an important role as they tend to form over inland areas rather than over the seas during the summer part of the year.

Keywords: surface solar radiation; global radiation; solar energy; satellite; Baltic area; coastline; cloud; convection; climate

---

1. Introduction

Globally, clouds are more prevalent over seas than over land [1,2]. Over mid-to-high latitudes in the Northern Hemisphere, however, surface-based synoptic weather reports show higher cloud amount over land than over seas [3]. In Sweden, the land–sea contrast in cloud behavior was studied already by Ångström in 1928 [4]. He analyzed measurements of bright sunshine duration across the Swedish west coast and found that Vinga, a lighthouse in the outermost archipelago, had 11% more sunshine than Kålltorp, which is located slightly east of the central parts of Gothenburg. The distance between these two locations is ca 27 km. Ångström further referred to similar results on the Swedish east coast reported by J. Westman. An early sunshine climatology over Finland, presented by Lunelund in 1941 [5], also indicates a land–sea contrast in the duration of bright sunshine, with relatively high values along the coastline and in the archipelago.

More recently, the subject was studied by Karlsson [6], who created a satellite-based Scandinavian cloud climatology using Advanced Very High Resolution Radiometer (AVHRR) measurements. He found that the Baltic sea is less cloudy than the surrounding inland areas from April to September, and that this land–sea contrast in the summer time cloudiness is most pronounced during the afternoon, indicating a role of convective clouds forming over the inland areas, caused by solar heating of the surface. In winter, on the other hand, Karlsson [6] found the cloudiness in Scandinavia to be generally rather high, without any distinct geographical feature.

These general features of the cloud climate do, of course, have an effect on the surface solar radiation (SSR; also called global radiation) and thereby also on the solar energy resource. Thus, a land–sea contrast in the climatological SSR can be expected, with higher radiation over the sea than over land. Indeed, Persson [7], who studied the solar radiation climate of Sweden, found that the SSR in Visby (Gotland) was 12% higher than at a location of similar latitude, but located inland. Apart from this, however, existing literature on the SSR climate (e.g., [8]) or how it translates into solar energy potential over Europe (e.g., [9]) has not paid much attention to how the land–sea contrast in the SSR behaves quantitatively. Therefore, from the perspective of solar resource assessment, it would be important to have better information on the SSR climatology in coastal areas.

The aim of this paper is to quantify the land–sea contrast in the SSR climate over the Baltic Sea and its surrounding areas. For this, we used two different satellite-retrieved SSR data sets from EUMETSAT's Climate Monitoring Satellite Application Facility (CM SAF), a coastline data base and ground-based measurements of the SSR from the Swedish and Finnish networks.

2. Materials and Methods

2.1. Satellite SSR

In this study, we used the following satellite-based SSR data records:

1. CLARA-A2 (CM SAF cLoud, Albedo and surface RAdiation dataset from AVHRR data—Edition 2; doi:10.5676/EUM_SAF_CM/CLARA_AVHRR/V002), a data record based on measurements by the AVHRR sensor onboard a series of polar-orbiting satellites
2. SARAH-2 (Surface Radiation Data Set-Heliosat (SARAH)—Edition 2; doi:10.5676/EUM_SAF_CM/SARAH/V002), a data record derived from satellite-observations of the visible channels of the Meteosat Visible Infra-Red Imager (MVIRI) and the Spinning Enhanced Visible and Infrared Imager (SEVIRI) instruments onboard the geostationary Meteosat satellites

Both data records cover more than 30 years, from the early 1980s up to 2015, and have been produced by EUMETSAT's CM SAF project. Note, however, that we here focus on the period 2003–2013 (see Section 2.5).

Riihelä et al. [10] found—using a previous version of these data records (i.e., SARAH-v001 and CLARA-A1)—that both CLARA and SARAH are capable of estimating the monthly mean SSR with an accuracy better than 10 W/m^2, when compared to ground-based measurements in Sweden and Finland. Another finding of [10] is of particular interest for the present study: they showed, that CLARA in general shows higher SSR values over the Baltic Sea than SARAH, and that CLARA is in better agreement with the ground-based measurements at Utö in the Finnish archipelago.

The CLARA-A2 data record and underlying algorithms are discussed by Karlsson et al. [11] and references therein. The update to version CLARA-A2 included major efforts to correct and homogenize the original satellite radiances measured by the AVHRR instruments. As regards the SSR record, the update substantially improved the spatial coverage. In CLARA-A2, SSR estimates are unavailable only over snow-covered surfaces. Monthly and daily SSR values are available on a regular $0.25° \times 0.25°$ global latitude-longitude grid.

The SARAH-2 data record and its algorithms are comprehensively presented in EUMETSAT's Algorithm Theoretical Baseline Document [12]. The update to version SARAH-2 included further improvements to the homogeneity of the data record, empirical correction of view-angle dependency of the cloud albedo and adjustment of the water vapour column based on surface elevation. The SARAH-2 products are available as monthly, daily, and 30 min values on a regular $0.05° \times 0.05°$ latitude–longitude grid. Note, however, that SARAH-2 is based on geostationary satellite measurements and therefore does not provide global coverage. The SARAH-2 SSR data, for example, leaves large parts of central and northern Finland uncovered (see, e.g., Figure 3-3 in [12]).

In this study, we used monthly values of the SSR as provided by CLARA-A2 and SARAH-2. In addition to the real-sky estimate of the SSR, both data sets also provide a clear-sky counterpart representing the SSR under the same atmospheric conditions, but assuming cloudless skies.

2.2. Coastline Information

The Global Self-consistent Hierarchical High-resolution Geography (GSHHG) database [13,14] contains coastline information constructed from hierarchically arranged closed polygons. The data are available in five different resolutions ranging from crude to full. In this study, we used the low resolution data, version 2.3.5 of GSHHG with some further modifications as explained in Section 2.4.

2.3. Ground-Based Pyranometer Measurements

Finally, we also used pyranometer measurements of the SSR (global radiation) of the Finnish and Swedish networks as a reference to which the satellite-retrieved results were compared. The stations, listed in Table 1, are operated by the Swedish Meteorological and Hydrological Institute and the Finnish Meteorological Institute. The same stations were included also in a previous study by Riihelä et al. [10], where more details on the instruments used and measurement data can be found.

Table 1. Solar radiation measurement stations included in this study arranged according to their distance to the coastline. Negative distances indicate locations off-shore.

Station	Distance [km]	Latitude [°N]	Longitude [°E]
Visby	−94	57.673	18.345
Utö	−80	59.784	21.368
Svenska Högarna	−41	59.442	19.502
Karlstad	−2	59.359	13.427
Luleå	1	65.544	22.111
Göteborg	3	57.688	11.980
Helsinki-Kumpula	3	60.203	24.961
Stockholm	4	59.353	18.063
Umeå	5	63.811	20.240
Norrköping	7	58.582	16.148
Lund	10	55.714	13.212
Helsinki-Vantaa	11	60.327	24.957
Jokioinen	54	60.814	23.498
Växjö	84	56.927	14.731
Borlänge	98	50.488	15.430

2.4. Distance to the Coastline

In order to scrutinize the land–sea contrast in the climatological SSR, we analyzed the behaviour of the climatological mean SSR as a function of the distance to the coastline. Figure 1 shows the region included in our analysis and the calculated distance to the coastline (see below for information on how the coastline is defined). The measurement stations in Finland and Sweden included in this study are shown as black filled squares.

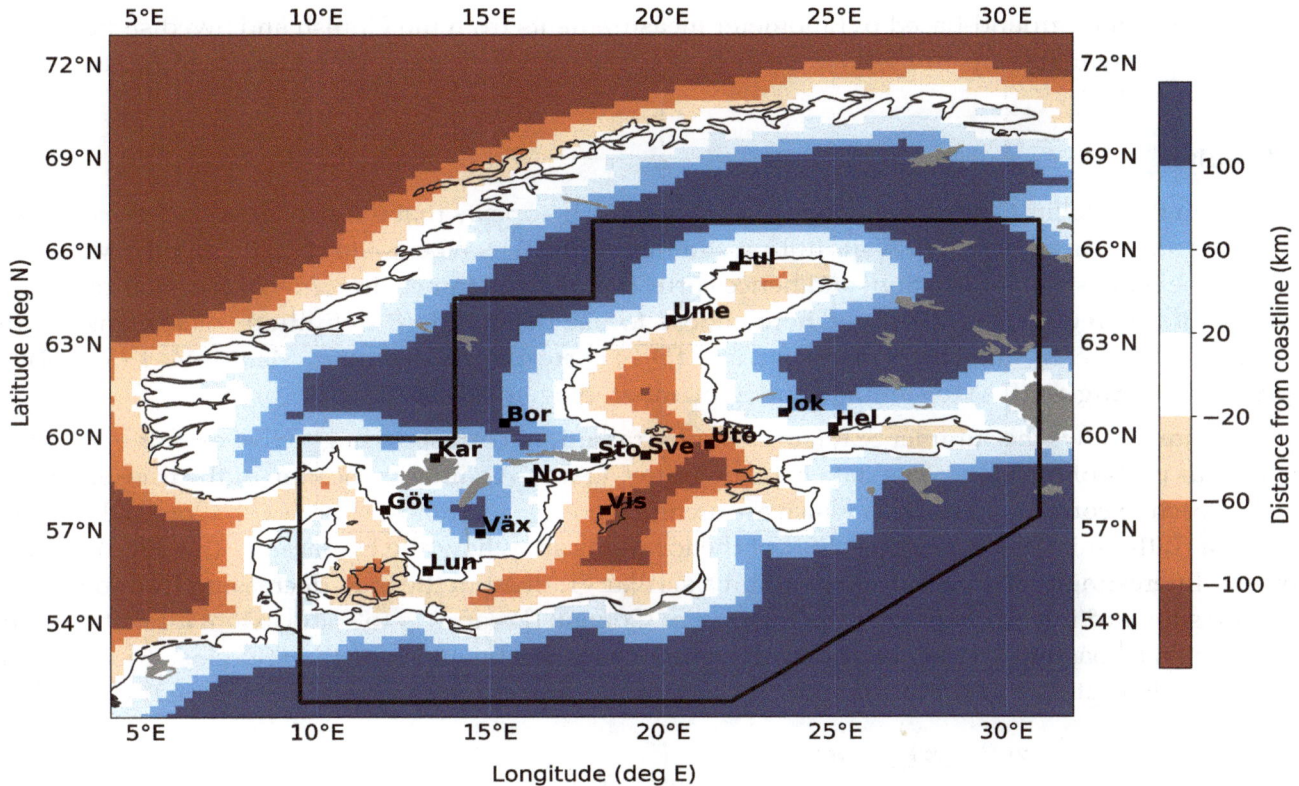

Figure 1. Map of the Baltic region included in this study, bounded by the black line. The color scale indicates the distance to the coastline of the CLARA satellite grid and the black filled squares represent the ground-based measurement stations.

The distance between the center of the satellite grid box and the coastline was calculated for both CLARA-A2 and SARAH-2 as well as for the ground-based measurement stations (see Figure 1 and Table 1). This was done using the haversine formula for the distance along a great circle on a spherical surface. For this, we used a slightly modified version of the low resolution GSHHG database. Our aim here was to obtain an appropriate level of detail in the coastline information, considering the resolution of the satellite data records and the expected distance over which a coastline gradient in the SSR climate can be seen. We wanted to avoid a too high level of detail, which may cause ambiguities in the interpretation of the coastline, in particular, in highly heterogeneous regions.

Starting from the low resolution GSHHG data, we decreased the distance between adjacent points by iteratively adding a point, positioned at the midpoint between the two neighboring points, until the distance between adjacent points became less than 5 km. We also excluded some lakes and islands from our analysis by including only polygons with an area larger than 0.8 square degrees and an area to circumference ratio larger than 0.21 degrees.

The resulting coastline information for the Baltic region is indicated by the distance to the coastline shown in Figure 1. It can be seen, for example, that the islands of Gotland and Själland are here interpreted as sea, which is also the case for Åland as well as other islands of the Baltic. Because of the discreteness of the coastline data used, the calculated distance to the coastline may have an error up to 2.5 km, with additional uncertainty introduced by differences between the true coastline with all its detail and the coastline as represented by our version of the GSHHG data.

2.5. Period of Analysis

Riihelä et al. [10] showed that the CLARA satellite SSR record shows better performance as compared with ground-based measurements during the more recent period, when a higher number of polar orbiting satellite overpasses are available for each day. Furthermore, Müller et al. [15]

argued that only 10 recent years should be used when estimating the solar energy resource. For these reasons, and since ground-based pyranometer measurements from the Finnish and Swedish networks were readily available up to the year 2013 (based on [10]) , we have here chosen to focus on the period 2003–2013.

3. Results

From the satellite cloud climatology of Karlsson [6], it appears that the strongest land–sea contrast in clouds and solar radiation prevails during summer. While it can be noted that the months May–July exhibit the highest solar radiation, and hence corresponds to 'summer' in terms of the solar radiation climate of this region (e.g., [16]), we here choose to use the standard definitions of seasons in our analysis. This means that December–February (DJF) corresponds to winter, March–May (MAM) to spring, June–August (JJA) to summer, and September–November (SON) to autumn.

Figure [2] shows the summer SSR (surface solar radiation) corresponding to five example transects in various parts of the Baltic. The data here are from CLARA-A2 and the location of the transects have been chosen to correspond to areas which are archipelago-free or at most exhibit a shallow archipelago. Palanga (Lithuania) has the highest SSR of the locations shown, while Haparanda (Sweden) has the lowest. Interestingly, all five transects exhibit a similar gradient in the SSR, decreasing from off-shore locations toward inland. Note that the transect of Gdansk (Poland) is to some extent almost parallel with the true coastline. This may have an influence on the behavior of the SSR close to the coast, where the first satellite grid box off-shore (at around −5 km) exhibits a relatively low SSR value.

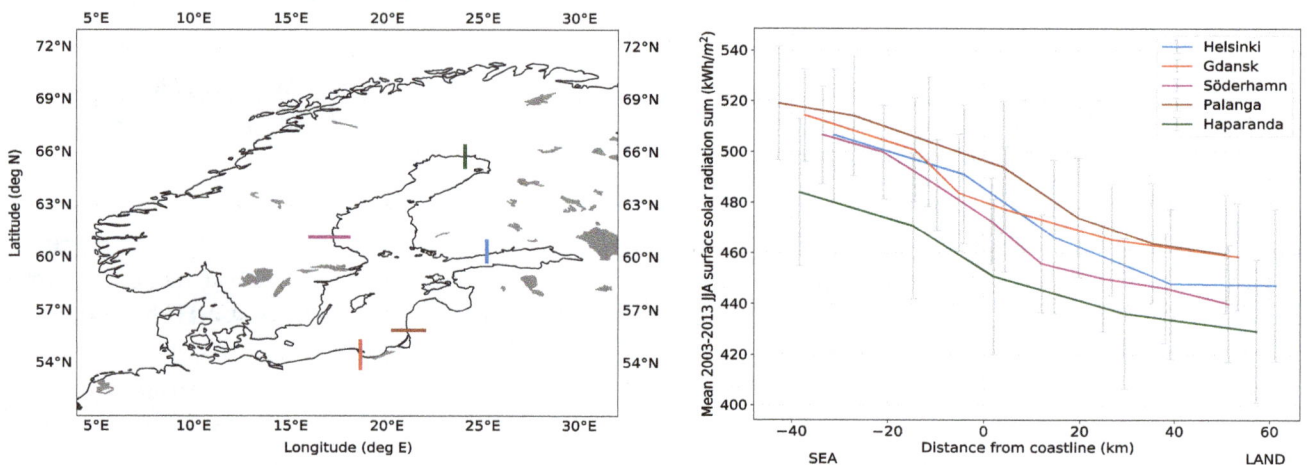

Figure 2. (left) Map showing the location of the transects for which **(right)** the climatological summer (JJA) SSR according to CLARA-A2 is presented as a function of distance to the coastline. The light grey error bars indicate the standard deviation of the annual summer values.

Figure 3 shows the seasonally averaged SSR for the Baltic region over the period 2003–2013 as a function of distance to the coastline. The figure shows both the SSR based on the two satellite records and as depicted by ground-based pyranometer measurements. Here, the data have been arranged according to the distance to the coastline into the following bins: ±5 km, 5–15 km, 15–25 km, and so on. The standard deviation of the satellite-based SSR shown in the figure is the median of the standard deviation over the grid boxes at the distance of interest. Similarly, when more than one ground-based measurement station is included in a specific distance bin, the standard deviation is presented as the median of the standard deviation over the stations of interest.

Table 2 shows the average SSR 20 km out on the sea (SSR_{-20km}), at the coastline (SSR_{0km}), and 20 km inland (SSR_{+20km}) together with the difference in the SSR between the sea and the land expressed as $\Delta SSR_{\pm 20km} = (SSR_{-20km} - SSR_{+20km})/SSR_{0km} \times 100\%$.

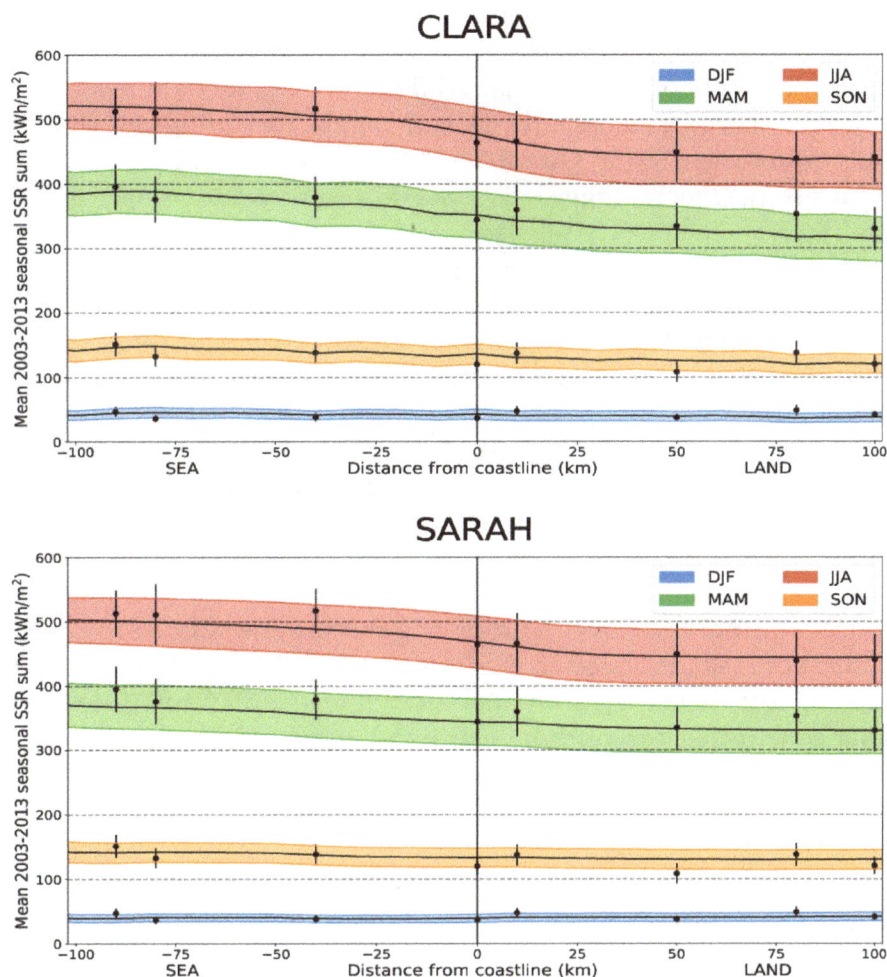

Figure 3. Climatological average (solid line) and standard deviation (colored shading) of the SSR for each season as a function of distance to the coastline for CLARA-A2 (**upper** panel) and SARAH-2 (**lower** panel). Ground-based pyranometer measurements of the SSR are shown as black filled circles with corresponding error bars denoting the standard deviation. See text for details.

Table 2. Comparison of the climatological SSR at 20 km out on the sea, at the coastline, and 20 km inland according to CLARA-A2 and SARAH-2.

| | $SSR_{-20km} [kWh/m^2]$ | | SSR_{0km} | | SSR_{+20km} | | $\Delta SSR_{\pm20km} [\%]$ | |
	CLARA	SARAH	CLARA	SARAH	CLARA	SARAH	CLARA	SARAH
DJF	43.3	38.5	42.4	39.8	41.2	40.6	4.8	−5.3
MAM	368.2	349.0	351.0	344.4	339.0	338.3	8.3	3.1
JJA	499.1	481.1	476.0	468.2	453.2	452.6	9.6	6.1
SON	137.5	134.2	134.6	133.5	129.9	131.9	5.6	1.7
YEAR	1042.0	997.0	998.3	980.1	957.8	957.9	8.4	4.0

Both satellite data records exhibit a land–sea contrast in the SSR, which is strongest during the summer period (JJA). Furthermore, springtime (MAM) SSR is higher over the seas than over land. In fact, a separate inspection of individual months reveals that the land–sea contrast in the SSR is strongest for the months May–July, hence coinciding with the months showing the highest climatological SSR. In autumn (SON), there is still a small distinguishable gradient in the SSR over the coastline in the CLARA-A2 data record, while SARAH-2 shows a rather flat behavior. In winter (DJF), SARAH-2 shows a slightly reversed gradient, with somewhat lower SSR over the sea than over land (see Table 2). We note, however, that winter time SSR from both SARAH-2 and CLARA-A2 suffer from larger uncertanties due to low sun and possibly snow-covered surfaces.

From Figure 3, it can be further noted, that while both satellite data records show reasonable agreement with the ground-based pyranometer measurements, the spring and summer time land–sea contrast in the SSR is better depicted in CLARA-A2. Both CLARA-A2 and the ground-based SSR measurements exhibit a somewhat stronger land–sea contrast than SARAH-2.

The Cloud Modification Factor (CMF), also referred to as the clear sky index, is a measure of the attenuating effect of clouds on solar radiation [17]. It is generally defined as the ratio of the SSR under all-sky conditions to that for otherwise the same conditions, but with a cloudless sky [18]. Both satellite data records enable the calculation of the CMF, since they provide, in addition to the real-sky estimate of the SSR, a clear-sky counterpart representing the SSR under cloudless skies. Figure 4 shows the monthly CMF of CLARA-A2 and SARAH-2, respectively. The CMF shows a rather similar land–sea contrast as the climatological SSR, with values decreasing from off-shore toward inland areas. Comparing Figure 4 with Figure 3 and Table 2 further reveals an approximate quantitative agreement between the behavior of the CMF and the SSR, supporting the straightforward hypothesis that most of the SSR gradient is caused by clouds. This conclusion is further corroborated by the results of Karlsson [6], who studied the diurnal behavior of the cloudiness in the Baltic area. He found a pronounced diurnal cycle over land during the summer (June–August), with maximum cloudiness in the afternoon. As discussed in the introduction, this indicates that convective clouds forming over the inland areas during the summer part of the year play an important role in the Baltic cloud climate. Similar conclusions can further be drawn indirectly from thunderstorm and lightning climatologies [19,20], showing significant land–sea contrast as well.

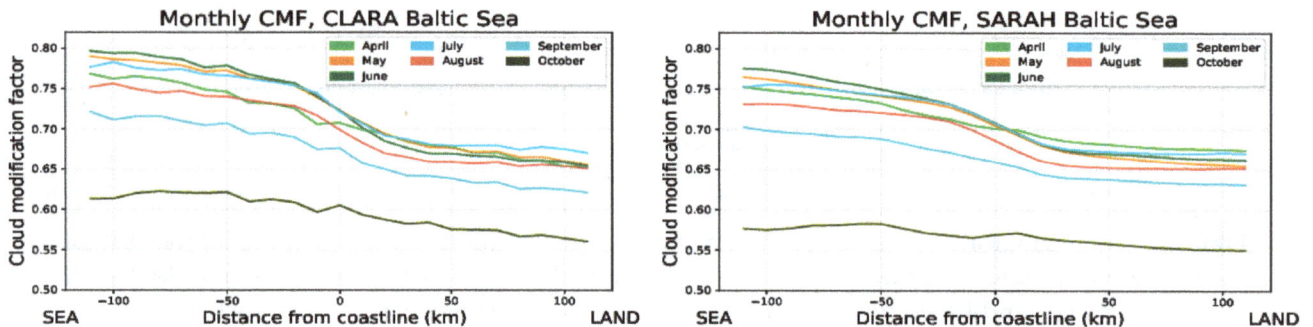

Figure 4. Climatological monthly Cloud Modification Factor (CMF) as a function of distance to the coastline for CLARA-A2 (**left** panel) and SARAH-2 (**right** panel).

4. Discussion

In this paper, we have studied the climatological behavior of the surface solar radiation (SSR). More specifically, we have used two satellite climate data records, CLARA-A2 and SARAH-2, together with ground-based pyranometer measurements of the SSR to study the land–sea contrast in the climatological SSR over the Baltic area. The results show that off-shore locations on average receive higher SSR than inland areas, in particular during the summer half of the year. For CLARA-A2, which shows better agreement with ground-based measurements, the annual SSR is 8% higher 20 km off the coastline than 20 km inland. SARAH-2 exhibits a smaller difference of 4%. We also show that the observed land–sea contrast in the SSR corresponds closely to the behavior of clouds.

The climatological SSR, a measure of the solar radiation received at a horizontal surface, is an important indicator of the solar energy production potential. As regards photovoltaic (PV) electricity production, however, also the temperature plays a role. The relative efficiency of PV cells typically decreases by 0.5% for 1°C increase in the cell temperature [21]. As temperatures of off-shore and coastal locations are typically cooler than those observed inland during the summer period in the

Baltic, these regions will gain an additional advantage through the temperature effect of PV cells. Quantification of this temperature effect is, however, left for future work.

An interesting question that remains is why CLARA-A2 shows better agreement than SARAH-2 with pyranometer measurements of the SSR in terms of the land–sea contrast. Although further studies are needed to understand the differences between CLARA-A2 and SARAH-2 in this context, the differences are likely linked to satellite observing geometry and retrieval algorithms. The comparably large viewing angles of the geostationary satellite, for example, require corrections to account for overestimation of the cloud optical thickness, in particular under broken cloud conditions. Such a correction has been applied in SARAH-2, however, a systematic underestimation of the SSR due to viewing geometry cannot be ruled out. As clouds behave differently over sea than over land (as shown in this paper), such an underestimation could have bearing on the results of the present study.

Another factor which may be of relevance relates to assumptions regarding the cloud transmissivity, which is one of the key parameters that determine the SSR. In the case of CLARA-A2, the cloud transmissivity is derived using the reflected top-of-the-atmosphere shortwave flux estimated from the satellite radiance measurements considering auxiliary information (e.g., surface albedo). For SARAH-2, the effective cloud albedo is estimated from the contrast of the cloud reflectivity to the clear-sky reflectivity, which might be larger over sea than over land (even for clouds of similar properties). Furthermore, the estimation of the shortwave reflected flux and the assignment to the cloud transmissivity in the CLARA-A2 algorithm depends also on the surface albedo and other local parameters. For both algorithms used here, effects induced by land–sea contrasts cannot be ruled out. The present study indicates that those effects that smoothen the land–sea contrast in SSR are smaller in CLARA-A2 than in SARAH-2, at least for the region studied, located at comparably high latitudes.

Author Contributions: A.V.L. designed the study and wrote the majority of the manuscript; A.H. and A.R. performed the analysis and contributed to the design of the study and writing of the manuscript; T.C., J.T. and R.M. contributed with ground-based and satellite data expertize and took part in writing of the manuscript. All authors have read and agreed to the published version of the manuscript.

Acknowledgments: We thank EUMETSAT's CMSAF project for providing the satellite-based SSR data used in this study.

References

1. Rossow, W.; Schiffer, R. Advances in understanding clouds from ISCCP. *Bull. Am. Meteorl. Soc.* **1999**, *80*, 2261–2284. [CrossRef]

2. Stubenrauch, C.J.; Chedin, A.; Rädel, G.; Scott, N.A.; Serrar, S. Cloud Properties and Their Seasonal and Diurnal Variability from TOVS Path-B. *J. Clim.* **2006**, *19*, 5531–5553. [CrossRef]

3. Hahn, C.; Warren, S. *A Gridded Climatology of Clouds over Land (1971–96) and Ocean (1954-97) from Surface Observations Worldwide*; Technical Report Numeric Data Product NDP-026E; Carbon Dioxide Information Analysis Center, Oak Ridge National Laboratory: Oak Ridge, TN, USA, 2007. [CrossRef]

4. Ångström, A. *Solstrålning Och Ljus På Den Svenska Västkusten*; Lundbergs Boktryckeri: Göteborg, Sweden, 1928.

5. Lunelund, H. Über die Sonnenscheindauer in Finnland. *Soc. Scient. Fenn. Comm. Phys. Math.* **1941**, *XI*, 1–14.

6. Karlsson, K. A 10 year cloud climatology over Scandinavia derived from NOAA advanced very high resolution radiometer imagery. *Int. J. Climatol.* **2003**, *23*, 1023–1044.10.1002/joc.916. [CrossRef]

7. Persson, T. Solar radiation climate in Sweden. *Phys. Chem. Earth Part B Hydrol. Ocean. Atmos.* **1999**, *24*, 275–279. [CrossRef]

8. Posselt, R.; Mueller, R.; Stöckli, R.; Trentmann, J. Remote sensing of solar surface radiation for climate monitoring—The CM-SAF retrieval in international comparison. *Remote Sens. Environ.* **2012**, *118*, 186–198. [CrossRef]

9. Šúri, M.; Huld, T.A.; Dunlop, E.D. PV-GIS: A web-based solar radiation database for the calculation of PV potential in Europe. *Int. J. Sustain. Energy* **2005**, *24*, 55–67. [CrossRef]

10. Riihelä, A.; Carlund, T.; Trentmann, J.; Müller, R.; Lindfors, A.V. Validation of CM SAF Surface Solar Radiation Datasets over Finland and Sweden. *Remote Sens.* **2015**, *7*, 6663–6682. [CrossRef]

11. Karlsson, K.G.; Anttila, K.; Trentmann, J.; Stengel, M.; Fokke Meirink, J.; Devasthale, A.; Hanschmann, T.; Kothe, S.; Jääskeläinen, E.; Sedlar, J.; et al. CLARA-A2: The second edition of the CM SAF cloud and radiation data record from 34 years of global AVHRR data. *Atmos. Chem. Phys.* **2017**, *17*, 5809–5828. [CrossRef]

12. Trentmann, J.; Pfeifroth, U. *Algorithm Theoretical Baseline Document, Meteosat Solar Surface Radiation and Effective Cloud Albedo, Climate Data Records—Heliosat, SARAH-2*; Technical Report Issue 2.2; EUMETSAT: Darmstadt, Germany, 2017.

13. Wessel, P.; Smith, W. A global, self-consistent, hierarchical, high-resolution shoreline database. *J. Geophys. Res. Solid Earth* **1996**, *101*, 8741–8743.10.1029/96JB00104. [CrossRef]

14. GSHHG. Global Self-Consistent Hierarchical High-Resolution Geography (GSHHG). Available online: http://www.soest.hawaii.edu/pwessel/gshhg/index.html (accessed on 16 October 2020).

15. Müller, B.; Wild, M.; Driesse, A.; Behrens, K. Rethinking solar resource assessments in the context of global dimming and brightening. *Sol. Energy* **2014**, *99*, 272–282. [CrossRef]

16. Tuononen, M.; O'Connor, E.J.; Sinclair, V.A. Evaluating solar radiation forecast uncertainty. *Atmos. Chem. Phys.* **2019**, *19*, 1985–2000. [CrossRef]

17. Calbo, J.; Pages, D.; Gonzalez, J.A. Empirical studies of cloud effects on UV radiation: A review. *Rev. Geophys.* **2005**, *43*. [CrossRef]

18. Lindfors, A.; Arola, A. On the wavelength-dependent attenuation of UV radiation by clouds. *Geophys. Res. Lett.* **2008**, *35*. [CrossRef]

19. Mäkelä, A.; Enno, S.E.; Haapalainen, J. Nordic Lightning Information System: Thunderstorm climate of Northern Europe for the period 2002–2011. *Atmos. Res.* **2014**, *139*, 46–61. [CrossRef]

20. Taszarek, M.; Allen, J.; Pucik, T.; Groenemeijer, P.; Czernecki, B.; Kolendowicz, L.; Lagouvardos, K.; Kotroni, V.; Schulz, W. A Climatology of Thunderstorms across Europe from a Synthesis of Multiple Data Sources. *J. Clim.* **2019**, *32*, 1813–1837. [CrossRef]

21. Radziemska, E. The effect of temperature on the power drop in crystalline silicon solar cells. *Renew. Energy* **2003**, *28*, 1–12. [CrossRef]

11

Multistep-Ahead Solar Radiation Forecasting Scheme Based on the Light Gradient Boosting Machine

Jinwoong Park, Jihoon Moon, Seungmin Jung and Eenjun Hwang *

School of Electrical Engineering, Korea University, 145 Anam-ro, Seongbuk-gu, Seoul 02841, Korea;
timeless@korea.ac.kr (J.P.); johnny89@korea.ac.kr (J.M.); jmkstcom@korea.ac.kr (S.J.)
* Correspondence: ehwang04@korea.ac.kr

Abstract: Smart islands have focused on renewable energy sources, such as solar and wind, to achieve energy self-sufficiency. Because solar photovoltaic (PV) power has the advantage of less noise and easier installation than wind power, it is more flexible in selecting a location for installation. A PV power system can be operated more efficiently by predicting the amount of global solar radiation for solar power generation. Thus far, most studies have addressed day-ahead probabilistic forecasting to predict global solar radiation. However, day-ahead probabilistic forecasting has limitations in responding quickly to sudden changes in the external environment. Although multistep-ahead (MSA) forecasting can be used for this purpose, traditional machine learning models are unsuitable because of the substantial training time. In this paper, we propose an accurate MSA global solar radiation forecasting model based on the light gradient boosting machine (LightGBM), which can handle the training-time problem and provide higher prediction performance compared to other boosting methods. To demonstrate the validity of the proposed model, we conducted a global solar radiation prediction for two regions on Jeju Island, the largest island in South Korea. The experiment results demonstrated that the proposed model can achieve better predictive performance than the tree-based ensemble and deep learning methods.

Keywords: smart island; solar energy; solar radiation forecasting; light gradient boosting machine; multistep-ahead prediction; feature importance

1. Introduction

Due to the serious problems caused by the use of fossil fuels, much attention has been focused on renewable energy sources (RESs) and smart grid technology to reduce greenhouse gas emissions [1,2]. Smart grid technology incorporates information and communication technology into the existing power grid using diverse smart sensors [3]. Smart grid technology can optimize the energy supply and demand by exchanging power production and consumption information between consumers and suppliers [4]. In particular, many countries, including smart islands, are replacing fossil fuels with RESs for energy self-sufficiency and carbon-free energy generation [5–7]. Two representative RESs are wind and global solar radiation. Although wind power has a smaller installation area and better power production than solar power, it suffers from higher maintenance costs and more noise. For example, due to various support policies of the Korean government related to renewable energies and smart grid technologies [8], the demand for photovoltaics (PV) is rapidly increasing in South Korea [9]. PV are best known as a method of generating electric power using solar cells to convert energy from the sun into a flow of electrons using the PV effect. Moreover, PV power system is based on an ecofriendly and infinite resource, and is cheaper to build than other power generation systems [10].

Various meteorological factors influence the PV system, and global solar radiation is the most crucial factor in the PV system [11,12]. Therefore, accurate global solar radiation forecasting is essential for the optimal operation of PV systems [13]. Recently, artificial neural network (ANN)-based global solar radiation forecasting models, such as the shallow neural network (SNN), deep neural network (DNN), and long short-term memory (LSTM) network, have been constructed to handle the nonlinearity and fluctuation of global solar radiation [14–20]. In addition, many studies have been conducted to predict global solar radiation accurately based on an ensemble learning technique that combines several weak models. For instance, in [21], the authors constructed two global solar radiation forecasting models based on the ANN and random forest (RF) methods. Then, they demonstrated that RF, which is an ensemble learning technique, exhibited better prediction performance than the ANN. In [22], the authors proposed four global solar radiation forecasting models based on the bagging and boosting techniques and analyzed the excellence and feature importance of the ensemble learning techniques.

Because global solar radiation is affected by diverse factors, such as season, time, and weather variables, predicting global solar radiation is challenging in the time domain [13]. The ensemble learning technique can avoid the overfitting problem and perform a more accurate prediction than the single model [23]. In this paper, we propose a novel forecasting model for multistep-ahead (MSA) global solar radiation predictions based on the light gradient boosting machine (LightGBM), which is a tree-based ensemble learning technique. The LightGBM can perform learning and prediction very quickly, which reduces the time needed for MSA prediction and performs more accurate predictions. Our forecasting model uses the meteorological information provided by the Korea Meteorological Administration (KMA) for global solar radiation prediction. In addition, to handle the uncertainty of PV scheduling, our MSA forecasting scheme makes hourly solar forecasts from 8 a.m. to 6 p.m. for 24 h from the current time. Usually, the farther the prediction point is from the learning point, the higher the probability that various changes will occur during the trend and pattern of the meteorological conditions and global solar radiation. To address this issue, we used time-series cross-validation (TSCV). We conducted rigorous experiments to compare the performance of LightGBM, various tree-based ensembles, and deep learning methods. Finally, we used the feature importance of the proposed model to provide interpretable forecasting results.

The contributions of this paper are as follows:

1. We proposed an MSA forecasting scheme for the efficient PV system operation.
2. We proposed an interpretable forecasting model based on feature importance analysis.
3. We increased the accuracy of global solar radiation forecasting using TSCV.

This paper is organized as follows. In Section 2, we describe the overall process for constructing a LightGBM-based forecasting model for MSA global solar radiation forecasting. In Section 3, we analyze the experimental results and describe the interpretable forecasting results of our proposed model. Lastly, we discuss in Section 4 the conclusions and some future research directions.

2. Materials and Methods

2.1. Data Collection and Preprocessing

In this paper, we used the date/time, meteorological data, and historical global solar radiation data provided by the KMA as input variables to construct a global solar radiation forecasting model. We considered two regions located on Jeju Island. Jeju is the largest island in South Korea and is implementing various measures to change into a smart island. For instance, it is enforcing diverse energy policies that encourage a shift from conventional fossil fuels to RESs. The two regions that we selected for validating prediction performance are Ildo-1 dong (latitude: 33.51411 and longitude: 126.52969) and Gosan-ri (latitude: 33.29382 and longitude: 126.16283). The data collection period is from 8 a.m. to 6 p.m. for a total of eight years from 2011 to 2018, and the collected data include temperature, humidity, wind speed, and global solar radiation. The meteorological observation data provided by the KMA include extra data, such as soil temperature, total cloud volume, ground-surface temperature, and sunshine amount. However, because the sky condition (also known as weather observation), temperature, humidity, and wind speed are provided by KMA's short-term weather forecasts, as shown in Figure 1, we only considered these factors [24].

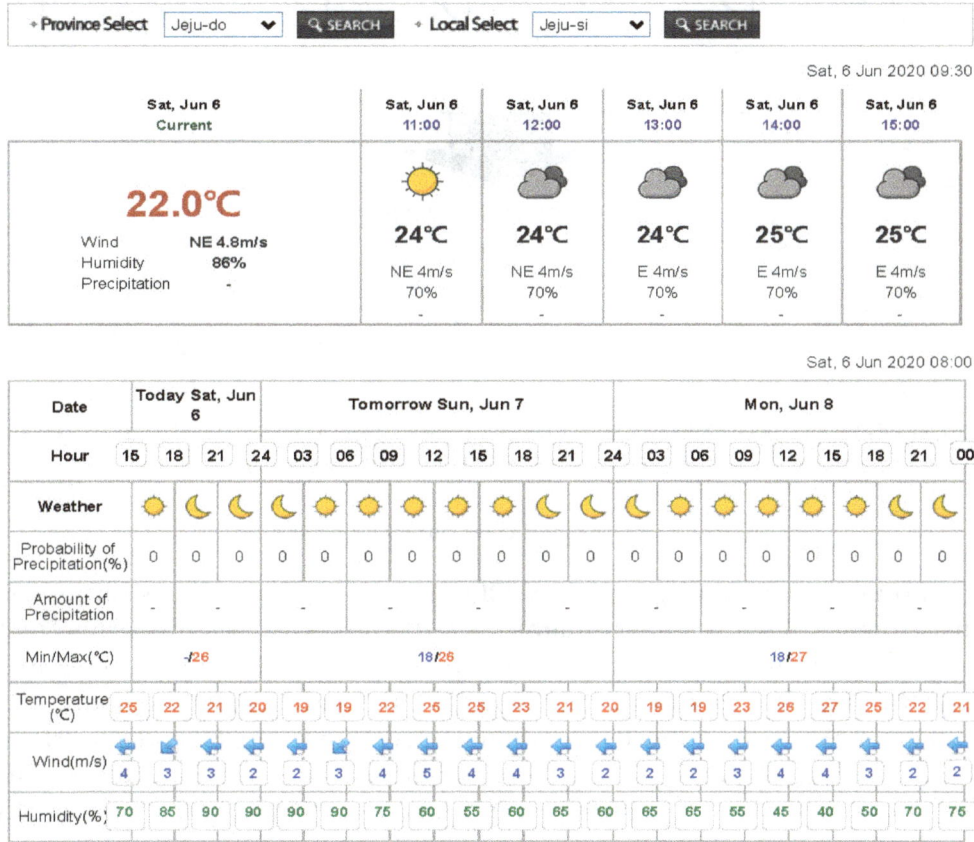

Figure 1. Short-term weather forecast by the Korea Meteorological Administration.

In the meteorological data we collected, about 0.1% of the total data for each category were missing, and the missing values were indicated as −1. Because the temperature, humidity, wind speed, and global solar radiation have continuous data characteristics, missing values can be estimated using linear interpolation. The sky condition data were presented as categorical values from 1 to 4, and missing values were approximated using logistic regression for similarity with the adjacent data.

For the date, to reflect the periodicity, one-dimensional data were augmented with continuous data in two-dimensional space using Equations (1) and (2) [25]. In the equations, end-of-month (*EoM*) indicates the last day of the month. The equations converted each Julian date into a value from 1 to 365. For instance, the Julian date of January 1 is converted to 1, and December 31 is converted to 365. In the case of leap years, 366 was used instead of 365 in the equations. Figure 2 illustrates an example of preprocessing the date data.

$$Date_X = \sin\left(360° \times \left(\sum_{1}^{Month-1} EoM + Day\right)/365\right) \tag{1}$$

$$Date_Y = \cos\left(360° \times \left(\sum_{1}^{Month-1} EoM + Day\right)/365\right) \tag{2}$$

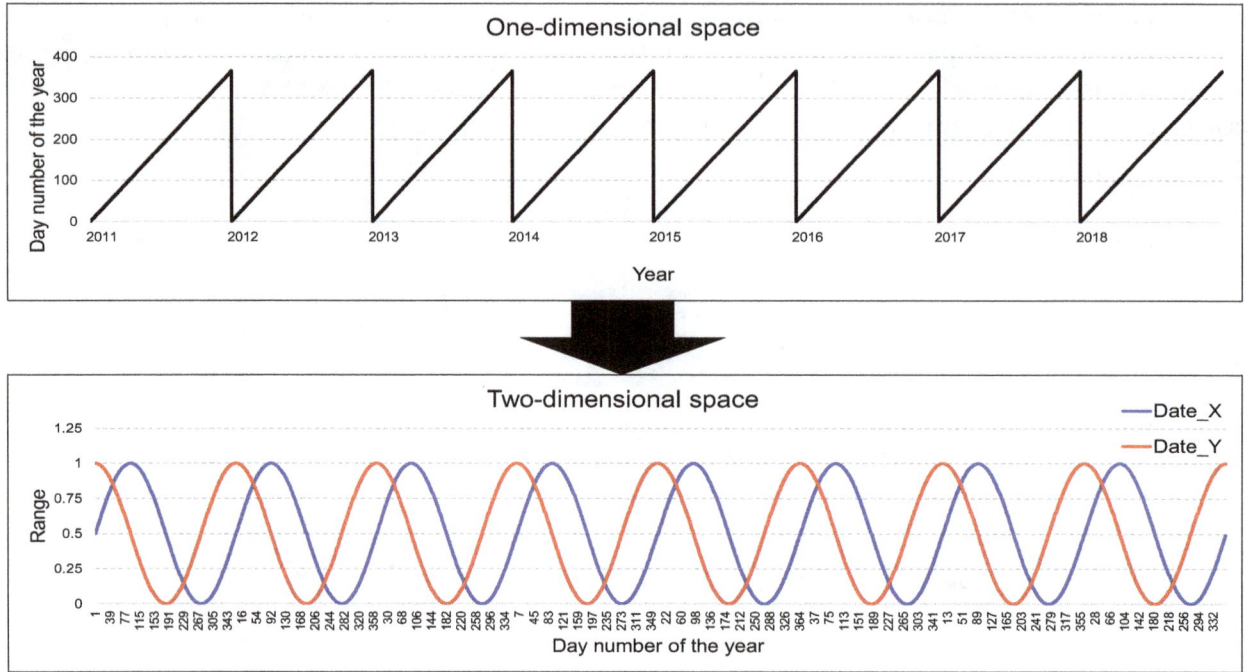

Figure 2. Example of date data preprocessing.

The cloud amount is provided by the KMA. Of the two popular methods for representing cloud amount, which are meteorology 1/8 and climatology 1/10, the KMA uses the second method. Hence, the cloud amount is represented by eleven scales (i.e., from 0 for a clear sky to 10 for an overcast sky). The sky condition data have four interval scales [26,27]: 1 for clear (0 ≤ cloud amount ≤ 2), 2 for partly cloudy (3 ≤ cloud amount ≤ 5), 3 for mostly cloudy (6 ≤ cloud amount ≤ 8), and 4 for cloudy (9 ≤ cloud amount ≤ 10). Because we represent the sky condition data using one-hot encoding, a value of 1 is placed in the binary variable for a specific sky condition, and 0 is used for the other sky conditions. Time data were also represented by interval scales. Global solar radiation is highest during the day from 12 to 2 p.m. To assess these variables more effectively, we used one-hot encoding to represent time intervals.

In addition, to reflect the recent trends in global solar radiation, we used the sky condition, temperature, humidity, wind speed, and global solar radiation of the day before the forecast point as input variables. We considered 30 input variables to construct our prediction model, as shown in Table 1. As our goal is to perform MSA (all time points for the next 24 h) forecasting, we needed all the input variables for 11 prediction time points. Therefore, we used 330 input variables (i.e., 30 input variables × 11 prediction time points) with 32,143 tuples for the MSA forecasting model construction, as shown in Figure 3.

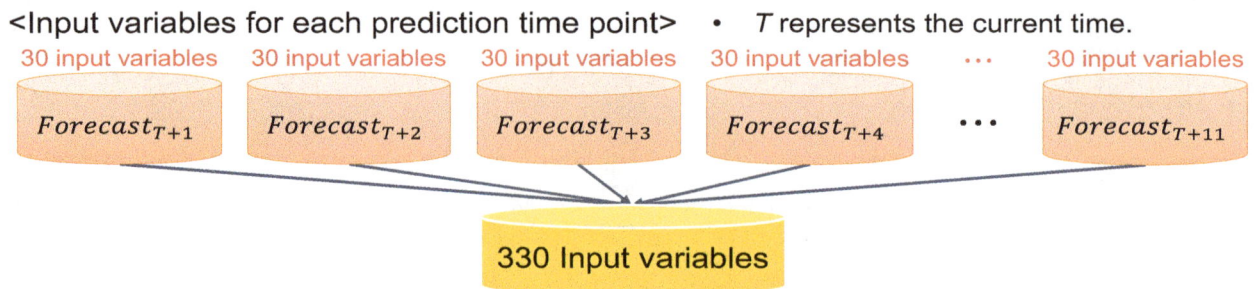

Figure 3. Input variable configuration for multistep-ahead (MSA) global solar radiation forecasting.

Table 1. List of input variables (IV) for the proposed model.

IV #	Input Variable (Feature)	IV #	Input Variable (Feature)
IV01	$Date_X$ (numeric)	IV16	Mostly cloudy (binary)
IV02	$Date_Y$ (numeric)	IV17	Cloudy (binary)
IV03	8 a.m. (binary)	IV18	Temperature (numeric)
IV04	9 a.m. (binary)	IV19	Humidity (numeric)
IV05	10 a.m. (binary)	IV20	Wind speed (numeric)
IV06	11 a.m. (binary)	IV21	$Date_X$ 1 day before (numeric)
IV07	12 p.m. (binary)	IV22	$Date_Y$ 1 day before (numeric)
IV08	1 p.m. (binary)	IV23	Clear 1 day before (binary)
IV09	2 p.m. (binary)	IV24	Partly cloudy 1 day before (binary)
IV10	3 p.m. (binary)	IV25	Mostly cloudy 1 day before (binary)
IV11	4 p.m. (binary)	IV26	Cloudy 1 day before (binary)
IV12	5 p.m. (binary)	IV27	Temperature 1 day before (numeric)
IV13	6 p.m. (binary)	IV28	Humidity 1 day before (numeric)
IV14	Clear (binary)	IV29	Wind speed 1 day before (numeric)
IV15	Partly cloudy (binary)	IV30	Global solar radiation 1 day before (numeric)

2.2. Forecasting Model Construction

The purpose of our model is to predict global solar radiation for the next 11 time points from the current time. To construct a global solar radiation forecasting model, we used LightGBM, a gradient boosting machine (GBM)-based model. The LightGBM model [28] is based on a gradient boosting decision tree (GBDT) applying gradient-based one-side sampling and exclusive feature bundling technologies. Unlike the conventional GBM tree splitting method, a leafwise method is used to create complex models to achieve higher accuracy; hence, it is useful for time-series forecasting. Because of the GBDT and leafwise method, LightGBM has the advantages of reduced memory usage and faster training speed. The LightGBM contains various hyperparameters to be tuned. Among them, the learning rate, number of iterations, and number of leaves are closely related to the prediction accuracy. In addition, overfitting can be prevented by adjusting the colsample by tree and subsample hyperparameters. Moreover, LightGBM also can use different algorithms for its learning iterations. In this paper, we constructed two LightGBM models using two boosting types: GBDT and dropouts meet multiple additive regression trees (DART) [29] for comparison. Both models perform predictions on multiple outputs using the *MultiOutputRegressor* module in scikit-learn (v. 0.22.1).

In general, to evaluate a forecasting method, we first divide a dataset into training and test sets. Then, we construct the forecasting model using the training set. Finally, we evaluate the performance of the forecasting model using the test set. A greater time interval between training and forecasting lowers the prediction performance [30]. To solve this problem, we applied TSCV, which is popularly used when data exhibit time-series characteristics and are focused on a single forecast of the dataset [6]. The TSCV uses all data before the prediction point as a training set and predicts the next forecasting point by setting it as a test set, iteratively.

However, if TSCV is performed at every point, it requires a considerable amount of time to train and forecast. To reduce this overhead, we conducted monthly TSCV, as shown in Figure 4. In addition, for interpretable global solar radiation forecasting, we analyzed the variable importance changes for the 30 input variables by obtaining the feature importance using LightGBM.

2.3. Baseline Models

To demonstrate the performance of our model, we constructed various forecasting models based on the tree-based ensemble and deep learning methods.

In the case of tree-based ensemble learning methods, because they combine several weak models effectively, they usually exhibit better prediction performance than a single model. In the experiment, we considered RF, GBM, and extreme gradient boosting (XGBoost) ensemble methods to construct MSA global solar radiation forecasting models. The RF method trains each tree independently by using a randomly selected sample of the data. As the RF method tends to reduce the correlation between trees, it provides a robust model for out-of-sample forecasting [31]. In addition, the GBM is a forward-learning ensemble method that obtains predictive results using gradually improved estimations [32]. The adjusted model is built by applying the residuals of the previous model, and this procedure is repeated N times to build a robust model. We constructed two GBM models by considering the quantile regression and Huber loss functions, respectively. The XGBoost method is an algorithm that can prevent overfitting by reducing the tree correlation using the shrinkage method [33]. Moreover, it can perform parallel processing by applying the column subsampling method. The XGBoost method constructs a weak model and evaluates the consistency using the training set. After that, the method constructs an adjusted prediction model with the explanatory variable for the gradient in the direction in which consistency increases using the gradient descent method. This procedure is repeated N times to build a robust model [34]. We constructed two XGBoost models by applying two boosting types (i.e., GBDT and DART). To predict multiple outputs, we constructed an RF model using the *MultivariateRandomForest* [35] package in R (v. 3.5.1) and the GBM and XGBoost models using the *MultiOutputRegressor* module in scikit-learn (v. 0.22.1) [36].

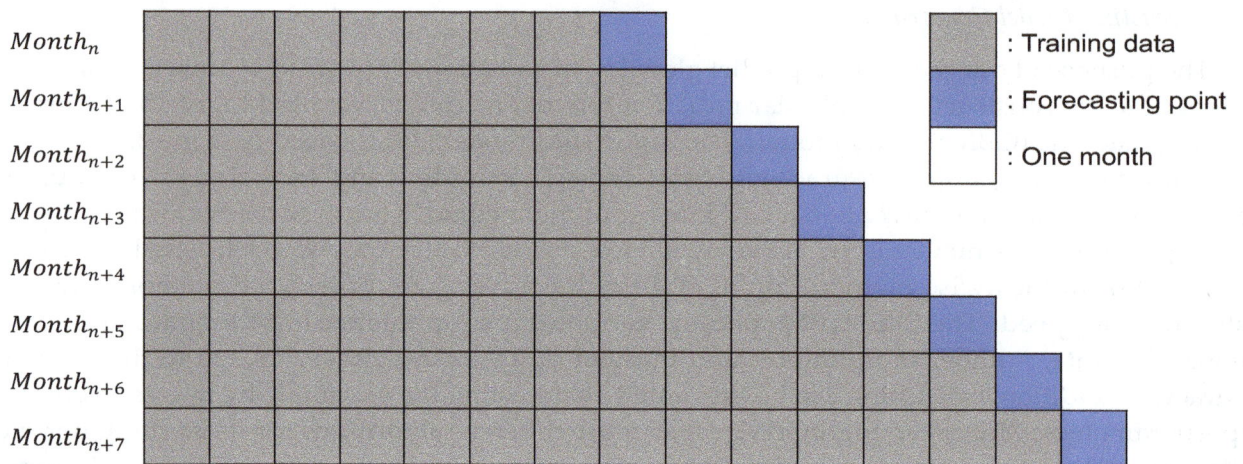

Figure 4. Example of monthly time-series cross-validation.

For deep learning-based MSA global solar radiation forecasting models, we considered the SNN, DNN, LSTM network, and attention-based LSTM (ATT-LSTM) network. These models require a sufficient amount of training data for accurate predictive performance, and the models can overfit the data if the training data are insufficient [37]. A typical ANN consists of an input layer, one or more hidden layers, and an output layer, and each layer consists of one or more hidden nodes [38,39]. The ANNs have various hyperparameters that affect prediction performance [38]. These hyperparameters include the number of hidden layers, number of hidden nodes, an activation function, and so on. In addition, the SNN has one hidden layer, and the DNN has two or more hidden layers [39]. The LSTM network [40] is a model that can solve the long-term dependency problem of the existing recurrent neural network. The LSTM network is useful for training sequence data in the time-series forecasting method. Nevertheless, although the length of the input variable is long, the forecasting accuracy of the sequence-to-sequence model suffers due to focusing on all input variables. To solve this problem, an attention mechanism [41] has been developed in the field of machine translation. The attention mechanism comprises an encoder that builds a vector from the input variable and a decoder that outputs a dependent variable using the vector output by the encoder as input. The decoder part performs the model training focused on data representing high similarity by indicating the similarity with the encoder as a value; hence, it can exhibit accurate forecasting performance. Applying the attention mechanism to the LSTM described above focuses the model on specific vectors so that it obtains more accurate forecasting results [42].

In our previous work [7], we constructed several deep learning models for MSA global solar radiation forecasting in the same experimental environment. We used the dropout method to control the weight of the hidden layers to prevent overfitting. To do this, we found optimal hyperparameter values for each deep learning model, as indicated in Table 2.

Table 2. Selected optimal hyperparameters for each deep learning model.

Models	Selected Hyperparameters
SNN	Number of hidden layers: 1 Number of hidden nodes: 14 Activation function: sigmoid Loss function: mean squared error Optimizer: Adam
DNN	Number of hidden layers: 7 Activation function: ReLU, SELU [43] Remaining hyperparameters are the same as those for the SNN model
LSTM network	Sequence length: 11 Number of hidden layers: 2 Activation function: ReLU, SELU [43] Loss function: Huber loss Optimizer: RMSProp Batch size: 11 Learning rate: 0.000001 Epoch: 5,000
ATT-LSTM network	Number of attention layers: 1 Remaining hyperparameters are the same as those for the LSTM network model

Notes: SNN: shallow neural network; DNN: deep neural network; LSTM: long short-term memory; ATT-LSTM: attention-based LSTM; ReLU: rectified linear unit; SELU: scaled exponential linear unit.

3. Results and Discussion

In the experiments, we used two global solar radiation datasets collected from two regions from 2011 to 2018. The two regions are Ildo-1 and Gosan-ri on Jeju island. We divided each dataset into two parts at a ratio of 75:25: a training set (in-sample) spanning 2011 to 2016, and a test set (out-of-sample) spanning 2017 to 2018. Table 3 lists various statistical analysis for the datasets by considering the training and test sets. The statistical analysis was performed by using Excel's Descriptive Statistics data analysis tool. Figure 5 represents the boxplots of the global solar radiation data for each region.

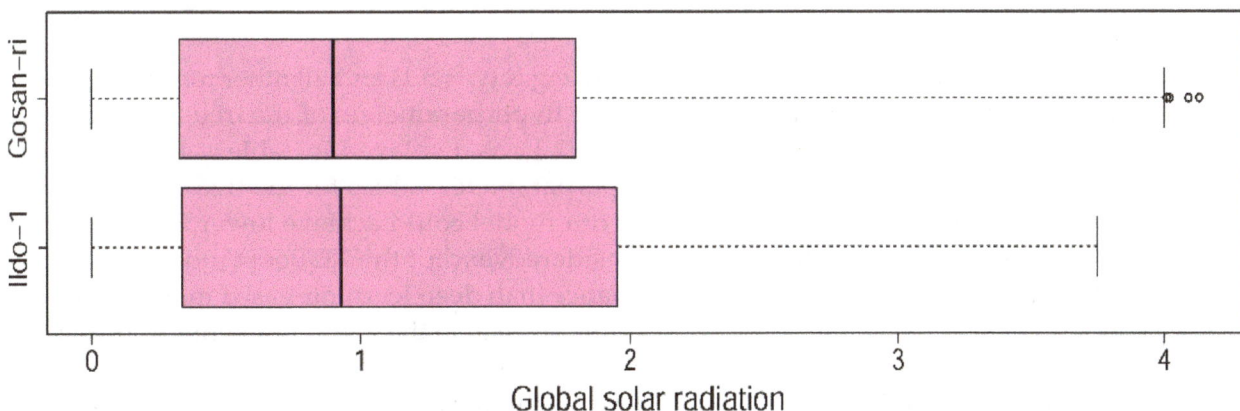

Figure 5. Boxplots by region (MJ/m^2).

Table 3. Statistical analysis of global solar radiation data by region (MJ/m^2).

Statistics	Ildo-1		Gosan-ri	
	Training Set	Test Set	Training Set	Test Set
Mean	1.188	1.258	1.179	1.044
Standard error	0.006	0.011	0.006	0.009
Median	0.910	1.010	0.910	0.840
Mode	0	0	0	0
Standard deviation	0.995	1.000	0.989	0.842
Sample variance	0.990	1.000	0.979	0.710
Kurtosis	−0.756	−0.869	−0.400	−0.419
Skewness	0.659	0.568	0.764	0.713
Range	3.750	3.720	4.130	3.550
Minimum	0	0	0	0
Maximum	3.750	3.720	4.130	3.550
Sum	28,656.9	10,097.9	28,422.8	8383.9
Count	24,112	8030	24,112	8030

For continuous data, such as humidity, wind speed, temperature, and historical global solar radiation, we performed standardization using Equation (3). In the equation, x_i and x denote the input variable and original data, respectively. In addition, μ and σ denote the average of the original data and the standard deviation, respectively.

$$x_i = \frac{x - \mu}{\sigma} \tag{3}$$

To evaluate the prediction performance of the models, we used four metrics: mean biased error (MBE), mean absolute error (MAE), root mean square error (RMSE), and normalized root mean square error (NRMSE), as shown in Equations (4)–(7). Here, A_t and F_t represent the actual and forecasted values, respectively, at time t, n indicates the number of observations, and $\overline{A}$ represents the average of the actual values.

$$MBE = \frac{1}{n} \sum_{t=1}^{n} (A_t - F_t) \tag{4}$$

$$MAE = \frac{1}{n} \sum_{t=1}^{n} |F_t - A_t| \tag{5}$$

$$RMSE = \sqrt{\frac{\sum_{t=1}^{n} (F_t - A_t)^2}{n}} \tag{6}$$

$$NRMSE = \frac{\sqrt{\frac{\sum_{t=1}^{n} (F_t - A_t)^2}{n}}}{\overline{A}} \times 100 \tag{7}$$

We implemented an RF-based forecasting model using R (v. 3.5.1) and all other forecasting models using Python (v. 3.6). We found optimal values for the hyperparameters of the tree-based ensemble learning models via *GridSearchCV* in scikit-learn (v. 0.22.1), as displayed in Table 4. Because the two regions are close together, we obtained the same hyperparameter values for the two regions.

Tables 5–13 and Figures 6–13 demonstrate that our model could achieve lower RMSE and MAE values than all other forecasting models that we considered, except the XGBoost model. In addition, tree-based ensemble models exhibited better performance than deep learning-based models. Moreover, the TSCV scheme demonstrated better prediction performance than the holdout scheme, as presented in Table 13. The XGBoost and LightGBM methods exhibited a similar prediction performance. However, regarding the aspect of the training and testing time, LightGBM took 220 s, whereas XGBoost took 3798 s. That is, LightGBM is 17 times faster than XGBoost. Hence, LightGBM has a clear advantage in terms of accuracy and time. In the forecasting results of LightGBM, we observed that the MAE and RMSE values were lowest at the first time point, and as the distance increased, these values increased.

Table 4. Selected hyperparameters for each ensemble learning model. Selected values are bold.

Models		Package or Module	Selected Hyperparameters
Random forest		*MultivariateRandomForest*	Number of trees: 128 [44] Number of features: 110 [44]
GBM	Quantile regression	*GradientBoostingRegressor* *GridSearchCV*	Learning rate: 0.01, **0.05**, 0.1 Number of iterations: 100, 250, **500** Maximum depth of the tree: 5, **10**
	Huber loss		Learning rate: 0.01, **0.05**, 0.1 Number of iterations: 100, 250, **500** Maximum depth of the tree: **5**, 10
XGBoost	GBDT	*XGBoost 1.0.2* *GridSearchCV*	Learning rate: **0.01**, 0.05, 0.1 Number of iterations: 250, **500**, 1000 Maximum depth of the tree: 6, **8**, 10 Subsample: 0.5, 0.75, **1.0** Colsample by tree: 0.5, 0.75, **1.0** Colsample by level: **0.5**, 0.75, 1.0 Colsample by node: 0.5, 0.75, **1.0**
	DART		Learning rate: **0.01**, 0.05, 0.1 Number of iterations: 250, 500, **1000** Maximum depth of the tree: 6, **8**, 10 Subsample: 0.5, 0.75, **1.0** Colsample by tree: 0.5, 0.75, **1.0** Colsample by level: **0.5**, 0.75, 1.0 Colsample by node: 0.5, 0.75, **1.0**
LightGBM	GBDT	*LightGBM 2.3.1* *GridSearchCV*	Learning rate: 0.01, **0.05**, 0.1 Number of iterations: **1000**, 1500 Number of leaves: **64** Subsample: **0.5** Colsample by tree: **1.0**
	DART (our model)		Learning rate: 0.01, 0.05, **0.1** Number of iterations: **1000**, 1500 Number of leaves: **64** Subsample: **0.5** Colsample by tree: **1.0**

Notes: GBM: gradient boosting machine; XGBoost: extreme gradient boosting; LightGBM: light GBM; GBDT: gradient boosting decision tree; DART: dropouts meet multiple additive regression trees.

Table 5. Mean bias error (MBE) distribution for each model for Ildo-1 (MJ/m^2).

Models	Points										
	1	2	3	4	5	6	7	8	9	10	11
SNN (Dropout O)	−0.05	−0.04	−0.03	−0.03	−0.03	−0.05	−0.06	−0.07	−0.07	−0.06	−0.04
SNN (Dropout X)	−0.04	−0.01	−0.04	−0.05	−0.05	−0.05	−0.04	−0.04	−0.04	−0.04	−0.04
DNN-ReLU (Dropout O)	−0.04	0	−0.06	−0.07	−0.08	−0.09	−0.10	−0.12	−0.13	−0.13	−0.14
DNN-ReLU (Dropout X)	−0.06	−0.06	−0.07	−0.07	−0.08	−0.08	−0.08	−0.08	−0.08	−0.07	−0.06
DNN-SELU (Dropout O)	0.05	0.08	0.02	−0.03	−0.04	−0.05	−0.03	0	−0.03	−0.02	−0.02
DNN-SELU (Dropout X)	−0.06	−0.06	−0.09	−0.10	−0.10	−0.10	−0.10	−0.10	−0.10	−0.10	−0.11
LSTM-ReLU (Dropout O)	−0.10	0.05	−0.02	−0.06	−0.06	−0.05	−0.04	−0.02	−0.02	−0.02	−0.23
LSTM-ReLU (Dropout X)	−0.19	−0.01	−0.02	−0.06	−0.10	−0.12	−0.12	−0.11	−0.10	−0.09	−0.08
LSTM-SELU (Dropout O)	−0.09	−0.02	−0.06	−0.07	−0.06	−0.04	−0.03	−0.03	−0.02	−0.03	−0.03
LSTM-SELU (Dropout X)	−0.10	−0.03	−0.06	−0.09	−0.10	−0.09	−0.09	−0.09	−0.09	−0.09	−0.09
ATT−LSTM-RELU (Dropout O)	−0.16	−0.14	−0.09	−0.07	−0.07	−0.01	0.02	0.03	0.01	−0.05	−0.09
ATT−LSTM-RELU (Dropout X)	−0.14	−0.11	−0.11	−0.06	0	0.09	0.07	0.09	0.12	0.01	−0.02
ATT−LSTM-SELU (Dropout O)	−0.04	0.02	−0.06	0.01	0.04	−0.06	−0.10	0	0.03	−0.02	−0.03
ATT−LSTM-SELU (Dropout X)	0.13	0.13	0.08	−0.05	−0.11	−0.18	−0.25	−0.18	−0.29	−0.30	−0.20
RF (TSCV)	−0.03	−0.04	−0.05	−0.06	−0.07	−0.07	−0.07	−0.07	−0.07	−0.06	−0.06
RF (Holdout)	−0.04	−0.05	−0.07	−0.08	−0.09	−0.09	−0.09	−0.09	−0.08	−0.08	−0.08
GBM-Huber (TSCV)	−0.02	−0.03	−0.04	−0.05	−0.05	−0.05	−0.05	−0.04	−0.04	−0.04	−0.03

Table 5. *Cont.*

Models	Points										
	1	2	3	4	5	6	7	8	9	10	11
GBM-Huber (Holdout)	−0.03	−0.04	−0.05	−0.07	−0.07	−0.07	−0.07	−0.06	−0.06	−0.05	−0.04
GBM-Quantile (TSCV)	0.24	0.32	0.35	0.38	0.39	0.41	0.42	0.43	0.45	0.45	0.46
GBM-Quantile (Holdout)	0.24	0.31	0.34	0.36	0.38	0.39	0.40	0.42	0.44	0.45	0.46
XGBoost-GDBT (TSCV)	−0.02	−0.03	−0.05	−0.05	−0.05	−0.05	−0.05	−0.05	−0.05	−0.04	−0.04
XGBoost-GDBT (Holdout)	−0.03	−0.05	−0.06	−0.07	−0.08	−0.07	−0.07	−0.07	−0.07	−0.06	−0.06
XGBoost-DART (TSCV)	−0.02	−0.03	−0.04	−0.05	−0.05	−0.05	−0.05	−0.05	−0.05	−0.04	−0.04
XGBoost-DART (Holdout)	−0.03	−0.05	−0.06	−0.07	−0.08	−0.07	−0.07	−0.07	−0.06	−0.06	−0.06
LightGBM-GDBT (TSCV)	−0.02	−0.03	−0.04	−0.04	−0.05	−0.05	−0.05	−0.05	−0.04	−0.05	−0.05
LightGBM-GDBT (Holdout)	−0.03	−0.05	−0.06	−0.07	−0.07	−0.08	−0.08	−0.07	−0.07	−0.08	−0.07
LightGBM-DART (TSCV)	−0.03	−0.04	−0.05	−0.06	−0.06	−0.06	−0.06	−0.05	−0.05	−0.05	−0.05
LightGBM-DART (Holdout)	−0.04	−0.06	−0.07	−0.08	−0.08	−0.08	−0.08	−0.07	−0.07	−0.07	−0.06

Table 6. Mean absolute error (MAE) distribution for each model for Ildo-1. A cooler color indicates a lower MAE value, whereas a warmer color indicates a higher MAE value (MJ/m^2).

Models	Points										
	1	2	3	4	5	6	7	8	9	10	11
SNN (Dropout O)	0.445	0.406	0.394	0.390	0.386	0.385	0.385	0.386	0.385	0.385	0.385
SNN (Dropout X)	0.413	0.390	0.393	0.392	0.390	0.388	0.387	0.387	0.388	0.391	0.395
DNN-ReLU (Dropout O)	0.419	0.384	0.389	0.386	0.382	0.379	0.380	0.382	0.385	0.388	0.391
DNN-ReLU (Dropout X)	0.445	0.401	0.398	0.387	0.384	0.385	0.383	0.381	0.381	0.381	0.381
DNN-SELU (Dropout O)	0.349	0.327	0.343	0.354	0.363	0.368	0.375	0.382	0.395	0.395	0.405
DNN-SELU (Dropout X)	0.408	0.383	0.380	0.378	0.376	0.374	0.373	0.373	0.374	0.376	0.378
LSTM-ReLU (Dropout O)	0.365	0.388	0.409	0.412	0.420	0.427	0.436	0.436	0.446	0.458	0.460
LSTM-ReLU (Dropout X)	0.380	0.399	0.417	0.426	0.431	0.442	0.456	0.467	0.468	0.470	0.470
LSTM-SELU (Dropout O)	0.318	0.332	0.350	0.360	0.372	0.377	0.371	0.373	0.378	0.379	0.370
LSTM-SELU (Dropout X)	0.357	0.379	0.400	0.412	0.413	0.438	0.446	0.467	0.488	0.501	0.517
ATT-LSTM-RELU (Dropout O)	0.291	0.324	0.329	0.340	0.347	0.348	0.357	0.363	0.368	0.382	0.394
ATT-LSTM-RELU (Dropout X)	0.272	0.299	0.324	0.332	0.339	0.351	0.349	0.360	0.375	0.378	0.392
ATT-LSTM-SELU (Dropout O)	0.238	0.291	0.311	0.330	0.333	0.346	0.368	0.348	0.358	0.371	0.381
ATT-LSTM-SELU (Dropout X)	0.261	0.301	0.307	0.319	0.343	0.376	0.414	0.386	0.439	0.452	0.415
RF (TSCV)	0.215	0.272	0.306	0.330	0.347	0.363	0.375	0.388	0.401	0.414	0.428
RF (Holdout)	0.223	0.280	0.315	0.340	0.358	0.371	0.383	0.396	0.409	0.423	0.435
GBM-Huber (TSCV)	0.186	0.251	0.292	0.318	0.337	0.351	0.359	0.368	0.374	0.382	0.385
GBM-Huber (Holdout)	0.189	0.257	0.300	0.330	0.350	0.359	0.370	0.377	0.383	0.392	0.394
GBM-Quantile (TSCV)	0.288	0.376	0.418	0.445	0.462	0.476	0.484	0.498	0.511	0.518	0.525
GBM-Quantile (Holdout)	0.287	0.375	0.413	0.445	0.461	0.470	0.482	0.497	0.510	0.524	0.530
XGBoost-GDBT (TSCV)	0.194	0.257	0.296	0.324	0.339	0.348	0.356	0.366	0.375	0.384	0.390
XGBoost-GDBT (Holdout)	0.197	0.263	0.305	0.333	0.349	0.358	0.367	0.374	0.383	0.390	0.396
XGBoost-DART (TSCV)	0.184	0.249	0.289	0.317	0.333	0.346	0.353	0.362	0.369	0.377	0.382
XGBoost-DART (Holdout)	0.188	0.255	0.298	0.328	0.346	0.357	0.364	0.372	0.379	0.386	0.393
LightGBM-GDBT (TSCV)	0.189	0.253	0.292	0.319	0.334	0.348	0.355	0.363	0.371	0.377	0.381
LightGBM-GDBT (Holdout)	0.193	0.261	0.305	0.331	0.346	0.359	0.368	0.373	0.382	0.391	0.394
LightGBM-DART (TSCV)	0.189	0.252	0.290	0.318	0.333	0.344	0.350	0.358	0.365	0.374	0.380
LightGBM-DART (Holdout)	0.193	0.257	0.300	0.328	0.344	0.355	0.359	0.369	0.377	0.384	0.389

Table 7. Root mean square error (RMSE) distribution for each model for Ildo-1. A cooler color indicates a lower RMSE value, whereas a warmer color indicates a higher RMSE value (MJ/m^2).

Models	Points										
	1	2	3	4	5	6	7	8	9	10	11
SNN (Dropout O)	0.588	0.546	0.537	0.534	0.528	0.526	0.524	0.525	0.524	0.525	0.528
SNN (Dropout X)	0.545	0.532	0.534	0.533	0.529	0.526	0.524	0.524	0.526	0.529	0.534
DNN-ReLU (Dropout O)	0.545	0.528	0.529	0.525	0.519	0.514	0.513	0.515	0.519	0.523	0.527
DNN-ReLU (Dropout X)	0.598	0.548	0.545	0.525	0.520	0.518	0.515	0.514	0.513	0.514	0.515
DNN-SELU (Dropout O)	0.392	0.441	0.465	0.479	0.487	0.499	0.506	0.508	0.522	0.519	0.532
DNN-SELU (Dropout X)	0.541	0.526	0.523	0.512	0.515	0.511	0.509	0.509	0.510	0.511	0.514
LSTM-ReLU (Dropout O)	0.544	0.547	0.545	0.541	0.540	0.540	0.547	0.551	0.559	0.570	0.636
LSTM-ReLU (Dropout X)	0.545	0.550	0.543	0.544	0.500	0.543	0.550	0.551	0.558	0.567	0.637
LSTM-SELU (Dropout O)	0.408	0.448	0.462	0.473	0.484	0.494	0.506	0.516	0.517	0.518	0.519
LSTM-SELU (Dropout X)	0.431	0.467	0.481	0.492	0.506	0.507	0.521	0.531	0.551	0.551	0.558
ATT-LSTM-RELU (Dropout O)	0.381	0.430	0.446	0.464	0.478	0.481	0.491	0.499	0.505	0.515	0.528
ATT-LSTM-RELU (Dropout X)	0.383	0.428	0.458	0.466	0.475	0.496	0.498	0.512	0.529	0.530	0.543
ATT-LSTM-SELU (Dropout O)	0.329	0.395	0.431	0.455	0.464	0.479	0.502	0.493	0.508	0.519	0.528
ATT-LSTM-SELU (Dropout X)	0.357	0.415	0.431	0.450	0.475	0.509	0.557	0.538	0.586	0.598	0.561
RF (TSCV)	0.302	0.378	0.421	0.450	0.471	0.490	0.504	0.516	0.528	0.540	0.554
RF (Holdout)	0.308	0.386	0.431	0.462	0.484	0.502	0.515	0.527	0.538	0.550	0.564
GBM-Huber (TSCV)	0.285	0.369	0.417	0.451	0.472	0.490	0.498	0.507	0.513	0.518	0.520
GBM-Huber (Holdout)	0.288	0.376	0.427	0.465	0.490	0.501	0.514	0.521	0.526	0.530	0.533
GBM-Quantile (TSCV)	0.412	0.527	0.594	0.638	0.663	0.682	0.695	0.714	0.732	0.739	0.749
GBM-Quantile (Holdout)	0.409	0.523	0.584	0.637	0.660	0.671	0.689	0.706	0.727	0.739	0.751
XGBoost-GDBT (TSCV)	0.289	0.371	0.417	0.452	0.471	0.484	0.494	0.504	0.511	0.517	0.521
XGBoost-GDBT (Holdout)	0.290	0.376	0.427	0.466	0.485	0.498	0.508	0.514	0.520	0.523	0.528
XGBoost-DART (TSCV)	0.280	0.363	0.410	0.444	0.466	0.481	0.491	0.499	0.504	0.509	0.514
XGBoost-DART (Holdout)	0.283	0.369	0.421	0.460	0.483	0.498	0.508	0.513	0.519	0.522	0.527
LightGBM-GDBT (TSCV)	0.285	0.368	0.415	0.449	0.469	0.488	0.496	0.507	0.512	0.515	0.520
LightGBM-GDBT (Holdout)	0.289	0.377	0.431	0.466	0.485	0.504	0.514	0.519	0.526	0.532	0.537
LightGBM-DART (TSCV)	0.284	0.366	0.411	0.444	0.464	0.479	0.487	0.496	0.502	0.508	0.514
LightGBM-DART (Holdout)	0.288	0.370	0.421	0.459	0.482	0.494	0.502	0.511	0.518	0.523	0.525

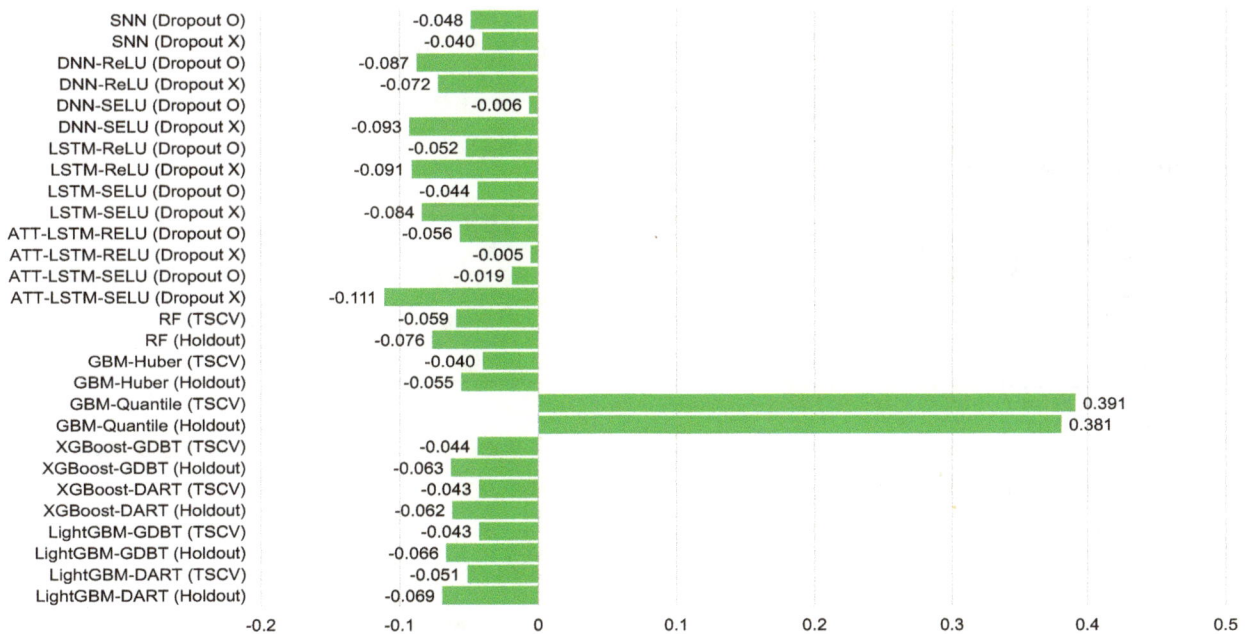

Model	Value
SNN (Dropout O)	-0.048
SNN (Dropout X)	-0.040
DNN-ReLU (Dropout O)	-0.087
DNN-ReLU (Dropout X)	-0.072
DNN-SELU (Dropout O)	-0.006
DNN-SELU (Dropout X)	-0.093
LSTM-ReLU (Dropout O)	-0.052
LSTM-ReLU (Dropout X)	-0.091
LSTM-SELU (Dropout O)	-0.044
LSTM-SELU (Dropout X)	-0.084
ATT-LSTM-RELU (Dropout O)	-0.056
ATT-LSTM-RELU (Dropout X)	-0.005
ATT-LSTM-SELU (Dropout O)	-0.019
ATT-LSTM-SELU (Dropout X)	-0.111
RF (TSCV)	-0.059
RF (Holdout)	-0.076
GBM-Huber (TSCV)	-0.040
GBM-Huber (Holdout)	-0.055
GBM-Quantile (TSCV)	0.391
GBM-Quantile (Holdout)	0.381
XGBoost-GDBT (TSCV)	-0.044
XGBoost-GDBT (Holdout)	-0.063
XGBoost-DART (TSCV)	-0.043
XGBoost-DART (Holdout)	-0.062
LightGBM-GDBT (TSCV)	-0.043
LightGBM-GDBT (Holdout)	-0.066
LightGBM-DART (TSCV)	-0.051
LightGBM-DART (Holdout)	-0.069

Figure 6. Average mean bias error for each model of Ildo-1 (MJ/m^2).

Table 8. Normalized root mean square error (NRMSE) distribution for each model for Ildo-1. A cooler color indicates a lower NRMSE value, whereas a warmer color indicates a higher NRMSE value (%).

Models	Points										
	1	2	3	4	5	6	7	8	9	10	11
SNN (Dropout O)	46.6	43.3	42.6	42.4	41.9	41.7	41.6	41.6	41.5	41.6	41.8
SNN (Dropout X)	42.9	41.7	41.4	41.1	40.8	40.5	40.3	40.3	40.4	40.5	40.7
DNN-ReLU (Dropout O)	43.6	41.9	42.0	41.6	41.1	40.8	40.7	40.9	41.1	41.4	41.8
DNN-ReLU (Dropout X)	47.4	43.5	42.4	41.7	41.2	41.0	40.8	40.7	40.7	40.7	40.8
DNN-SELU (Dropout O)	31.2	35.0	36.9	38.0	38.7	39.7	40.2	40.4	41.5	41.3	42.2
DNN-SELU (Dropout X)	43.2	42.2	42.4	42.2	42.0	41.7	41.5	41.5	41.7	41.9	42.4
LSTM-ReLU (Dropout O)	48.2	43.5	42.4	42.1	41.8	41.4	41.1	40.9	40.8	41.0	41.2
LSTM-ReLU (Dropout X)	55.9	48.7	47.2	46.5	46.1	45.9	45.5	45.8	45.3	45.4	45.6
LSTM-SELU (Dropout O)	46.7	43.1	42.3	41.6	41.2	40.9	40.8	40.9	41.1	41.3	41.5
LSTM-SELU (Dropout X)	50.5	45.0	44.3	44.0	43.9	43.7	43.1	43.3	43.5	43.7	43.9
ATT-LSTM-RELU (Dropout O)	30.2	34.2	35.4	36.9	37.9	38.2	39.0	39.6	40.1	40.9	41.9
ATT-LSTM-RELU (Dropout X)	30.4	34.0	36.3	37.0	37.7	39.4	39.5	40.7	42.0	42.1	43.1
ATT-LSTM-SELU (Dropout O)	26.1	31.4	34.2	36.1	36.8	38.0	39.8	39.1	40.3	41.2	42.0
ATT-LSTM-SELU (Dropout X)	28.3	32.9	34.2	35.8	37.7	40.4	44.2	42.7	46.6	47.6	44.5
RF (TSCV)	24.0	30.0	33.4	35.7	37.4	38.9	40.1	41.0	42.0	43.0	44.1
RF (Holdout)	24.5	30.7	34.3	36.7	38.5	39.9	41.0	41.9	42.8	43.8	44.8
GBM-Huber (TSCV)	22.6	29.3	33.1	35.8	37.5	38.9	39.6	40.3	40.8	41.2	41.4
GBM-Huber (Holdout)	22.8	29.9	33.9	36.9	39.0	39.8	40.9	41.4	41.8	42.1	42.3
GBM-Quantile (TSCV)	32.7	41.9	47.2	50.7	52.7	54.2	55.2	56.7	58.2	58.8	59.5
GBM-Quantile (Holdout)	32.5	41.6	46.4	50.6	52.4	53.4	54.8	56.2	57.8	58.8	59.7
XGBoost-GDBT (TSCV)	22.9	29.4	33.1	35.9	37.4	38.5	39.3	40.1	40.6	41.1	41.4
XGBoost-GDBT (Holdout)	23.1	29.9	33.9	37.0	38.5	39.6	40.3	40.8	41.4	41.6	42.0
XGBoost-DART (TSCV)	22.3	28.9	32.6	35.3	37.0	38.2	39.0	39.7	40.1	40.5	40.8
XGBoost-DART (Holdout)	22.5	29.3	33.5	36.5	38.4	39.6	40.4	40.8	41.3	41.5	41.9
LightGBM-GDBT (TSCV)	22.7	29.3	33.0	35.7	37.2	38.8	39.5	40.3	40.7	40.9	41.4
LightGBM-GDBT (Holdout)	23.0	29.9	34.2	37.0	38.5	40.0	40.9	41.2	41.8	42.3	42.7
LightGBM-DART (TSCV)	22.5	29.1	32.7	35.3	36.9	38.1	38.7	39.4	39.9	40.4	40.9
LightGBM-DART (Holdout)	22.9	29.4	33.5	36.5	38.3	39.3	39.9	40.6	41.1	41.6	41.8

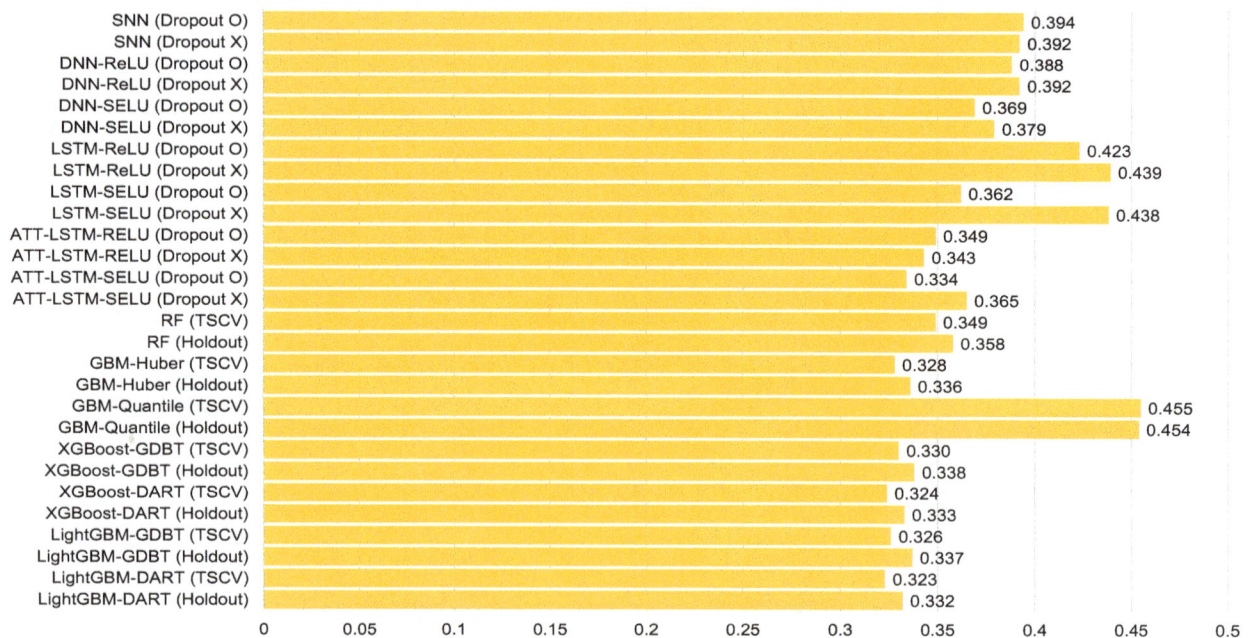

Figure 7. Average mean absolute error for each model of Ildo-1 (MJ/m^2).

Model	Value
SNN (Dropout O)	0.394
SNN (Dropout X)	0.392
DNN-ReLU (Dropout O)	0.388
DNN-ReLU (Dropout X)	0.392
DNN-SELU (Dropout O)	0.369
DNN-SELU (Dropout X)	0.379
LSTM-ReLU (Dropout O)	0.423
LSTM-ReLU (Dropout X)	0.439
LSTM-SELU (Dropout O)	0.362
LSTM-SELU (Dropout X)	0.438
ATT-LSTM-RELU (Dropout O)	0.349
ATT-LSTM-RELU (Dropout X)	0.343
ATT-LSTM-SELU (Dropout O)	0.334
ATT-LSTM-SELU (Dropout X)	0.365
RF (TSCV)	0.349
RF (Holdout)	0.358
GBM-Huber (TSCV)	0.328
GBM-Huber (Holdout)	0.336
GBM-Quantile (TSCV)	0.455
GBM-Quantile (Holdout)	0.454
XGBoost-GDBT (TSCV)	0.330
XGBoost-GDBT (Holdout)	0.338
XGBoost-DART (TSCV)	0.324
XGBoost-DART (Holdout)	0.333
LightGBM-GDBT (TSCV)	0.326
LightGBM-GDBT (Holdout)	0.337
LightGBM-DART (TSCV)	0.323
LightGBM-DART (Holdout)	0.332

Table 9. Mean bias error (MBE) distribution for each model of Gosan-ri (MJ/m^2).

Models	Points										
	1	**2**	**3**	**4**	**5**	**6**	**7**	**8**	**9**	**10**	**11**
SNN (Dropout O)	−0.04	−0.01	−0.02	−0.02	−0.02	−0.02	−0.02	−0.02	−0.02	−0.02	−0.02
SNN (Dropout X)	0.01	0.01	−0.01	−0.02	−0.02	−0.02	−0.01	0	0	0.01	0.02
DNN-ReLU (Dropout O)	−0.01	0	0.01	0.02	0.02	0.02	0.02	0.02	0.02	0.03	0.03
DNN-ReLU (Dropout X)	0.04	0.07	0.05	0.04	0.03	0.03	0.04	0.04	0.04	0.04	0.04
DNN-SELU (Dropout O)	0.04	0.06	0.07	0.06	0.04	0.05	0.05	0.06	0.06	−0.01	−0.01
DNN-SELU (Dropout X)	−0.02	0.02	0.01	−0.02	−0.03	−0.04	−0.04	−0.04	−0.05	−0.05	−0.05
LSTM-ReLU (Dropout O)	−0.13	0.08	0.08	0.05	0.02	−0.01	−0.01	−0.01	0.01	0.03	0.04
LSTM-ReLU (Dropout X)	−0.16	0.02	0.03	0	−0.03	−0.05	−0.05	−0.05	−0.40	−0.03	−0.02
LSTM-SELU (Dropout O)	0	0.03	0	−0.02	−0.04	−0.04	−0.04	−0.04	−0.04	−0.04	−0.04
LSTM-SELU (Dropout X)	−0.05	0.04	0.02	0	−0.02	−0.03	−0.03	−0.03	−0.02	−0.02	−0.02
ATT-LSTM-RELU (Dropout O)	0.02	−0.07	−0.01	0.03	0.04	0.13	0.16	0.10	−0.02	0.06	0
ATT-LSTM-RELU (Dropout X)	−0.05	−0.09	−0.08	−0.16	−0.24	−0.20	−0.25	−0.15	−0.24	−0.28	−0.25
ATT-LSTM-SELU (Dropout O)	0.05	0.05	0.01	−0.02	−0.01	0.01	−0.08	−0.11	−0.14	−0.06	−0.02
ATT-LSTM-SELU (Dropout X)	−0.04	−0.06	−0.01	−0.02	−0.04	−0.02	−0.02	0	−0.01	0.03	−0.01
RF (TSCV)	0	0	0	0	0	0	0	0.01	0.01	0.02	0.02
RF (Holdout)	0	0	0	0	0	0	0.01	0.01	0.02	0.02	0.03
GBM-Huber (TSCV)	−0.01	−0.01	−0.01	−0.01	−0.01	0	0	0	0	0	0
GBM-Huber (Holdout)	−0.01	−0.01	−0.01	−0.01	−0.01	0	0.01	0.01	0.01	0.01	0.01
GBM-Quantile (TSCV)	0.37	0.46	0.53	0.58	0.60	0.62	0.63	0.65	0.66	0.67	0.68
GBM-Quantile (Holdout)	0.26	0.32	0.36	0.39	0.42	0.41	0.43	0.44	0.45	0.47	0.46
XGBoost-GDBT (TSCV)	−0.01	−0.01	−0.01	−0.01	−0.01	−0.01	−0.01	0	−0.01	0	0
XGBoost-GDBT (Holdout)	−0.01	−0.01	−0.01	−0.01	−0.01	−0.01	0	0	0	0	0
XGBoost-DART (TSCV)	0	−0.01	−0.01	−0.01	0	0	0	0	0	0	0
XGBoost-DART (Holdout)	0	−0.01	0	0	0	0.01	0.01	0.01	0.01	0.01	0.01
LightGBM-GDBT (TSCV)	−0.01	−0.01	0	0	0	0	0	0	0	0	0
LightGBM-GDBT (Holdout)	0	−0.01	0	0	0.01	0.01	0.01	0.01	0.01	0.01	0.01
LightGBM-DART (TSCV)	−0.01	−0.02	−0.01	−0.01	−0.01	−0.01	−0.01	0	−0.01	0	0
LightGBM-DART (Holdout)	−0.01	−0.02	−0.01	−0.01	−0.01	0	0.01	0	0	0	0

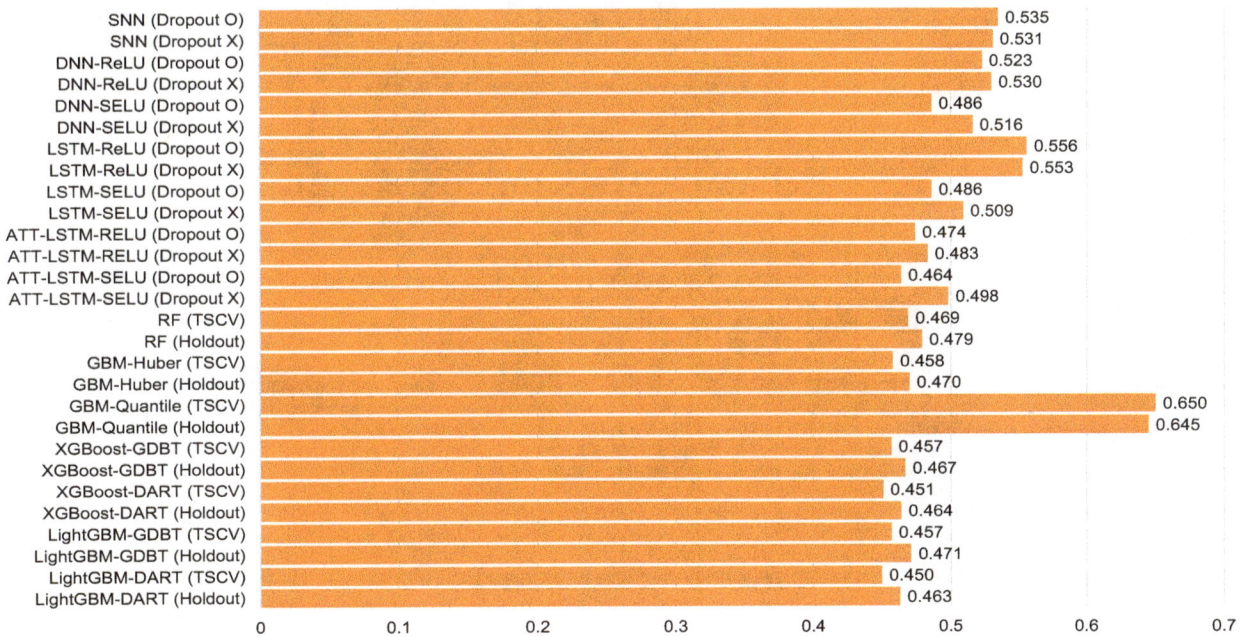

Figure 8. Average root mean square error for each model of Ildo-1 (MJ/m^2).

Table 10. Mean absolute error (MAE) distribution for each model of Gosan-ri. A cooler color indicates a lower MAE value, whereas a warmer color indicates a higher MAE value (MJ/m^2).

Models	Points										
	1	2	3	4	5	6	7	8	9	10	11
SNN (Dropout O)	0.426	0.399	0.391	0.387	0.381	0.377	0.373	0.374	0.375	0.374	0.373
SNN (Dropout X)	0.388	0.375	0.374	0.370	0.364	0.360	0.358	0.357	0.357	0.356	0.355
DNN-ReLU (Dropout O)	0.409	0.381	0.377	0.373	0.370	0.369	0.367	0.367	0.369	0.367	0.374
DNN-ReLU (Dropout X)	0.421	0.399	0.387	0.394	0.384	0.379	0.371	0.388	0.401	0.399	0.389
DNN-SELU (Dropout O)	0.278	0.306	0.325	0.343	0.355	0.365	0.367	0.379	0.388	0.387	0.397
DNN-SELU (Dropout X)	0.409	0.381	0.380	0.375	0.370	0.367	0.364	0.363	0.365	0.367	0.367
LSTM-ReLU (Dropout O)	0.345	0.358	0.359	0.362	0.370	0.377	0.386	0.376	0.376	0.378	0.380
LSTM-ReLU (Dropout X)	0.377	0.399	0.401	0.408	0.415	0.420	0.422	0.421	0.421	0.421	0.420
LSTM-SELU (Dropout O)	0.298	0.312	0.320	0.340	0.352	0.367	0.371	0.383	0.388	0.389	0.390
LSTM-SELU (Dropout X)	0.340	0.351	0.352	0.357	0.360	0.361	0.369	0.372	0.371	0.371	0.371
ATT-LSTM-RELU (Dropout O)	0.221	0.265	0.289	0.317	0.332	0.365	0.377	0.365	0.366	0.371	0.373
ATT-LSTM-RELU (Dropout X)	0.240	0.295	0.311	0.346	0.385	0.383	0.409	0.384	0.416	0.445	0.442
ATT-LSTM-SELU (Dropout O)	0.219	0.262	0.289	0.316	0.327	0.340	0.354	0.363	0.376	0.378	0.383
ATT-LSTM-SELU (Dropout X)	0.231	0.273	0.297	0.310	0.327	0.338	0.347	0.359	0.367	0.376	0.380
RF (TSCV)	0.176	0.227	0.256	0.277	0.293	0.307	0.319	0.331	0.343	0.353	0.363
RF (Holdout)	0.180	0.230	0.260	0.280	0.297	0.310	0.324	0.335	0.346	0.356	0.364
GBM-Huber (TSCV)	0.164	0.224	0.259	0.284	0.302	0.314	0.323	0.330	0.340	0.347	0.352
GBM-Huber (Holdout)	0.165	0.226	0.264	0.290	0.306	0.316	0.324	0.334	0.340	0.350	0.355
GBM-Quantile (TSCV)	0.289	0.359	0.410	0.453	0.468	0.481	0.492	0.505	0.510	0.522	0.530
GBM-Quantile (Holdout)	0.297	0.367	0.418	0.457	0.484	0.483	0.504	0.511	0.523	0.541	0.534
XGBoost-GDBT (TSCV)	0.168	0.222	0.254	0.280	0.295	0.306	0.314	0.324	0.334	0.341	0.346
XGBoost-GDBT (Holdout)	0.170	0.224	0.258	0.285	0.299	0.308	0.316	0.325	0.334	0.343	0.349
XGBoost-DART (TSCV)	0.159	0.216	0.249	0.275	0.291	0.302	0.312	0.321	0.329	0.339	0.345
XGBoost-DART (Holdout)	0.161	0.220	0.255	0.281	0.296	0.307	0.316	0.325	0.333	0.341	0.348
LightGBM-GDBT (TSCV)	0.167	0.225	0.259	0.283	0.300	0.312	0.320	0.331	0.336	0.343	0.353
LightGBM-GDBT (Holdout)	0.169	0.228	0.265	0.290	0.305	0.316	0.325	0.336	0.343	0.348	0.355
LightGBM-DART (TSCV)	0.163	0.218	0.252	0.276	0.293	0.304	0.314	0.322	0.332	0.341	0.347
LightGBM-DART (Holdout)	0.165	0.222	0.259	0.283	0.297	0.311	0.318	0.328	0.336	0.344	0.352

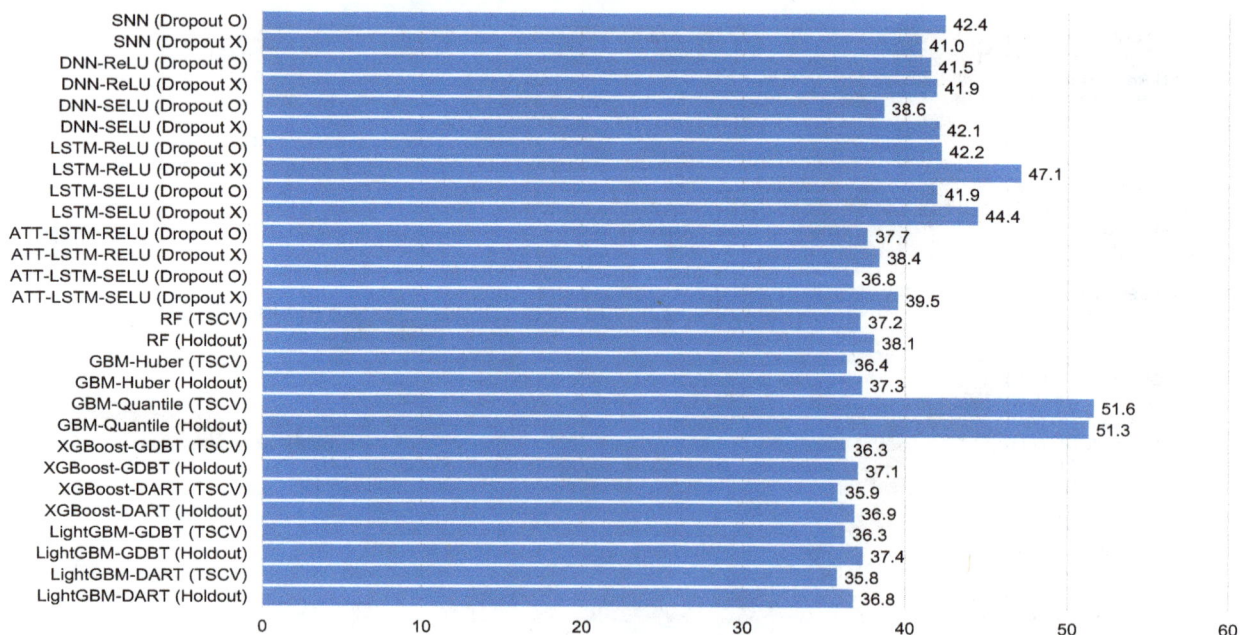

Figure 9. Average normalized root mean square error for each model of Ildo-1 (%).

Table 11. Root mean square error (RMSE) distribution for each model for Gosan-ri. A cooler color indicates a lower RMSE value, whereas a warmer color indicates a higher RMSE value (MJ/m^2).

Models	Points										
	1	2	3	4	5	6	7	8	9	10	11
SNN (Dropout O)	0.559	0.533	0.528	0.531	0.530	0.527	0.525	0.526	0.523	0.522	0.521
SNN (Dropout X)	0.518	0.518	0.520	0.517	0.512	0.506	0.502	0.501	0.499	0.497	0.497
DNN-ReLU (Dropout O)	0.522	0.519	0.519	0.512	0.506	0.500	0.496	0.495	0.495	0.496	0.497
DNN-ReLU (Dropout X)	0.552	0.522	0.523	0.522	0.520	0.520	0.521	0.521	0.521	0.522	0.526
DNN-SELU (Dropout O)	0.366	0.407	0.431	0.452	0.469	0.484	0.490	0.503	0.516	0.514	0.524
DNN-SELU (Dropout X)	0.532	0.519	0.521	0.517	0.514	0.511	0.509	0.509	0.510	0.511	0.511
LSTM-ReLU (Dropout O)	0.400	0.420	0.437	0.442	0.451	0.462	0.473	0.488	0.501	0.511	0.511
LSTM-ReLU (Dropout X)	0.455	0.462	0.477	0.484	0.491	0.500	0.511	0.531	0.547	0.567	0.580
LSTM-SELU (Dropout O)	0.375	0.380	0.410	0.444	0.467	0.486	0.492	0.510	0.526	0.533	0.538
LSTM-SELU (Dropout X)	0.406	0.417	0.451	0.471	0.490	0.496	0.500	0.508	0.522	0.547	0.555
ATT-LSTM-RELU (Dropout O)	0.306	0.370	0.403	0.435	0.454	0.485	0.497	0.495	0.508	0.515	0.517
ATT-LSTM-RELU (Dropout X)	0.338	0.416	0.445	0.490	0.541	0.539	0.569	0.527	0.566	0.586	0.581
ATT-LSTM-SELU (Dropout O)	0.297	0.360	0.401	0.436	0.452	0.473	0.500	0.515	0.529	0.524	0.526
ATT-LSTM-SELU (Dropout X)	0.322	0.380	0.415	0.432	0.456	0.470	0.485	0.499	0.511	0.522	0.524
RF (TSCV)	0.257	0.322	0.359	0.387	0.409	0.426	0.441	0.457	0.470	0.481	0.490
RF (Holdout)	0.260	0.326	0.364	0.391	0.413	0.431	0.449	0.464	0.475	0.485	0.494
GBM-Huber (TSCV)	0.251	0.324	0.368	0.400	0.425	0.442	0.457	0.457	0.466	0.477	0.491
GBM-Huber (Holdout)	0.252	0.327	0.373	0.408	0.430	0.446	0.457	0.471	0.479	0.489	0.491
GBM-Quantile (TSCV)	0.388	0.480	0.549	0.601	0.625	0.644	0.659	0.682	0.687	0.704	0.711
GBM-Quantile (Holdout)	0.397	0.488	0.556	0.607	0.641	0.648	0.673	0.685	0.706	0.723	0.720
XGBoost-GDBT (TSCV)	0.252	0.320	0.362	0.396	0.416	0.430	0.445	0.458	0.469	0.496	0.480
XGBoost-GDBT (Holdout)	0.253	0.321	0.367	0.403	0.423	0.435	0.448	0.461	0.472	0.480	0.483
XGBoost-DART (TSCV)	0.244	0.314	0.357	0.391	0.414	0.428	0.443	0.457	0.465	0.474	0.480
XGBoost-DART (Holdout)	0.246	0.318	0.364	0.399	0.419	0.434	0.448	0.462	0.472	0.479	0.484
LightGBM-GDBT (TSCV)	0.251	0.323	0.367	0.399	0.422	0.440	0.454	0.468	0.474	0.482	0.493
LightGBM-GDBT (Holdout)	0.254	0.328	0.375	0.408	0.429	0.448	0.461	0.477	0.486	0.489	0.495
LightGBM-DART (TSCV)	0.247	0.318	0.360	0.391	0.414	0.431	0.445	0.458	0.470	0.477	0.482
LightGBM-DART (Holdout)	0.249	0.322	0.367	0.400	0.420	0.438	0.452	0.465	0.475	0.484	0.490

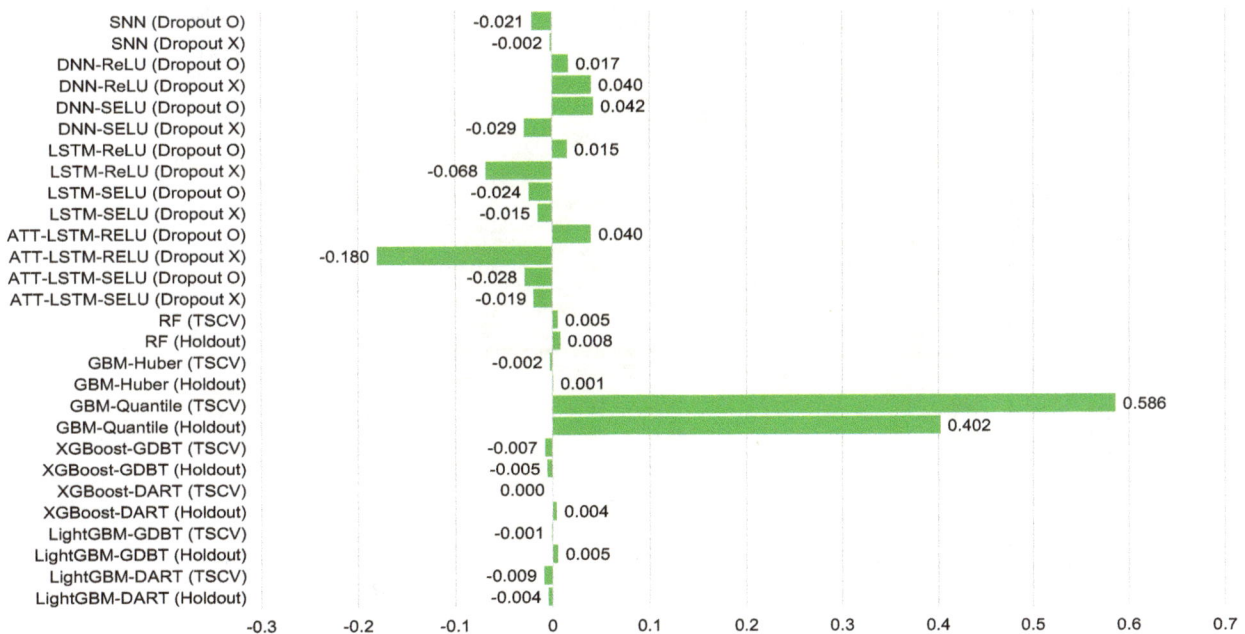

Figure 10. Average mean bias error for each model for Gosan-ri (MJ/m^2).

Table 12. Normalized root mean square error (NRMSE) distribution for each model for Gosan-ri. A cooler color indicates a lower NRMSE value, whereas a warmer color indicates a higher NRMSE value (%).

Models	Points										
	1	**2**	**3**	**4**	**5**	**6**	**7**	**8**	**9**	**10**	**11**
SNN (Dropout O)	53.4	51.0	50.5	50.7	50.7	50.4	50.2	50.3	49.9	49.9	49.8
SNN (Dropout X)	49.5	49.5	49.7	49.4	48.9	48.4	48.0	47.8	47.7	47.5	47.4
DNN-ReLU (Dropout O)	52.7	49.9	50.0	49.9	49.7	49.7	49.8	49.7	49.7	49.8	50.2
DNN-ReLU (Dropout X)	49.9	49.7	49.6	48.9	48.3	47.8	47.4	47.3	47.3	47.3	47.4
DNN-SELU (Dropout O)	35.1	39.0	41.3	43.2	44.9	46.3	46.9	48.2	49.4	49.2	50.2
DNN-SELU (Dropout X)	50.8	49.6	49.7	49.3	49.1	48.9	48.6	48.6	48.7	48.8	48.7
LSTM-ReLU (Dropout O)	56.6	50.6	50.6	49.7	48.9	48.7	48.4	48.3	48.1	48.1	48.0
LSTM-ReLU (Dropout X)	60.6	53.7	54.0	53.1	52.4	52.4	52.2	52.5	51.8	52.0	52.0
LSTM-SELU (Dropout O)	49.7	49.1	48.9	49.0	49.2	49.4	49.6	49.8	49.8	49.7	49.6
LSTM-SELU (Dropout X)	54.4	51.3	51.6	51.5	51.2	51.4	51.4	51.5	51.1	51.0	51.2
ATT-LSTM-RELU (Dropout O)	29.3	35.4	38.6	41.6	43.5	46.4	47.6	47.4	48.6	49.3	49.5
ATT-LSTM-RELU (Dropout X)	26.9	33.1	35.3	38.9	43.0	42.8	45.2	41.9	45.0	46.6	46.2
ATT-LSTM-SELU (Dropout O)	28.5	34.5	38.4	41.7	43.3	45.3	47.8	49.3	50.7	50.1	50.3
ATT-LSTM-SELU (Dropout X)	30.8	36.4	39.8	41.3	43.6	45.0	46.4	47.7	48.8	50.0	50.1
RF (TSCV)	24.6	30.8	34.4	37.1	39.1	40.8	42.3	43.8	45.0	46.1	47.0
RF (Holdout)	24.9	31.2	34.8	37.4	39.6	41.3	43.0	44.4	45.5	46.5	47.3
GBM-Huber (TSCV)	24.0	31.0	35.2	38.3	40.7	42.3	43.7	44.6	45.7	46.5	47.0
GBM-Huber (Holdout)	24.1	31.3	35.8	39.0	41.2	42.7	43.7	45.1	45.8	46.9	47.1
GBM-Quantile (TSCV)	37.2	46.0	52.6	57.6	59.9	61.7	63.1	65.3	65.8	67.5	68.2
GBM-Quantile (Holdout)	38.1	46.7	53.2	58.1	61.4	62.0	64.4	65.6	67.6	69.2	68.9
XGBoost-GDBT (TSCV)	24.1	30.6	34.6	37.9	39.9	41.2	42.6	43.9	44.9	45.6	46.0
XGBoost-GDBT (Holdout)	24.2	30.8	35.1	38.6	40.5	41.7	42.9	44.2	45.2	45.9	46.3
XGBoost-DART (TSCV)	23.4	30.1	34.2	37.5	39.6	41.0	42.4	43.7	44.6	45.4	46.0
XGBoost-DART (Holdout)	23.6	30.4	34.8	38.2	40.1	41.6	42.9	44.2	45.2	45.8	46.4
LightGBM-GDBT (TSCV)	24.1	31.0	35.2	38.2	40.4	42.2	43.5	44.8	45.4	46.2	47.2
LightGBM-GDBT (Holdout)	24.3	31.4	35.9	39.1	41.1	42.9	44.1	45.6	46.6	46.8	47.4
LightGBM-DART (TSCV)	23.6	30.4	34.5	37.4	39.6	41.2	42.6	43.8	45.0	45.7	46.1
LightGBM-DART (Holdout)	23.8	30.9	35.2	38.3	40.2	42.0	43.3	44.5	45.5	46.3	46.9

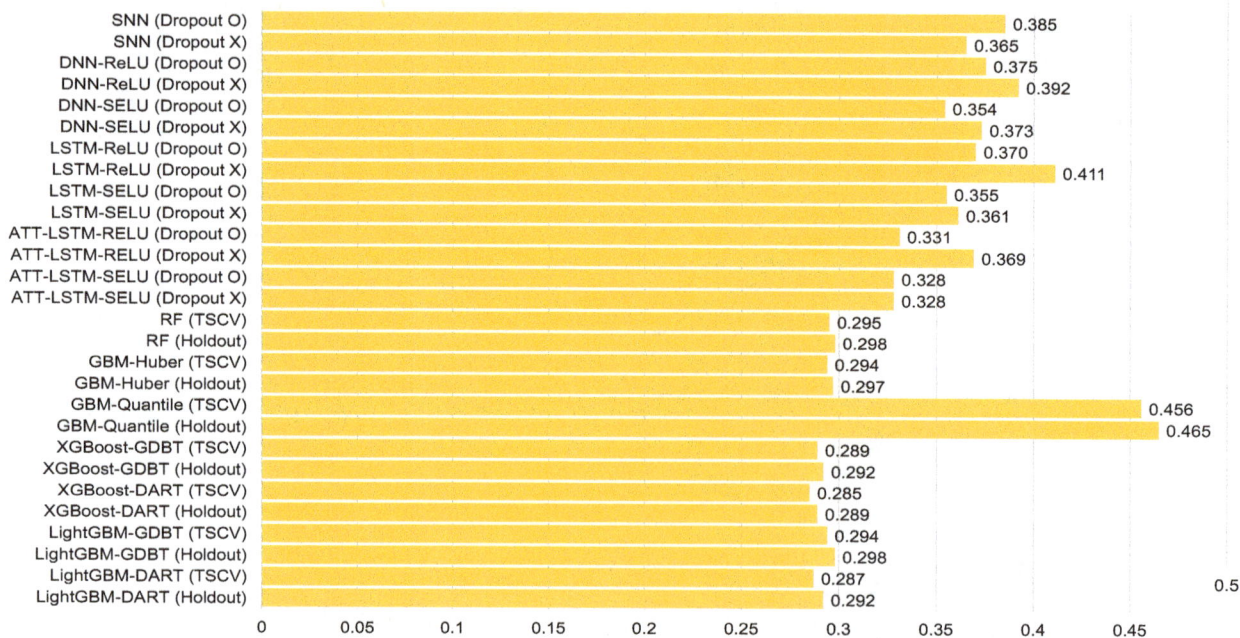

Figure 11. Average mean absolute error for each model for Gosan-ri (MJ/m^2).

Table 13. Average mean bias error, mean absolute error, root mean square error, and normalized root mean square error comparison according to the forecasting models.

Models	Ildo-1				Gosan-ri			
	MBE	MAE	RMSE	NRMSE	MBE	MAE	RMSE	NRMSE
SNN (Dropout O)	−0.048	0.394	0.535	42.4	−0.021	0.530	0.385	50.6
SNN (Dropout X)	−0.040	0.392	0.531	41.0	−0.002	0.508	0.365	48.5
DNN-ReLU (Dropout O)	−0.087	0.388	0.523	41.5	0.017	0.505	0.375	50.1
DNN-ReLU (Dropout X)	−0.072	0.392	0.530	41.9	0.040	0.525	0.392	48.3
DNN-SELU (Dropout O)	−0.006	0.369	0.486	38.6	0.042	0.469	0.354	44.9
DNN-SELU (Dropout X)	−0.093	0.379	0.516	42.1	−0.029	0.515	0.373	49.2
LSTM-ReLU (Dropout O)	−0.052	0.423	0.556	42.2	0.015	0.463	0.370	49.6
LSTM-ReLU (Dropout X)	−0.091	0.439	0.553	47.1	−0.068	0.510	0.411	53.3
LSTM-SELU (Dropout O)	−0.044	0.362	0.486	41.9	−0.024	0.469	0.355	49.4
LSTM-SELU (Dropout X)	−0.084	0.438	0.509	44.4	−0.015	0.488	0.361	51.6
ATT-LSTM-RELU (Dropout O)	−0.056	0.349	0.474	37.7	0.040	0.453	0.331	43.4
ATT-LSTM-RELU (Dropout X)	−0.005	0.343	0.483	38.4	−0.180	0.509	0.369	40.4
ATT-LSTM-SELU (Dropout O)	−0.019	0.334	0.464	36.8	−0.028	0.456	0.328	43.6
ATT-LSTM-SELU (Dropout X)	−0.111	0.365	0.498	39.5	−0.019	0.456	0.328	43.6
RF (TSCV)	−0.059	0.349	0.469	37.2	0.005	0.409	0.295	39.2
RF (Holdout)	−0.076	0.358	0.479	38.1	0.008	0.414	0.298	39.6
GBM-Huber (TSCV)	−0.040	0.328	0.458	36.4	−0.002	0.417	0.294	39.9
GBM-Huber (Holdout)	−0.055	0.336	0.470	37.3	0.001	0.420	0.297	40.2
GBM-Quantile (TSCV)	0.391	0.455	0.650	51.6	0.586	0.612	0.456	58.6
GBM-Quantile (Holdout)	0.381	0.454	0.645	51.3	0.402	0.622	0.465	59.6
XGBoost-GDBT (TSCV)	−0.044	0.330	0.457	36.3	−0.007	0.411	0.289	39.2
XGBoost-GDBT (Holdout)	−0.063	0.338	0.467	37.1	−0.005	0.413	0.292	39.6
XGBoost-DART (TSCV)	−0.043	0.324	0.451	35.9	0.000	0.406	0.285	38.9
XGBoost-DART (Holdout)	−0.062	0.333	0.464	36.9	0.004	0.411	0.289	39.4
LightGBM-GDBT (TSCV)	−0.043	0.326	0.457	36.3	−0.001	0.416	0.294	39.8
LightGBM-GDBT (Holdout)	−0.066	0.337	0.471	37.4	0.005	0.423	0.298	40.5
LightGBM-DART (TSCV)	−0.051	0.323	0.450	35.8	−0.009	0.408	0.287	39.1
LightGBM-DART (Holdout)	−0.069	0.332	0.463	36.8	−0.004	0.415	0.292	39.7

Notes: MBE: mean bias error; MAE: mean absolute error; RMSE: root mean square error; NRMSE: normalized RMSE.

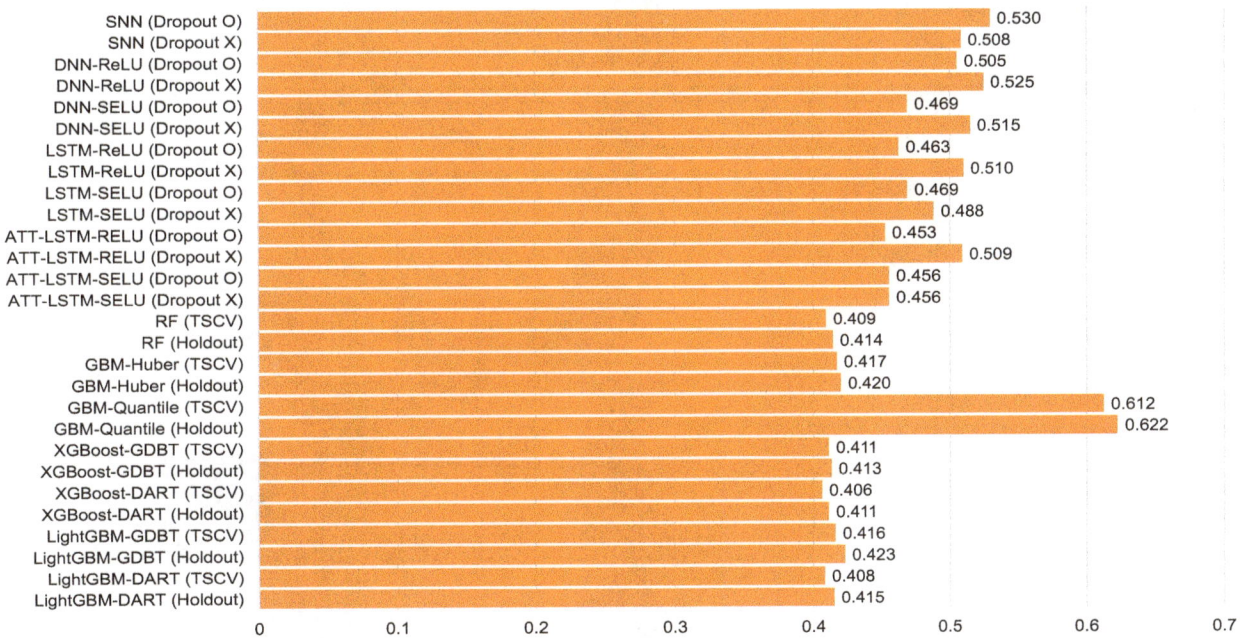

Figure 12. Average root mean square error for each model for Gosan-ri (MJ/m^2).

Feature importance is a measure of variable importance when data have obtained a subset of all features. Feature importance can be determined from logistic regression or tree-based models. We determined the feature importance of our model, LightGBM-DART (TSCV), at each test point (one month) according to the TSCV cycle. Figures 14 and 15 present a heat map graph that reveals the feature importance of the input variables mentioned in Table 2 for both regions. The variable importance values are exhibited in the range of 0 to 1 using minimum–maximum normalization to help readers understand. From the table, we confirmed that the day number of the year ($Date_X$ and $Date_y$) consistently exhibited high feature importance, and the temperature, humidity, and wind speed, among the meteorological information, presented high feature importance. In particular, the importance of humidity increased over time.

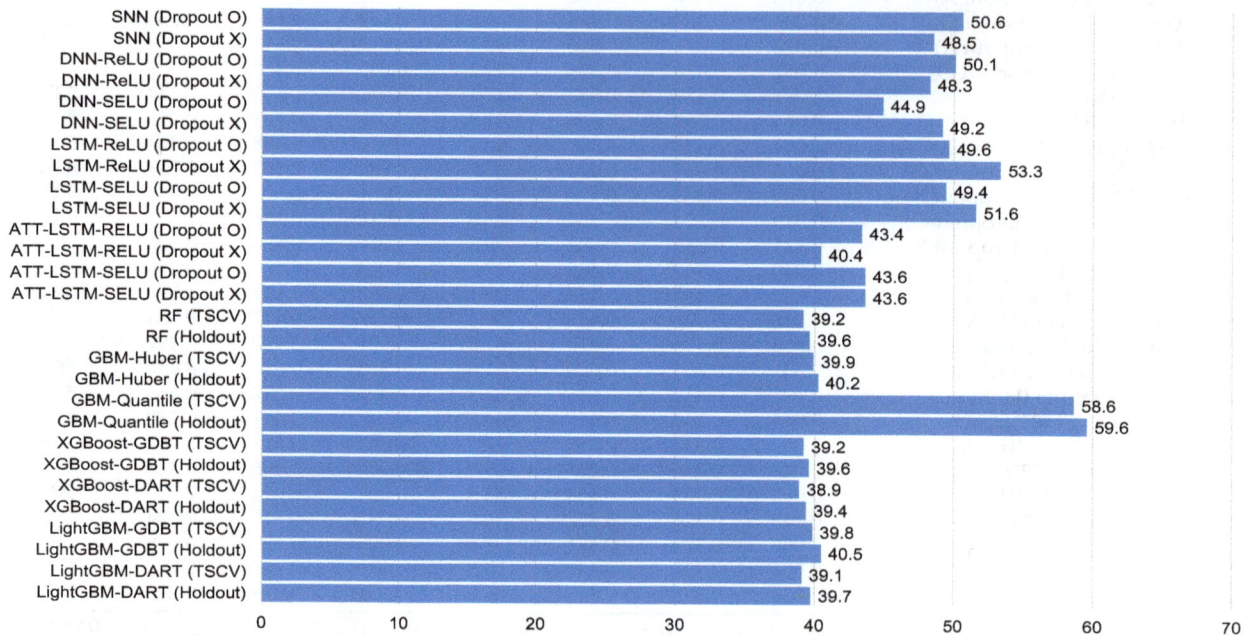

Figure 13. Average normalized root mean square error for each model for Gosan-ri (%).

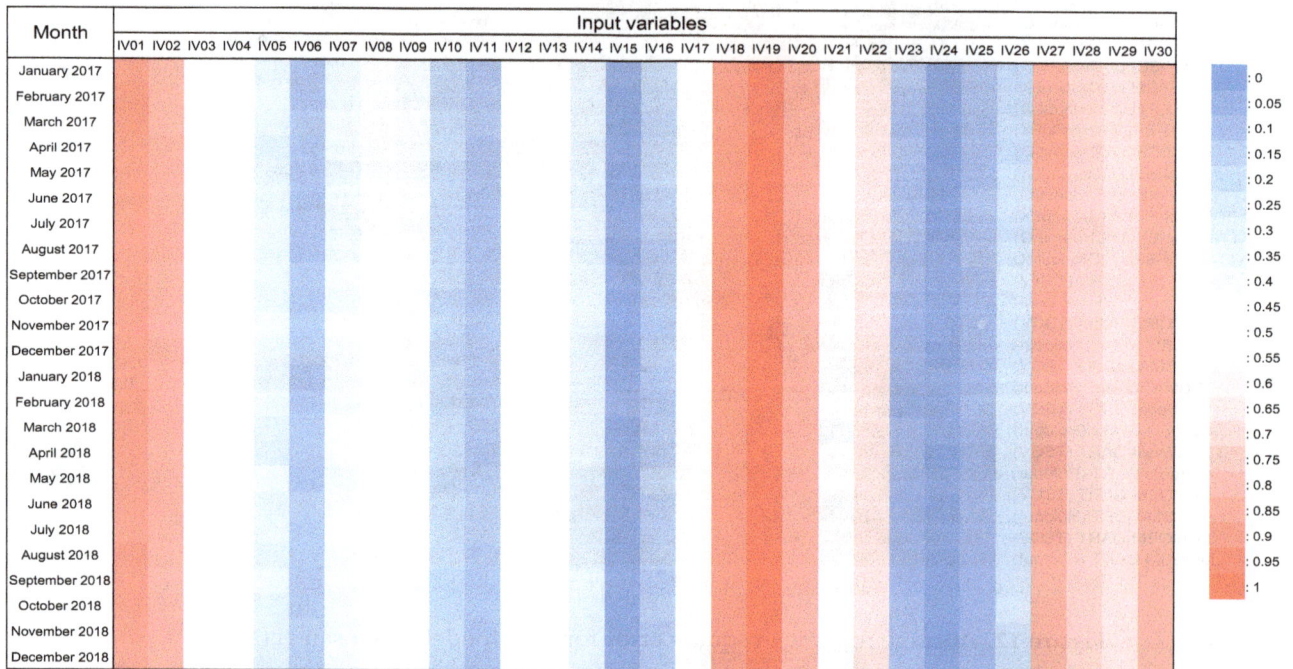

Figure 14. Result of feature importance via time-series cross-validation using the input variables in Table 2 for Ildo-1. A cooler color indicates a lower feature importance value, whereas a warmer color indicates a higher feature importance value.

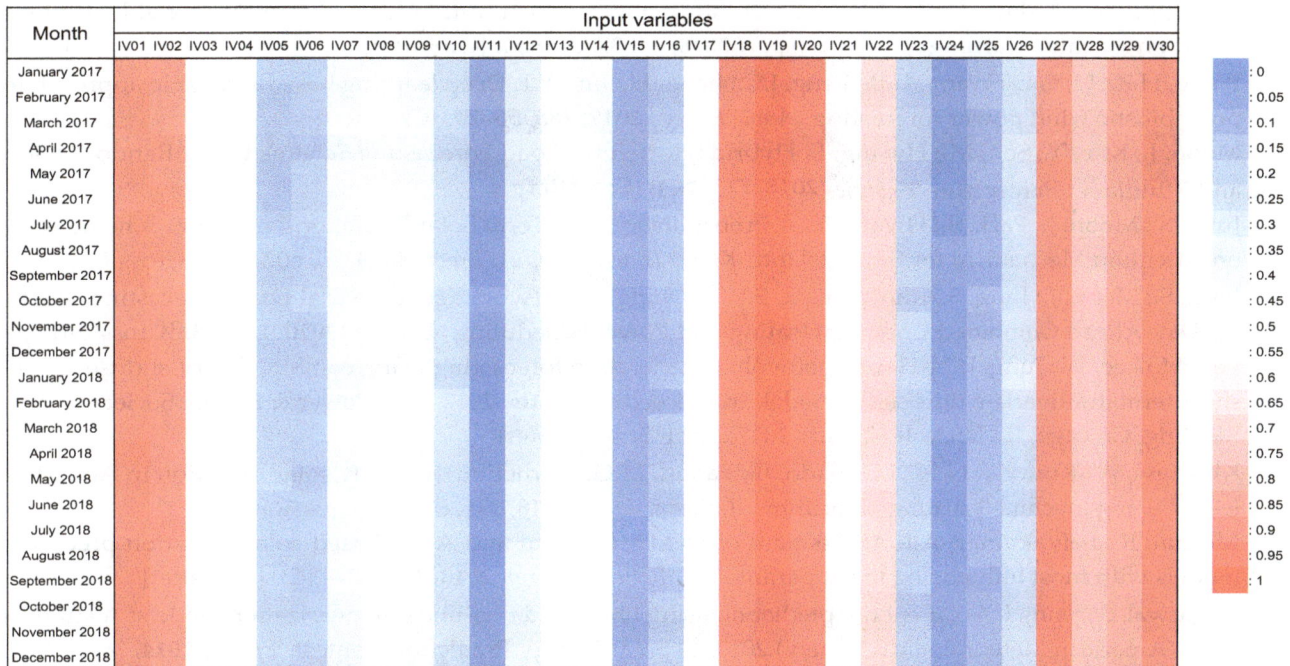

Figure 15. Result of feature importance via time-series cross-validation using the input variables in Table 2 for Gosan-ri. A cooler color indicates a lower feature importance value, whereas a warmer color indicates a higher feature importance value.

4. Conclusions

In this paper, we proposed an MSA global solar radiation forecasting method based on LightGBM. To do this, we first configured 330 input variables considering the time and weather information provided by KMA to forecast the global solar radiation at multiple time points over the next 24 h. Then, we constructed a LightGBM-based forecasting model with DART boosting. To evaluate the performance of our model, we implemented diverse ensemble-based models and deep learning-based models and compared their performance using global solar radiation data from Jeju Island. From the comparison, we confirmed that our model exhibited better forecasting performance than other methods. We plan to conduct a forecasting model using only historical global solar radiation data in the future to provide accurate global solar radiation forecasting in regions where meteorological information is not provided. We will also conduct smart grid scheduling based on photovoltaic forecasting.

Author Contributions: Conceptualization, J.P. and J.M.; methodology, J.P.; software, J.P. and S.J.; validation, J.P. and S.J.; formal analysis, J.M.; investigation, S.J.; data curation, J.P. and S.J.; writing—original draft preparation, J.P. and J.M.; writing—review and editing, E.H.; visualization, J.M.; supervision, E.H.; project administration, E.H.; funding acquisition, E.H. All authors have read and agreed to the published version of the manuscript.

References

1. Moon, J.; Kim, K.-H.; Kim, Y.; Hwang, E. A Short-Term Electric Load Forecasting Scheme Using 2-Stage Predictive Analytics. In Proceedings of the IEEE International Conference on Big Data and Smart Computing (BigComp), Shanghai, China, 15–17 January 2018; pp. 219–226. [CrossRef]
2. Abdel-Nasser, M.; Mahmoud, K. Accurate photovoltaic power forecasting models using deep LSTM-RNN. *Neural Comput. Appl.* **2019**, *31*, 2727–2740. [CrossRef]
3. Son, M.; Moon, J.; Jung, S.; Hwang, E. A Short-Term Load Forecasting Scheme Based on Auto-Encoder and Random Forest. In Proceedings of the International Conference on Applied Physics, System Science and Computers, Dubrovnik, Croatia, 26–28 September 2018; pp. 138–144. [CrossRef]

4. Moon, J.; Park, J.; Hwang, E.; Jun, S. Forecasting power consumption for higher educational institutions based on machine learning. *J. Supercomput.* **2018**, *74*, 3778–3800. [CrossRef]
5. Wang, H.Z.; Li, G.Q.; Wang, G.B.; Peng, J.C.; Jiang, H.; Liu, Y.T. Deep learning based ensemble approach for probabilistic wind power forecasting. *Appl. Energy* **2017**, *188*, 56–70. [CrossRef]
6. Moon, J.; Kim, Y.; Son, M.; Hwang, E. Hybrid Short-Term Load Forecasting Scheme Using Random Forest and Multilayer Perceptron. *Energies* **2018**, *11*, 3283. [CrossRef]
7. Jung, S.; Moon, J.; Park, S.; Hwang, E. A Probabilistic Short-Term Solar Radiation Prediction Scheme Based on Attention Mechanism for Smart Island. *KIISE Trans. Comput. Pract.* **2019**, *25*, 602–609. [CrossRef]
8. Park, S.; Moon, J.; Jung, S.; Rho, S.; Baik, S.W.; Hwang, E. A Two-Stage Industrial Load Forecasting Scheme for Day-Ahead Combined Cooling, Heating and Power Scheduling. *Energies* **2020**, *13*, 443. [CrossRef]
9. Lee, M.; Lee, W.; Jung, J. 24-Hour photovoltaic generation forecasting using combined very-short-term and short-term multivariate time series model. In Proceedings of the 2017 IEEE Power & Energy Society General Meeting, Chicago, IL, USA, 16–20 July 2017; pp. 1–5. [CrossRef]
10. Khosravi, A.; Koury, R.N.N.; Machado, L.; Pabon, J.J.G. Prediction of hourly solar radiation in Abu Musa Island using machine learning algorithms. *J. Clean. Prod.* **2018**, *176*, 63–75. [CrossRef]
11. Meenal, R.; Selvakumar, A.I. Assessment of SVM, empirical and ANN based solar radiation prediction models with most influencing input parameters. *Renew. Energy* **2018**, *121*, 324–343. [CrossRef]
12. Aggarwal, S.; Saini, L. Solar energy prediction using linear and non-linear regularization models: A study on AMS (American Meteorological Society) 2013–14 Solar Energy Prediction Contest. *Energy* **2014**, *78*, 247–256. [CrossRef]
13. Yadav, A.K.; Chandel, S. Solar radiation prediction using Artificial Neural Network techniques: A review. *Renew. Sustain. Energy Rev.* **2014**, *33*, 772–781. [CrossRef]
14. Rao, K.D.V.S.K.; Premalatha, M.; Naveen, C. Analysis of different combinations of meteorological parameters in predicting the horizontal global solar radiation with ANN approach: A case study. *Renew. Sustain. Energy Rev.* **2018**, *91*, 248–258. [CrossRef]
15. Cornaro, C.; Pierro, M.; Bucci, F. Master optimization process based on neural networks ensemble for 24-h solar irradiance forecast. *Sol. Energy* **2015**, *111*, 297–312. [CrossRef]
16. Dahmani, K.; Dizene, R.; Notton, G.; Paoli, C.; Voyant, C.; Nivet, M.L. Estimation of 5-min time-step data of tilted solar global irradiation using ANN (Artificial Neural Network) model. *Energy* **2014**, *70*, 374–381. [CrossRef]
17. Leva, S.; Dolara, A.; Grimaccia, F.; Mussetta, M.; Sahin, E. Analysis and validation of 24 hours ahead neural network forecasting for photovoltaic output power. *Math. Comput. Simul.* **2017**, *131*, 88–100. [CrossRef]
18. Amrouche, B.; le Pivert, X. Artificial neural network based daily local forecasting for global solar radiation. *Appl. Energy* **2014**, *130*, 333–341. [CrossRef]
19. Kaba, K.; Sarıgül, M.; Avcı, M.; Kandırmaz, H.M. Estimation of daily global solar radiation using deep learning model. *Energy* **2018**, *162*, 126–135. [CrossRef]
20. Qing, X.; Niu, Y. Hourly day-ahead solar irradiance prediction using weather forecasts by LSTM. *Energy* **2018**, *148*, 461–468. [CrossRef]
21. Benali, L.; Notton, G.; Fouilloy, A.; Voyant, C.; Dizene, R. Solar radiation forecasting using artificial neural network and random forest methods: Application to normal beam, horizontal diffuse and global components. *Renew. Energy* **2019**, *132*, 871–884. [CrossRef]
22. Lee, J.; Wang, W.; Harrou, F.; Sun, Y. Reliable solar irradiance prediction using ensemble learning-based models: A comparative study. *Energy Conv. Manag.* **2020**, *208*, 112582. [CrossRef]
23. Rew, J.; Cho, Y.; Moon, J.; Hwang, E. Habitat Suitability Estimation Using a Two-Stage Ensemble Approach. *Remote Sens.* **2020**, *12*, 1475. [CrossRef]
24. Kim, J.; Moon, J.; Hwang, E.; Kang, P. Recurrent inception convolution neural network for multi short-term load forecasting. *Energy Build.* **2019**, *194*, 328–341. [CrossRef]
25. Jung, S.; Moon, J.; Park, S.; Rho, S.; Baik, S.W.; Hwang, E. Bagging Ensemble of Multilayer Perceptrons for Missing Electricity Consumption Data Imputation. *Sensors* **2020**, *20*, 1772. [CrossRef]
26. Kim, K.H.; Oh, J.K.-W.; Jeong, W. Study on Solar Radiation Models in South Korea for Improving Office Building Energy Performance Analysis. *Sustainability* **2016**, *8*, 589. [CrossRef]

27. Lee, M.; Park, J.; Na, S.-I.; Choi, H.S.; Bu, B.-S.; Kim, J. An Analysis of Battery Degradation in the Integrated Energy Storage System with Solar Photovoltaic Generation. *Electronics* **2020**, *9*, 701. [CrossRef]

28. Ke, G.; Meng, Q.; Finley, T.; Wang, T.; Chen, W.; Ma, W.; Ye, Q.; Liu, T. LightGBM: A highly efficient gradient boosting decision tree. In *Advances in Neural Information Processing Systems*; Morgan Kaufmann Publishers: San Mateo, CA, USA, 2017; pp. 3148–3156.

29. Rashmi, K.V.; Gilad-Bachrach, R. DART: Dropouts meet Multiple Additive Regression Trees. In Proceedings of the AISTATS, San Diego, CA, USA, 9–12 May 2015.

30. De Livera, A.M.; Hyndman, R.J.; Snyder, R.D. Forecasting time series with complex seasonal patterns using exponential smoothing. *J. Am. Stat. Assoc.* **2011**, *106*, 1513–1527. [CrossRef]

31. Breiman, L.E.O. Random Forests. *Mach. Learn.* **2001**, *45*, 5–32. [CrossRef]

32. Natekin, A.; Knoll, A. Gradient Boosting Machines: A Tutorial. *Front. Neurorobot.* **2013**, *7*, 21. [CrossRef]

33. Chen, T.; Guestrin, C. XGBoost: A scalable tree boosting system. In Proceedings of the 22nd ACM SIGKDD International Conference on Knowledge Discovery and Data Mining, San Francisco, CA, USA, 13–17 August 2016.

34. Park, S.; Moon, J.; Hwang, E. 2-Stage Electric Load Forecasting Scheme for Day-Ahead CCHP Scheduling. In Proceedings of the IEEE International Conference on Power Electronics and Drive System (PEDS), Toulouse, France, 9–12 July 2019. [CrossRef]

35. Rahman, R.; Otridge, J.; Pal, R. IntegratedMRF: Random forest-based framework for integrating prediction from different data types. *Bioinformatics* **2017**, *33*, 1407–1410. [CrossRef] [PubMed]

36. Pedregosa, F.; Varoquaux, G.; Gramfort, A.; Michel, V.; Thirion, B.; Grisel, O.; Blondel, M.; Prettenhofer, P.; Weiss, R.; Dubourg, V. Scikit-learn: Machine learning in Python. *J. Mach. Learn. Res.* **2011**, *12*, 2825–2830.

37. Feng, C.; Cui, M.; Hodge, B.-M.; Zhang, J. A data-driven multi-model methodology with deep feature selection for short-term wind forecasting. *Appl. Energy* **2017**, *190*, 1245–1257. [CrossRef]

38. Moon, J.; Park, S.; Rho, S.; Hwang, E. A comparative analysis of artificial neural network architectures for building energy consumption forecasting. *Int. J. Distrib. Sens. Netw.* **2019**, *15*. [CrossRef]

39. Moon, J.; Jung, S.; Rew, J.; Rho, S.; Hwang, E. Combination of short-term load forecasting models based on a stacking ensemble approach. *Energy Build.* **2020**, *216*, 109921. [CrossRef]

40. Hochreiter, S.; Schmidhuber, J. Long short-term memory. *Neural Comput.* **1997**, *9*, 1735–1780. [CrossRef]

41. Bahdanau, D.; Cho, K.; Bengio, Y. Neural machine translation by jointly learning to align and translate. In Proceedings of the International Conference on Learning Representations, San Diego, CA, USA, 7–9 May 2015.

42. Li, H.; Shen, Y.; Zhu, Y. Stock Price Prediction Using Attention-based Multi-Input LSTM. In Proceedings of the 10th Asian Conference on Machine Learning (ACML 2018), Beijing, China, 14–16 November 2018; pp. 454–469.

43. Klambauer, G.; Unterthiner, T.; Mayr, A.; Hochreiter, S. Self-Normalizing Neural Networks. In Proceedings of the 31st International Conference on Neural Information Processing Systems (NIPS 2017), Long Beach, CA, USA, 4–9 December 2017; pp. 971–980.

44. Moon, J.; Kim, J.; Kang, P.; Hwang, E. Solving the Cold-Start Problem in Short-Term Load Forecasting Using Tree-Based Methods. *Energies* **2020**, *13*, 886. [CrossRef]

12

Characterizing Geological Heterogeneities for Geothermal Purposes through Combined Geophysical Prospecting Methods

Cristina Sáez Blázquez *, Pedro Carrasco García, Ignacio Martín Nieto, Miguel Ángel Maté-González, Arturo Farfán Martín and Diego González-Aguilera

Department of Cartographic and Land Engineering, University of Salamanca, Higher Polytechnic School of Avila, Hornos Caleros 50, 05003 Avila, Spain; retep81@usal.es (P.C.G.); nachomartin@usal.es (I.M.N.); mategonzalez@usal.es (M.Á.M.-G.); afarfan@usal.es (A.F.M.); daguilera@usal.es (D.G.-A.)
* Correspondence: u107596@usal.es

Abstract: Geothermal energy is becoming essential to deal with the catastrophic effect of climate change. Although the totality of the Earth's crust allows the exploitation of shallow geothermal resources, it is important to identify those areas with higher thermal possibilities. In this sense, geophysical prospecting plays a vital role in the recognition and estimation of potential geothermal resources. This research evaluates the geothermal conditions of a certain area located in the center of Spain. The evaluation is mainly based on geological and geophysical studies and, in particular, the Time Domain Electromagnetic Method and the Electrical Resistivity Tomography. Once we analyzed the geology and the historical thermal evidence near the study area, our geophysical results were used to define the geothermal possibilities from a double perspective. In relation to anomalous heat gradient, the identification of a fault and the contact with impermeable granitic materials at the depth of 180 m denotes a potential location for the extraction of groundwater. Regarding the common ground-source heat-pump uses, the analysis has allowed the determination of the most appropriate area for the location of the geothermal well field. Finally, the importance of accurately defining the position of the drillings was confirmed by using software GES-CAL.

Keywords: geothermal energy; geophysical prospecting; time domain electromagnetic method; electrical resistivity tomography; potential well field location; GES-CAL software

---

1. Introduction

The fight against climate change and its catastrophic effects is one of the main challenges that currently enrolls the whole world. Efforts are therefore focused on the exploration of renewable and clean energy sources that contribute to the gradual transition and reduction of fossil energies. Within the broad group of environmentally friendly resources, geothermal energy constitutes a versatile and excellent solution for electricity generation and other direct uses. The origin of the Earth's thermal energy is linked to the internal structure of the planet and the physical processes occurring there. The existence of this heat has been proved through the rocks' temperature, which increases with depth (gradient commonly averages 3 °C/100 m of depth). However, gradients above the average can be found in areas with particular geological conditions. Armstead [1] divided the Earth's crust into non-thermal and thermal areas, considering that the last ones are characterized by temperature gradients greater than 40 °C/km depth. Focusing on very low-enthalpy geothermal resources, they can be practically found at any point of the crust, thanks to the constant ground temperature from depths of 8–10 m. In these systems, heat can be extracted for heating and cooling applications, using geothermal heat pumps. From a certain depth, the ground can store the heat even seasonally, so that ground temperature is almost constant throughout the year. Within the common exploitation of these resources, some areas, with specific geological and stratigraphic conditions, are especially appropriate for the implementation of these geothermal systems.

The identification of potential geothermal areas is, nowadays, a challenging and costly task. The principal geological parameters of a geothermal reservoir to be defined are the tectonic structures (faults), permeability, lithology, temperature, and stress field. The most accurate way of determining the above factors is by in situ measurements in a borehole [2], which are frequently discarded because of technical and/or economic reasons. In this way, it is required the implementation of alternative solutions that allow an estimation of the geothermal potential. Fortunately, some of the mentioned parameters can be estimated from the surface, mainly by the application of geophysical methods. These techniques represent a primary tool for investigating the surface and are applicable to a wide range of issues. The principal application of geophysics is in prospecting for natural resources, but it is also used in geological surveying, in engineering, or archaeological-site investigations. Since these methods allow interpreting the ground stratigraphic and structural details at scales from tens to thousands of meters, they constitute a useful tool in hydrogeological and geothermal prospecting [3,4]. There are numerous methods within the term geophysics: Seismic, magnetic, gravimetric, thermal, or electric and electromagnetic techniques are some of the major geophysical methods used for geothermal exploration. Each one of these procedures presents a series of assets and limitations that must be analyzed before their selection [5,6].

Numerous studies have addressed the geothermal prospecting from the application of geophysics [7]. The most widespread methods are the seismic, the electric, and the magnetic ones [8–15], but many others can be found in the current literature. As an example of the large geophysical implementation with geothermal purposes, the following published research is worth highlighting. Abubakar et al. [16] used improved remote sensing techniques to identify hydrothermal alterations in the Yankari Park (Nigeria). Arzate et al. [17] deduced the geothermal field model of Los Humeros (Mexico) from the use of magneto-telluric soundings. Along these lines, Long et al. [18] and Volpi et al. [19] applied similar methods to thermally define a certain area of Oregon and Italy, respectively. Hermans et al. [20] analyzed the shallow geothermal possibilities of a sandy aquifer by using electric resistivity tomography. As can be deduced from the above, geophysics has been deeply implemented in the geothermal context. However, most of the existing works are mainly focused on analyzing anomalous geothermal possibilities, without defining the final and real uses that a certain area may have as ground-source heat-pump system. This reason, together with the fact that there is an alarming lack of geothermal systems in the area considered here, has contributed to the development of this work. Thus, this research is not exclusively focused on studying the anomalous geothermal resources; it also addresses the importance of using geophysics, although low-enthalpy uses are expected.

This research firstly aims to determine the lithological composition of the ground materials by using two different geophysical prospecting techniques. From this information, the possible ground water geothermal use is evaluated, to finally define the most suitable area (within a certain perimeter) for exploration of future geothermal applications. Considering the particular characteristics of the area under study, these applications are limited to low-enthalpy geothermal uses (trough heat-pump systems), but the possibility of finding a promising thermal anomaly in the form of a hot spring could also be analyzable. Within the large number of geophysical techniques, this work includes the Time Domain Electromagnetic Method (TDEM) and the Electrical Resistivity Tomography (ERT). Beyond the analysis of anomalous geothermal activity, results of this work are also used to determine the exact location of the well field as part of a ground-source heat-pump system. Furthermore, a specific geothermal software is used to highlight the importance of an accurate well-field design. The following subsections accordingly describe the geological composition of the study area, the fundamentals of the geophysical methods implemented in the work, and the results obtained from them. Finally, discussion and conclusion sections establish the possible geothermal applications in the mentioned area.

2. Materials and Methods

The principal objective of this section was to determine the characteristics of the study area, both from a geological and a thermal point of view. After this preliminary analysis, the geophysical methods implemented in this work are thoroughly described.

2.1. Characterization of the Area under Study

The area evaluated in this work is located in the province of Segovia, at the center of Spain. As shown in Figure 1, it is geologically constituted by mainly tertiary age materials in the Duero Basin. The most significant lithological units are described below.

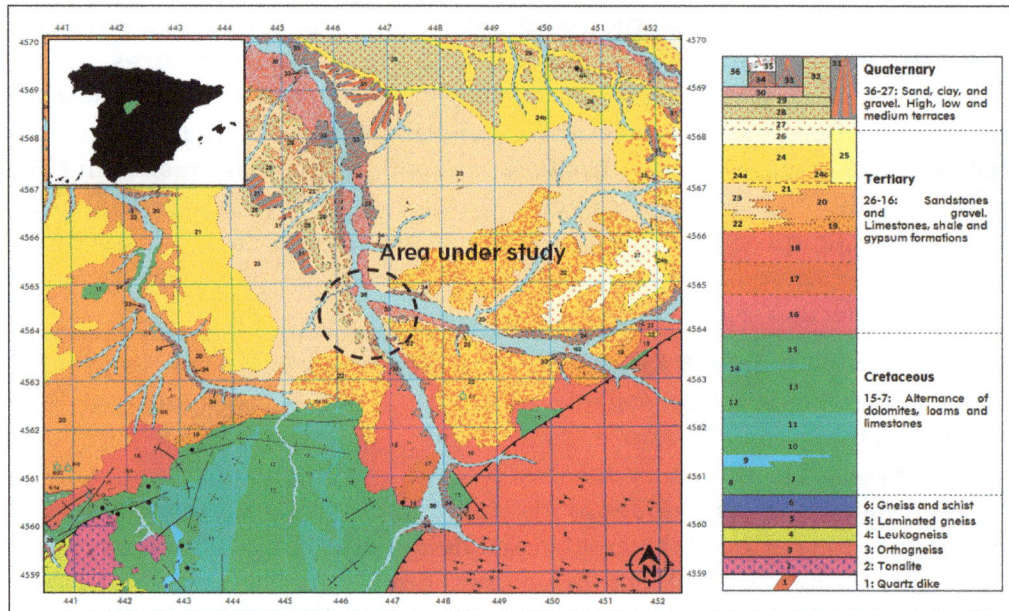

Figure 1. Geological composition of the area under study [21].

2.1.1. Tertiary

It is predominantly characterized by an alternation of clays, silts, sand, and sandstones. Sands are mineralogical constituted by quartz, feldspars, and micas, but also by igneous and metamorphic fragments (shales and granites). Sands are organized as sedimentary sections of lenticular and tabular geometry of variable thicknesses, with great variability in the vertical and horizontal layout. Clays and silts are included in the sand levels, behaving as isolation, to a greater or lesser degree, depending on the location.

Since it is a continental basin, in which the river sedimentation plays an important role, sands are arranged in lenticular layers with poor level of lateral continuity. Sections are distinguished by the frequency of the sandy layers and their permeability, but mainly by the permeability of the global matrix. Lenticular layers of sands and gravels encompassed in a semipermeable matrix behave as a large, heterogeneous, and anisotropic aquifer (confined or semi-confined, according to the different areas).

2.1.2. Calcareous Mesozoic

After the Tertiary, the area considered here is constituted by an alternation of dolomites, limestones, marls, and sandstones. Only the dolomitic and limestone–dolomitic sections of the Upper Cretaceous create aquifers of certain importance. The remaining areas are principally constituted by low-permeability materials.

2.1.3. Paleozoic

Gneisses and granites, considered as impermeable, are expected in those low deep areas with high levels of erosion and fragmentation.

Once we analyzed the prevailing geology, it was also required to evaluate the evidence of existing thermal waters in the area of study. The Geological and Mining Institute of Spain [22] allows consulting the Spanish mineral and thermal waters, areas with proven evidence of thermal water, bottling plants, or officially declared waters. In Figure 2, it is possible to observe the different mineral and thermal waters close to the area under study.

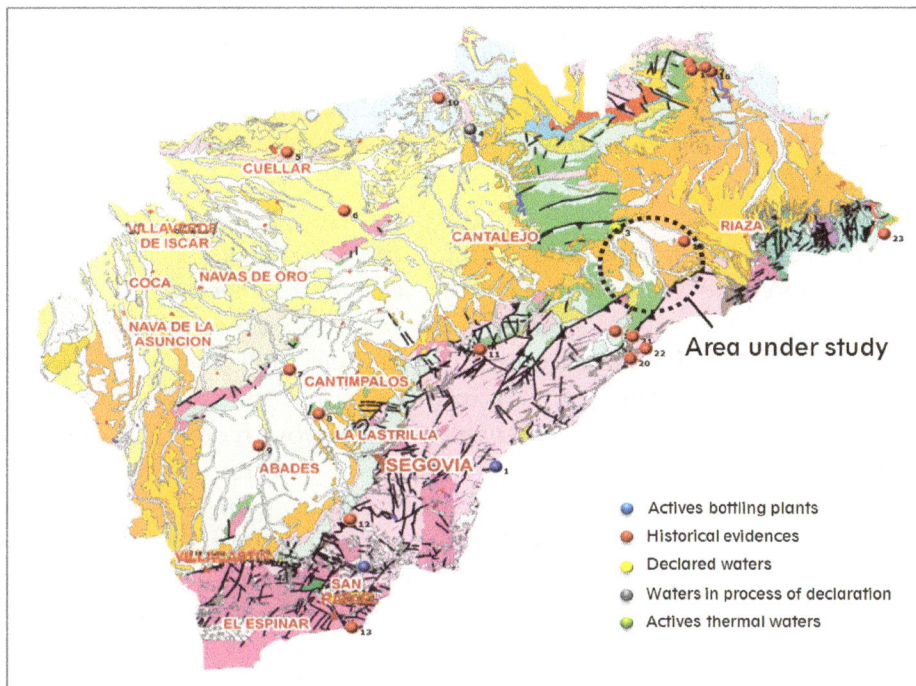

Figure 2. Thermal and mineral waters in Segovia, the Spanish province where the research is focused [22].

As can be graphically seen in Figure 2, there are several historical evidences near the studied area but also declared water. More information about these waters is presented in Table 1.

Table 1. Description of the water evidences close to the study area.

Classification	Description
18-Historical evidence	180 m deep spring water drilling Lithology: clays, sands, and gravels Water temperature: 14.7 °C
3-Declared water	Natural spring Lithology: Cretaceous limestones Water temperature: 20.8 °C
20-Historical evidence	Unavailable additional information
21-Historical evidence	Natural spring Lithology: carbonated Cretaceous Water temperature: 11.7 °C
22-Historical evidence	Unavailable additional information
23-Historical evidence	Unavailable additional information

From all of these data, it is especially remarkable the declared water number 3, in which an anomalous thermal gradient (of above 20 °C) was found. The proximity of these evidences to the studied area and the geological similarity among them mean an important starting point to justify the geophysical prospecting tests performed at later stages.

2.2. Geophysical Prospecting

Before addressing how the field tests were carried out in the area considered here, it is important to briefly describe the fundamentals of the geophysical methods selected in this work.

2.2.1. TDEM

At the beginning of the 1980s, the Time Domain Electromagnetic Method emerged as a very relevant innovation in the geophysical field. The subsequent application of this method in numerous works has allowed us to accumulate a remarkable experience in different hydrological and mining research, underground environmental pollution, archaeology, or in the location of structures and complex subsurface anomalies [23–25]. As this is a widespread method already presented in a large number of published research studies, only a brief description is provided below.

TDEM is a geophysical exploration technique used to measure the electrical resistivity or conductivity of the subsoil. The common array consists of a transmitter unit connected to a loop that receives and sends the signal to a receiver unit. By injecting a constant current into the transmitter loop, a stable primary magnetic field is generated in the ground. When this current is instantly stopped (also stopping the existing magnetic field), an electromagnetic induction of electrical currents is produced in the subsoil because of the Faraday's Law. These currents pass through closed paths in the ground and migrate in depth and laterally, while their intensity decreases with time, also generating a decreasing transient secondary magnetic field on the surface. This secondary field induces a time-varying voltage at the receiver. The way in which the voltage drops contains the information about the ground resistivity, since the magnitude and distribution of the induced currents depend on this property. In this context, short-time voltages provide information about the shallow resistivity, while the long-time voltages are linked to deeper resistivities. The principal phases explained above are synthesized in Figure 3.

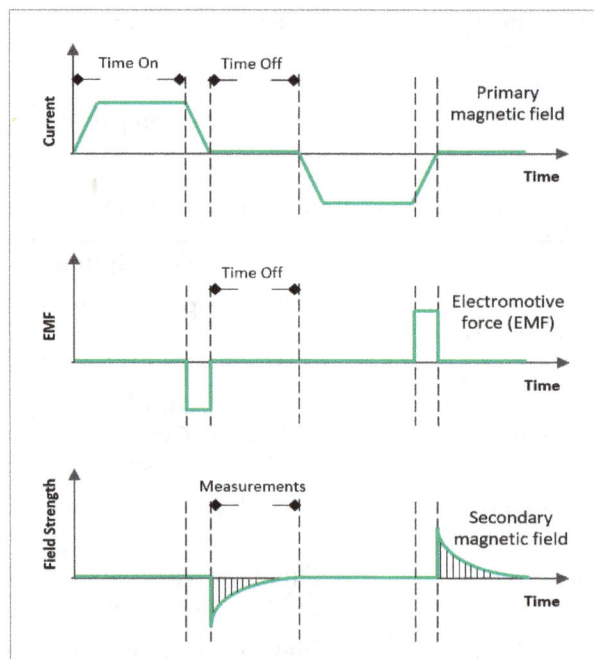

Figure 3. Time diagram in which the measurements are made by the receiver.

The advantage of TDEM regarding other electrical and electromagnetic methods is its low sensitivity to the separation between transmitter (T) and receiver (R). In this way, TDEM is the only electrical method that can be applied with a separation T/R lower than the depth of the structure that pretends to be found. This fact allows for the improvement of the lateral resolution of the method. However, longer distances are required for deeper prospecting to deal with noise effects [26].

In any case, the depth of prospecting in TDEM is determined, not by the T/R separation, but by the time, since the transmitter stops emitting at the associated magnetic moment. For greater depths, it is therefore necessary to collect the signal at later stages. It is obvious that, with short times, currents are concentrated in the superficial layers. The first electromotive force (EMF) measurements will be consequently more sensitive to the resistivity of the upper layers. As time passes, the current intensity reaches greater depths, and the measured EMF is more influenced by these depths. Furthermore, the current density decreases in the upper layers so that the electrical resistivity of these layers has low influence on the EMF measured over long periods of time. This fact contributes to eliminate the effect of near-surface resistivity variations, commonly the reason of losing quality in the final data obtained by other electrical prospecting methods.

In relation to the TDEM data processing, EMF, measured as a function of time, is converted into apparent resistivity. This resistivity is then introduced in an inversion tool which calculates the stratification of apparent conductivities by using the Spiker algorithm (adjusting in the best possible way to the curve of observed apparent resistivities). Specifications of the TDEM device can be found at Appendix A, Figure A1.

2.2.2. ERT

Two-dimensional electrical resistivity tomography is a technique widely used for the characterization of the subsoil with multiple purposes and applications. In general, its aim is the location of subsurface complex structures and anomalies (both geological and anthropic) [27–30].

ERT consists of measuring the apparent resistivity of the ground by using a tetra-electrode device with a constant separation "a" among electrodes. Distances between the couples of electrodes (transmitter–receiver) are then varied by multiples of a value "n". The basis of this method is injecting a constant current in the ground through the transmitter electrodes and measuring the potential difference between the receiver ones. The final result is an apparent resistivity section for several levels in depth "n". Data are subsequently processed by inversion mathematical algorithms. The inversion results in real depths and resistivities image that must be verified with the geological information, existing drillings, or geochemical or hydrological data. Through the results interpretation, a final diagnosis is obtained.

With the purpose of finally converting the distribution of the ground real resistivity into a geological structure, it is necessary to know and consider the typical resistivities for the different subsoil materials and the geology in the area of study [31–33].

Data processing of this research was performed in the inversion software RES2DINV. This tool is based on the least square's inversion technique with smoothing restriction, using Equation (1) [34–36].

$$(J^T J + uF)d = J^T g \tag{1}$$

$$F = f_x f_x{}^T + f_z f_z{}^T \tag{2}$$

where f_x = horizontal flattening filter; f_z = vertical flattening filter; J = partial derivates matrix; $J^T = J$ transposed matrix; u = softening factor; d = disturbance model vector; and g = discrepancy vector.

The 2D model implemented in this software divides the ground into a certain number of rectangular blocks. The objective is to determine the real resistivity of the rectangular blocks that would produce a pseudo-section of apparent resistivities as the ones measured in the field. Depending on the device used, the thickness of the first layer of blocks is variable (0.5 times the space among electrodes for Wenner and Schlumberger devices, 0.9 for the Pole–Pole, 0.3 for the Dipole–Dipole, or 0.6 for the

Pole–Dipole). For the next deeper layer, the thickness will be increased in 10% to 25%. Layers' depths can also be defined by the user. Specifications of the ERT device can be found at Appendix A, Figure A2.

2.3. Field Work

Regarding TDEM prospecting, ten tests were carried out in the perimeter of the area under study. All TDEMs were performed with a loop of 200 × 200 m and implemented the coincident loop mode. The UTM coordinates of each TDEM is included in Table 2. Coordinates were obtained by using a GNSS (Global Navigation Satellite System) with an accuracy of ±1 cm in planimetry. The reference system is the European Terrestrial Reference System 1989 (ETRS 89). The use of this system allows a perfect geo-location of the tests.

Table 2. UTM coordinates (time zone 30T) of each TDEM made in the area considered in this research.

Test	X	Y
TDEM-1	447,377	4,564,903
TDEM-2	447,289	4,564,603
TDEM-3	447,165	4,564,251
TDEM-4	446,949	4,564,139
TDEM-5	446,641	4,563,959
TDEM-6	446,353	4,563,763
TDEM-7	446,277	4,563,283
TDEM-8	446,637	4,562,667
TDEM-9	446,257	4,564,295
TDEM-10	448,085	4,562,743

For each TDEM, at least one register was made, using a staking of 1000 repetitions of the measurement for each channel. In each of the mentioned registers, 73 channels were measured. However, for the size of the loop, from channels 35 to 40, measurements were affected by the background noise, higher than the signal to be measured. First channels were also affected by the slope, so that first and last channels were discarded in the corresponding data processing. Figure 4 shows the TDEM equipment during the tests in the area under study.

Figure 4. TDEM equipment in the area under study.

Relative to the implementation of ERT, 725 m was measured in a profile distributed in the area under study. Pole–Dipole and Schlumberger devices were used in this prospecting. Pole–Dipole was selected considering the great depth of penetration (around 250 m in this case) and the high resolution. The separation among electrodes was of 25 m. The beginning and ending (in UTM coordinates) of the ERT profile are shown in Table 3. These coordinates were obtained with the previously mentioned GNSS system.

Table 3. UTM coordinates (time zone 30T) of each TDEM made in the area considered in this research.

Profile 1	X	Y
Starting point	446,708	4,564,016
Ending point	447,334	4,564,353

In Figure 5, it is possible to observe the location of the TDEM and ERT prospecting in the area considered in this research.

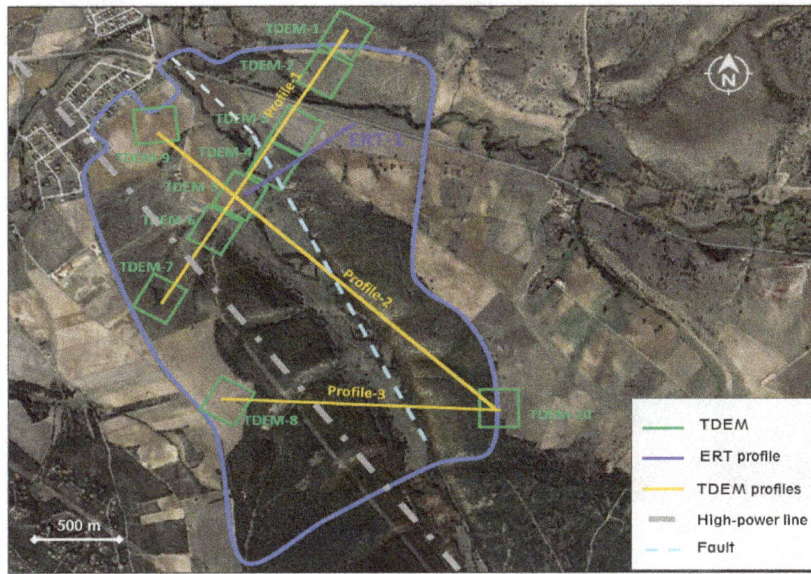

Figure 5. Location of the geophysical tests (TDEM and ERT) performed in the area under study.

3. Results

Results derived from the 2D interpretation of the electromagnetic tests (TDEM) are shown in Figures 6–8. Sections of iso-resistivities curves reflect the spatial variation (2D) of the apparent resistivity in each TDEM profile. This variation is mainly due to lithological alterations corresponding to different levels of ground materials. However, it is also possible to find sections with different resistivities in the same lithological formation because of other external factors.

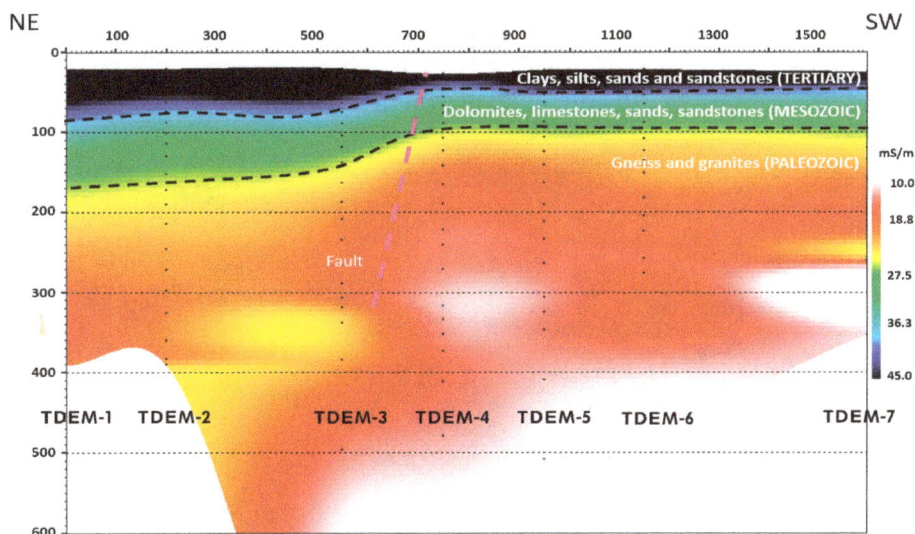

Figure 6. Two-dimensional interpretation of electromagnetic prospecting, TDEM-profile 1.

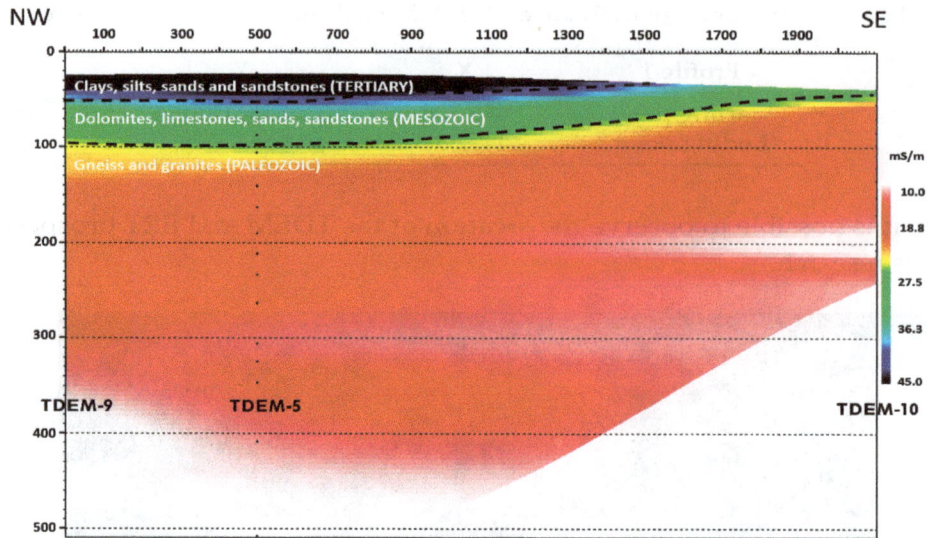

Figure 7. Two-dimensional interpretation of electromagnetic prospecting, TDEM-profile 2.

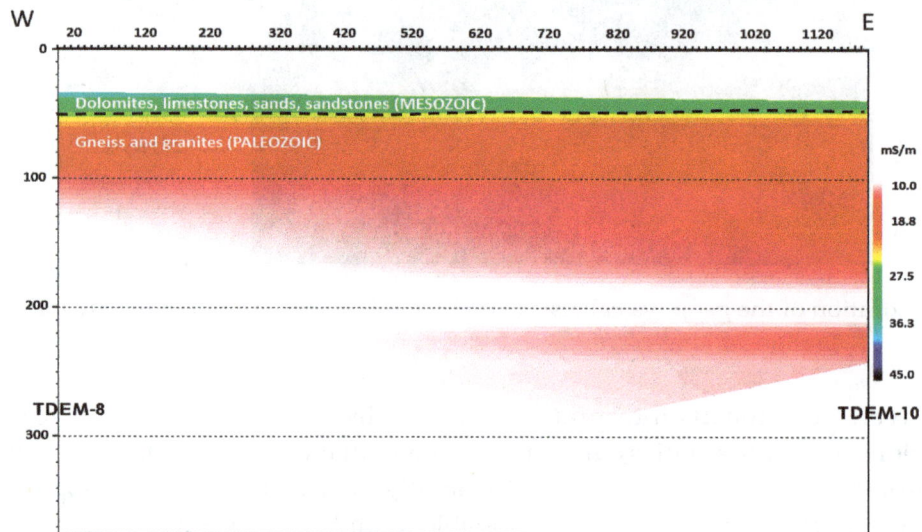

Figure 8. Two-dimensional interpretation of electromagnetic prospecting, TDEM-profile 3.

Field results from ERT prospecting were processed with the aim of building 2D geo-electric sections (sections of resistivities and depths) for the different profiles. The variations of resistivities derive from lithological changes in the lateral direction and in depth. The ERT section of this work is represented in Figure 9.

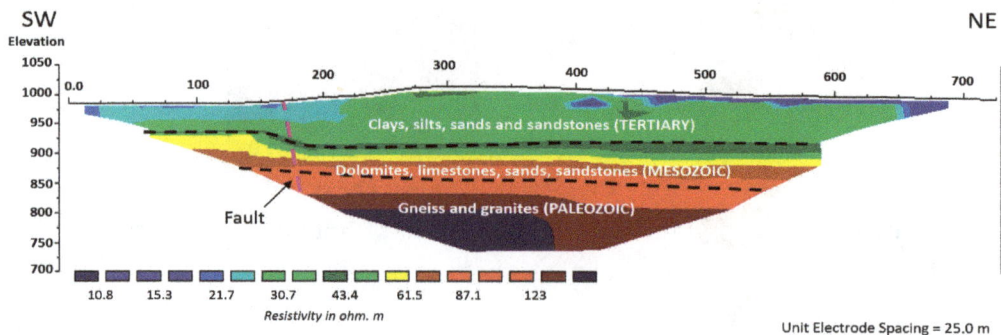

Figure 9. Two-dimensional resistivity section from the ERT profile in the area under study.

In Figures 6–9, above, the different geo-electric unities and the materials associated are represented. Geophysical data (resistivities) are converted into lithological information based on a series of correlation criteria that consider previous studies in the area (geology) and previous fieldwork in similar geological scenarios. The analysis of the geophysical prospecting brings to light the existence of three main unities: Tertiary (clays, silts, sands, and sandstones), Mesozoic (dolomites, limestones, sands, and sandstones), and Paleozoic (gneisses and granites). Results also evince the presence of a fault associated to the Duratón River.

The determination of each of the geological formations is mainly based on knowing the geology in the area (already described in Section 2.1) and the standard values of the resistivity/conductivity typically associated to each material. In this way, standard resistivity or conductivity values (obtained from different official databases) are the basis to connect each layer (with a specific resistivity) with a geological formation. However, beyond this information, the experience of the geophysicist is essential to accurately define materials of the ground. This technician is finally the key factor to achieve a reliable interpretation of the subsoil.

4. Discussion

The area under study is located in the Duero Basin, one of the stables areas of the Earth's crust characterized by normal geothermal gradient (3 °C/100 m). In this way, the geothermal possibilities are focused on the existence, at the appropriate depth, of permeable materials capable of containing and allowing the movement of fluids to extract the heat from the rock. Based on previous studies from the Geological and Mining Institute of Spain (IGME), the temperatures of the aquifers in the detrital Tertiary of the Duero Basin (area of this research) in the depth of 0–200 m vary in the range of 12–20 °C.

4.1. Anomalous Geothermal Possibilities

As mentioned above, the possibility of achieving an especial use of geothermal resources (beyond the normal use as ground-source heat-pump systems) in the area of study is limited to the existence of particular geological formations. The geophysical tests performed here reveal that, around the depth of 180 m, it is possible to find the contact with the impermeable granitic materials from the Paleozoic. In order to estimate the temperature in the aforementioned level, it is convenient to consider the average annual temperature in the area under study. This information can be found in Figure A3, from Appendix B. As Figure A3 shows, the temperature in the area of this research is around 11.7 °C for the period considered. From this value, and considering the normal geothermal gradient, the temperate at the depth of 180 m would be of about 17.1 °C. This temperature could be higher, since the contact is constituted by granitic materials (characterized by high thermal conductivities) that could provide a higher amount of heat to the groundwater.

From the depth of 180 m, it would be interesting (from the geothermal point of view) to perform a drilling in the location of the fault represented in Figures 6 and 9. In this position, it could be possible to find meteorized permeable granitic materials accumulating groundwaters. Additionally, it might be the case that deeper water flows reach the fault with higher temperatures. It should also be remembered the documented natural spring (Table 1) in the vicinity of the study area with temperatures of above 20 °C and placed in similar lithology. In the case that geological formations in depth are permeable or fractured, and if there is groundwater circulation, this water is capable of capturing the heat from the rocks and reaching the surface through crevices or faults. Once in the surface, it could lead to the generation of thermal waters or geysers (depending on the geothermal gradient, the groundwater temperature). In this sense, the existence of a fault in the area under study is not in itself an indication of anomalous geothermal activity, but if this was the case, the fault constitutes the way of using the geothermal resource in the surface.

Based on the results obtained throughout this research, Figure 10, below, shows the location of the area recommended for the collection of groundwater, with high possibilities of reaching an anomalous thermal gradient. Figure 10 also includes the exact position of the drilling to carry out the exploitation of the geothermal resource.

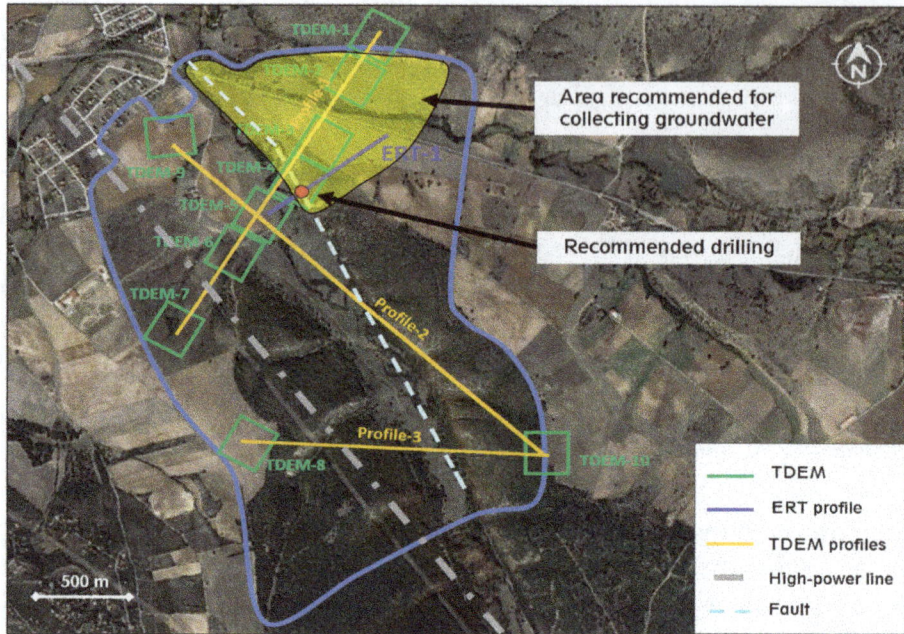

Figure 10. Location of the specific area and drilling recommended for the collection of groundwater.

UTM coordinates of the recommended drilling (previously indicated in Figure 10) are included in Table 4.

Table 4. UTM coordinates (time zone 30T) of the drilling suggested for the exploitation of possible geothermal resources.

UTM Coordinates	X	Y
Recommended drilling	446,872	4,564,070

In the end, the existence of geothermal anomalies is judged in this work by three main factors: the existent thermal evidence in the nearby areas, the results of the geophysical prospecting, and the estimation of the geothermal gradient in the area under study. The subsoil materials (known from the geophysics results) are a preliminary indication of the possible geothermal activity. However, this information cannot be used as final evidence. In this way, the existing thermal waters with thermal gradient above the mean value and the estimation of the geothermal gradient in the specific area of study are also useful and necessary to make a global evaluation. It must also be mentioned that the presence of granite formations does not guarantee the existence of geothermal anomalies. Despite this fact, the existence of these materials is favorable for finding geothermal activity.

4.2. Ground-Source Heat-Pump Uses

Low-enthalpy geothermal energy can be used at any point of the Earth's crust by the use of heat pumps as ground-source or groundwater heat-pump systems. However, the geological conditions of an area highly influence the global geothermal design and, hence, the final investment of the system.

The site under study is constituted by materials with high thermal conductivities located a few meters from the surface, especially in the south and northwest areas. These geological formations (limestones, dolomites, granites, etc.) make this place an ideal location for the implementation of ground-source heat-pump systems for heating and cooling purposes.

With the aim of highlighting the importance of the ground characterization to define the specific drilling area, the software GES-CAL [37], designed for the calculation of low-enthalpy geothermal systems, was used in this work. Thus, this tool is useful to compare the differences, in technical and economic terms, between a system placed in the most appropriate area (according to the geophysical results) and in an aleatory one.

Derived from the abovementioned information, Table 5 includes the results of GES-CAL software in the design of three ground-source heat-pump systems in the area under study. Case 1 is planned to be placed in the most favorable conditions in which the granite and gneiss formations are found from the depth 50 m (profile 3 of Figure 8). Case 2 is, in turn, located in the most extended area where the Paleozoic begins at the depth of 100 m. Finally, Case 3 considers the most unfavorable situation in which the consolidated formations are found from the depth of around 180 m (Figure 6). When performing the analysis of each assumption with GES-CAL, identical initial conditions are introduced in the software, except for the ground thermal conductivity. In function of the stratigraphic column prevailing in each case, an average value of the ground thermal conductivity was obtained by considering the length of each layer of material and its thermal conductivity from the surface to the depth of 150 m (common depth of the wells in shallow geothermal systems).

Table 5. Drilling length and initial investment required in the geothermal system of each case, according to the global ground thermal conductivity.

Cases	Ground Thermal Conductivity (W/mK)	Total Drilling Length * (m)	Initial Investment (€)
Case 1	2.178	136	22,993.38
Case 2	1.920	152	25,019.23
Case 3	1.286	218	29,729.28

* Associated to a vertical double-U heat-exchanger design.

It is important to mention that the average of the ground thermal conductivity of each case (included in Table 5) was obtained from the officially accepted thermal conductivity values for each material, according to the "Technical Building Code" (CTE) [38].

As can be noted from the results of Table 5, the selection of an appropriate location of the geothermal well field involves significant reductions of the global drilling length required in the system. Additionally, the economic module of GES-CAL allows for the comparison of the initial investment associated to each assumption. Observing Table 5 again, we see Case 1 requires an investment of around 23% lower than the one required by Case 3.

Beyond the economic side, it is also important to consider the technical factors that could compromise the performance of the geothermal installation. In this sense, the drilling method selected in the geothermal well field is completely influenced by the geological formations constituting the underground. Reverse circulation methods (associated to loosen materials) are frequently avoided in geothermal systems due to the difficulties of holding the materials during the drilling process without casing. In relation to the cases analyzed in this research, the stratigraphic column of Cases 2 and 3 is mainly constituted by non-consolidated materials (especially in Case 3, in which all the column is made up of this kind of material). This fact would oblige us to use the reverse-circulation technique when carrying out the geothermal wells of these cases, complicating the global process, but also raising the price of the initial investment. Regarding Case 1, the presence of consolidated materials at a more superficial level allows the implementation of rotary percussive drilling techniques. These methods are ideal for the geothermal drilling, because casing is not needed, and the general cost is lower than the reverse-circulation ones. It is important to clarify that GES-CAL does not provide information about the most suitable location of the wells; this location is specifically defined from the geophysical prospecting (distribution of the geological formations in the subsoil).

5. Conclusions

This research has particularly exposed the applicability of the geophysical techniques, TDEM and ERT, in the identification of potential areas for the exploitation of shallow geothermal resources. Geophysical prospecting results have revealed the exact lithology of the ground in the area under study. Based on this knowledge, the following statements were deduced from this work:

- The investigated area is located on the SE edge of the tertiary depression belonging to the Duero Basin. The most superficial geological unities are principally detrital materials: clays, sands, gravels, silts, and sandstones. More in depth, dolomites, limestones, and sandstones are found to finally reach the Paleozoic level (gneiss and granites). Results derived from the implementation of geophysics have allowed us to define the thickness of each layer of materials in the entire area under study.
- Regarding the exploitation of groundwater resources, the highest possibilities of locating thermal waters were detected in the NE side of the study area. More specifically, the fault associated to the Duratón River is considered the most favorable location to exploit the mentioned resource.
- In addition to the possible use of thermal waters, the implementation of ground-source heat-pump systems was analyzed in this research. Even though, in all the area under study, it is possible to install this kind of energy, the information obtained from the geophysical tests has also allowed us to define the most appropriate location of the geothermal well field. The GES-CAL tool was used to compare the design of the shallow geothermal system in different locations of the area. Through this analysis, it has verified that the precise location of the geothermal wells could mean significant economic savings, being also important to avoid possible technical problems during the drilling process.

In conclusion, this research has proved that geophysical prospecting methods, as the ones selected here, constitute a useful tool to firstly define the underground geological characterization, and then to analyze the potential areas for geothermal exploitation. All of this includes the detection of possible thermal water resources, as well as the establishment of the most suitable areas for the location of the well field (in the case of ground-source or groundwater heat-pump systems).

Author Contributions: Conceptualization, C.S.B. and P.C.G.; data curation, I.M.N. and M.Á.M.-G.; formal analysis, C.S.B., P.C.G., and I.M.N.; investigation, C.S.B.; methodology, C.S.B., P.C.G., I.M.N., and M.Á.M.-G.; resources, C.S.B., P.C.G., A.F.M., and D.G.-A.; supervision, A.F.M. and D.G.-A.; validation, C.S.B., P.C.G., I.M.N., M.Á.M.-G., A.F.M., and D.G.-A.; visualization, C.S.B.; writing—original draft, C.S.B. All authors have read and agreed to the published version of the manuscript.

Acknowledgments: We would like to thank the Department of Cartographic and Land Engineering of the Higher Polytechnic School of Avila, University of Salamanca, for allowing us to use their facilities and for their collaboration during the experimental phase of this research. We also want to thank the University of Salamanca and Santander Bank for providing a pre-doctoral grant (Training of University Teachers Grant) to the corresponding author of this paper; the grant has made possible the realization of the present work.

Appendix A

This Appendix specifies the technical characteristics of the device used when performing the corresponding geophysical prospecting of this research.

- TDEM equipment

TerraTEM is the device used at the electromagnetic prospecting. It is a new transient electromagnetic survey system that incorporates a 10 Amp transmitter and a true simultaneous 500 kHz three-component receiver. The unit is powered by an external 24 V battery pack system allowing 6–8 h of continuous operation. The GPS, which is mounted on the front panel, allows for geolocation information to be automatically recorded with soundings. The user interface comprises a 15″ color LCD panel and a touchscreen. System parameters are stored automatically with each sounding, for post-survey quality assurance. More information about the specifications of this device are shown in Table A1. Additionally, Figure A1 includes the mentioned device.

Table A1. Specifications of TerraTEM device.

TerraTEM	
Transmitter Output	10 Amps. (max.)
Receivers	1 Channel
High-Resolution Sampling Rates	500 kHz
Data Visualization and Processing in Field	Standard Software
Storage Device	1 GB Flash Disk
GPS Receiver°	12 Channels
Communications	USB and RS-232 Standard
Extra Stacking Options and Gain Functions	10 Selectable Gain Settings from 1 to 8.000
Operating Temperature	−10–40 °C
Resolution	23 nV
Transmitter Current	50 A at 6 V through to 120 V (6 kW)

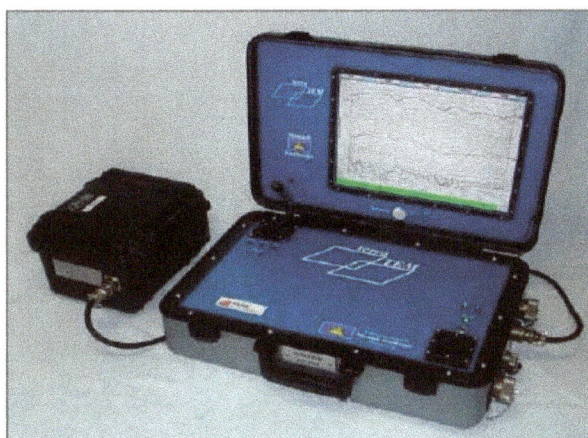

Figure A1. TerraTEM device used in the electromagnetic survey of this research.

- ERT equipment

Regarding the ERT tests, Syscal Pro was the device selected. It is an all-in-one multi-node resistivity and induce polarization sounding and profiling system. Syscal Pro gathers a 10-channels receiver and a 250 W internal transmitter, making it the more powerful system of the Syscal range. In Figure A2 and Table A2, it is possible to observe the mentioned equipment and its principal specifications.

Table A2. Specifications of Syscal Pro device.

Syscal Pro	
Transmitter max. voltage	800 V
Transmitter max. current	2.5 A, accuracy 0.2%
Transmitter max. power	250 W
Receiver max. voltage	15 V
Receiver resolution	1 microV
Electrodes	Up to 4000 can be used
Data flash memory	More than 21,000 readings
Serial link	RS-232 data download
Power supply	Two internal rechargeable 12 V, 7.2 Ah Optional external 12 V batteries
Casing	Shock resistant fiber-glass case
Operating Temperature	−20–70 °C

Figure A2. Syscal Pro device used when carrying out the 2D electrical resistivity tomography of this research.

Appendix B

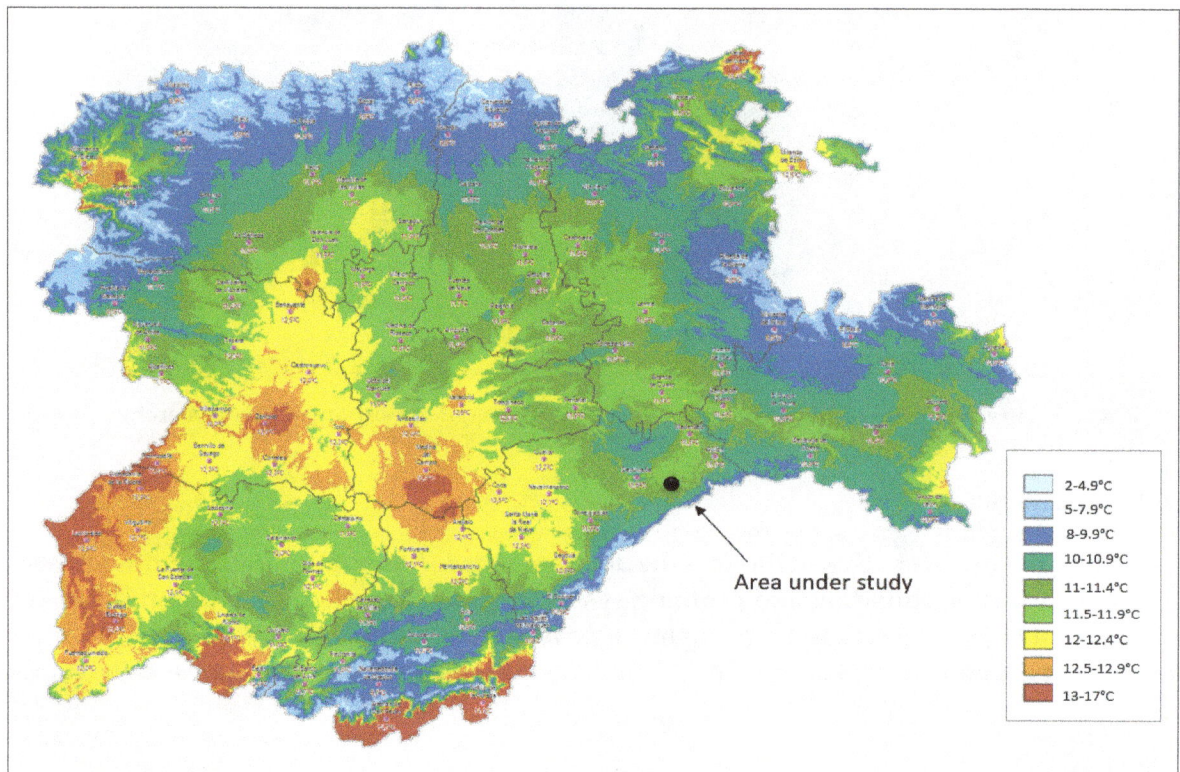

Figure A3. Average annual temperature in the area considered in this research for the period 1981–2010 [39].

References

1. Armstead, H.C. *Geothermal Energy: Its Past, Present and Future Contribution to the Energy Needs of Man*, 2nd ed.; E. & F.N Spon: New York, NY, USA; London, UK, 1983.

2. Bodvarsson, G. Evaluation of geothermal prospects and the objectives of geo-thermal exploration. *Geoexploration* **1970**, *8*, 7–17. [CrossRef]

3. Sena-Lozoya, E.B.; González-Escobar, M.; Gómez-Arias, E.; González-Fernández, A.; Gómez-Ávila, M. Seismic exploration survey northeast of the Tres Virgenes Geothermal Field, Baja California Sur, Mexico: A new Geothermal prospect. *Geothermics* **2020**, *84*, 101743. [CrossRef]

4.   Kearey, P.; Brooks, M.; Hill, I. *An Introduction to Geophysical Exploration*; John Wiley & Sons: Hoboken, NJ, USA, 2013.

5.   Aretouyap, Z.; Nouck, P.N.; Nouayou, R. A discussion of major geophysical methods used for geothermal exploration in Africa. *Renew. Sustain. Energy Rev.* **2016**, *58*, 775–781. [CrossRef]

6.   Mariita, N.O. Strengths and weaknesses of gravity and magnetics as exploration tools for geothermal energy. In Proceedings of the Short Course V on Exploration for Geothermal Resources, Naivasha, Kenya, 29 October–19 November 2010.

7.   DomraKana, J.; Djongyang, N.; Raïdandi, D.; Nouck, P.N.; Dadjé, A. A review of geophysical methods for geothermal exploration. *Renew. Sustain. Energy Rev.* **2015**, *44*, 87–95. [CrossRef]

8.   Bibby, H.M.; Risk, G.F.; Caldwell, T.G.; Bennie, S.L. Misinterpretation of electrical resistivity data in geothermal prospecting: A case study from the Taupo Volcanic Zone. In *Geological and Nuclear Sciences, Proceedings of the World Geothermal Congress*; World Geothermal Congress: Antalya, Turkey, 2005; pp. 1–8.

9.   Deckert, H.; Bauer, W.; Abe, S.; Horowitz, F.G.; Schneider, U. Geophysical greenfield exploration in the permo-carboniferous Saar–Nahe basin—The Wiesbaden Geothermal Project, Germany. *Geophys. Prospect.* **2018**, *66*, 144–160. [CrossRef]

10.   Subasinghe, N.D.; Nimalsiri, T.B.; Suriyaarachchi, N.B.; Hobbs, B.; Fonseka, M.; Dissanayake, C. *Study of Thermal Water Resources in Sri Lanka Using Time Domain Electromagnetics (TDEM). En Advanced Materials Research*; Trans Tech Publications Ltd.: Freienbach, Switzerland, 2014; pp. 3198–3201.

11.   Fadillah, T.; Sulistijo, B.; Notosiswoyo, S.; Kristanto, A.; Yushantarti, A. The Resistivity Structure of Aluvial in Geothermal Prospect. Using Time Domain Electromagnetic Methode (TDEM) Survey. In Proceedings of the World Geothermal Congress 2015, Melbourne, Australia, 19–25 April 2015.

12.   Sáez Blázquez, C.; Martín, A.F.; García, P.C.; González-Aguilera, D. Thermal conductivity characterization of three geological formations by the implementation of geophysical methods. *Geothermics* **2018**, *72*, 101–111. [CrossRef]

13.   Martín Nieto, I.; Farfán Martín, A.; Sáez Blázquez, C.; González-Aguilera, D.; Carrasco García, P.; Farfán Vasco, E.; Carrasco García, J. Use of 3D electrical resistivity tomography to improve the design of low enthalpy geothermal systems. *Geothermics* **2019**, *79*, 1–13. [CrossRef]

14.   Bibby, H.M.; Dawson, G.B.; Rayner, H.H.; Bennie, S.L.; Bromley, C.J. Electrical resistivity and magnetic investigations of the geothermal systems in the Rotorua area, New Zealand. *Geothermics* **1992**, *21*, 43–64. [CrossRef]

15.   Tang, X.; Zhang, J.; Pang, Z.; Hu, S.; Tian, J.; Bao, S. The eastern Tibetan Plateau geothermal belt, western China: Geology, geophysics, genesis, and hydrothermal system. *Tectonophysics* **2017**, *717*, 433–448. [CrossRef]

16.   Abubakar, A.J.A.; Hashim, M.; Pour, A.B. Remote sensing satellite imagery for prospecting geothermal systems in an aseismic geologic setting: Yankari Park, Nigeria. *Int. J. Appl. Earth Obs. Geoinf.* **2019**, *80*, 157–172. [CrossRef]

17.   Arzate, J.; Corbo-Camargo, F.; Carrasco-Núñez, G.; Hernández, J.; Yutsis, V. The Los Humeros (Mexico) geothermal field model deduced from new geophysical and geological data. *Geothermics* **2018**, *71*, 200–211. [CrossRef]

18.   Long, C.L.; Kaufmann, H.E. Reconnaissance geophysics of a known geothermal resource area, Weiser, Idaho and Vale, Oregon. *Geophysics* **1980**, *45*, 312–322. [CrossRef]

19.   Volpi, G.; Manzella, A.; Fiordelisi, A. Investigation of geothermal structures by magnetotellurics (MT): An example from the Mt. Amiata area, Italy. *Geothermics* **2003**, *32*, 131–145. [CrossRef]

20.   Hermans, T.; Vandenbohede, A.; Lebbe, L.; Nguyen, A. shallow geothermal experiment in a sandy aquifer monitored using electric resistivity tomography. *Geophysics* **2012**, *77*, 11–21. [CrossRef]

21.   Magna 50. *Mapa Geológico de España 1:50.000*; Instituto Geológico y Minero de España: Sepúlveda, Spain, 2003; hoja 431.

22.   Aguas Minerales y Termales de España. IGME, Instituto Geológico y Minero de España. Available online: http://aguasmineralesytermales.igme.es/inicio.aspx (accessed on 1 January 2020).

23.   Hapsoro, C.A.; Srigutomo, W.; Purqon, A. *3-D Modeling of Layered Earth Structure in the Geothermal Systems Using Time Domain Electromagnetics (TDEM) Method. IOP Conference Series: Earth and Environmental Science*; IOP Publishing: Bristol, UK, 2019; Volume 318.

24.   Goldman, M.; Rabinovich, B.; Rabinovich, M.; Gilad, D.; Gev, I.; Schirov, M. Application of the integrated NMR-TDEM method in groundwater exploration in Israel. *J. Appl. Geophys.* **1994**, *31*, 27–52. [CrossRef]

25. Descloitres, M.; Chalikakis, K.; Legchenko, A.; Moussa, A.M.; Genthon, P.; Favreau, G.; Le Coz, M.; Boucher, M.; Oï, M. Investigation of groundwater resources in the Komadugu Yobe Valley (Lake Chad Basin, Niger) using MRS and TDEM methods. *J. Afr. Earth Sci.* **2013**, *87*, 71–85. [CrossRef]

26. Fitterman, D.V.; Stewart, M.T. Transient Electromagnetic Sounding for Groundwater. *Geophysics* **1986**, *51*, 889–1033. [CrossRef]

27. Kumar, D.; Thiagarajan, S.; Rai, S.N. Deciphering geothermal resources in Deccan Trap region using electrical resistivity tomography technique. *J. Geol. Soc. India* **2011**, *78*, 541–548. [CrossRef]

28. Carrier, A.; Lupi, M.; Fishanger, F.; Nawratil de Bono, C. Deep-reaching electrical resistivity tomography (ERT) methods for middle-enthalpy geothermal prospection in the Geneva Basin, Switzerland. In Proceedings of the EGU General Assembly Conference Abstracts, Vienna, Austria, 8–13 April 2018; p. 20.

29. Chabaane, A.; Redhaounia, B.; Gabtni, H. Combined application of vertical electrical sounding and 2D electrical resistivity imaging for geothermal groundwater characterization: Hammam Sayala hot spring case study (NW Tunisia). *J. Afr. Earth Sci.* **2017**, *134*, 292–298. [CrossRef]

30. Maté-González, M.Á.; Sánchez-Aparicio, L.J.; Sáez Blázquez, C.; Carrasco García, P.; Álvarez-Alonso, D.; de Andrés-Herrero, M.; García-Davalillo, J.C.; González-Aguilera, D.; Herández Ruiz, M.; Jordá Bordehore, L.; et al. On the Combination of Remote Sensing and Geophysical Methods for the Digitalization of the San Lázaro Middle Paleolithic Rock Shelter (Segovia, Central Iberia, Spain). *Remote Sens.* **2019**, *11*, 2035. [CrossRef]

31. Daniels, F.; Alberty, R.A. *Physical Chemistry*; Mir: Moscow, Russia, 1966; p. 291.

32. Mainoo, P.A.; Manu, E.; Yidana, S.M.; Agyekum, W.A.; Stigter, T.; Duah, A.A.; Preko, K. Application of 2D-Electrical resistivity tomography in delineating groundwater potential zones: Case study from the voltaian super group of Ghana. *J. Afr. Earth Sci.* **2019**, *160*, 103618. [CrossRef]

33. Zhou, B.; Moosoo, W. Multi-Parameter Tomographic Inversion for Imaging 2D Electrical Resistivity Anisotropy. 1st Conference on Geophysics for Geothermal-Energy Utilization and Renewable-Energy Storage. *Eur. Assoc. Geosci. Eng.* **2019**, *1*, 1–5.

34. deGroot-Hedlin, C.; Constable, S. Occam's inversion to generate smooth, two-dimensional models from magnetotelluric data. *Geophysics* **1990**, *55*, 1613–1624. [CrossRef]

35. Rajesh, R.; Tiwari, R.K. FORTRAN code to convert resistivity Vertical Electrical Sounding data to RES2DINV format. *Site Characterisation Using Multi-Channel Anal. Surf. Waves Var. Locat. Kumaon Himalayas India* **2018**, *22*, 359–363.

36. Istiqomah, A.N. *Analisis Komparasi Inversi Software Res2Dinv dan Ipi2Win Pada Metode Vertical Electrical Sounding (VES)*; SKRIPSI Mahasiswa UM: Kuala Lumpur, Malaysia, 2019.

37. Sáez Blázquez, C.; Martín Nieto, I.; Mora, R.; Farfán Martín, A.; González-Aguilera, D. GES-CAL: A new computer program for the design of closed-loop geothermal energy systems. *Geothermics* **2020**, *87*, 101852. [CrossRef]

38. Structural solutions file. In *Technical Building Code, CTE*; Institute of Structural Sciences of Eduardo Torroja and Structural Institute of Castilla y León: Castilla y León, Spain, 2007.

39. Agencia Estatal de Meteorología (AEMET). *Series of Annual Average Temperatures of AEMET Stations (Spain). Observations of the Thirty-Year Period 1981–2010, Completed, Refined and Homogenized*; Agencia Estatal de Meteorología (AEMET): Madrid, Spain, 2010.

Permissions

List of Contributors

Joaquín Alonso-Montesinos
Department of Chemistry and Physics, University of Almería, 04120 Almería, Spain
CIESOL, Joint Centre of the University of Almería-CIEMAT, 04120 Almería, Spain

Guojiang Xiong, Xufeng Yuan and Jing Zhang
Guizhou Key Laboratory of Intelligent Technology in Power System, College of Electrical Engineering, Guizhou University, Guiyang 550025, China

Dongyuan Shi
State Key Laboratory of Advanced Electromagnetic Engineering and Technology, Huazhong University of Science and Technology, Wuhan 430074, China

Lin Zhu
Department of Electrical Engineering and Computer Science, University of Tennessee, Knoxville, TN 37996, USA

Gang Yao
Guizhou Electric Power Grid Dispatching and Control Center, Guiyang 550002, China

Susumu Shimada and Tetsuya Kogaki
Renewable Energy Research Center, National Institute of Advanced Industrial Science and Technology, Koriyama 963-0298, Japan

Jay Prakash Goit
Renewable Energy Research Center, National Institute of Advanced Industrial Science and Technology, Koriyama 963-0298, Japan
Department of Mechanical Engineering, Kindai University, Higashi-Hiroshima, Hiroshima 739-2116, Japan

Teruo Ohsawa
Graduate School of Maritime Sciences, Kobe University, Kobe 658-0022, Japan

Satoshi Nakamura
Coastal and Estuarin Environment Division, Port and Airport Research Institute, Yokosuka 239-0826, Japan

André R. Gonçalves, Silvia V. Pereira, Rodrigo S. Costa, Marcelo P. Pes, Francisco J. L. Lima and Enio B. Pereira
National Institute for Space Research, Av. dos Astronautas, 1758, São José dos Campos SP 12227-010, Brazil

Arcilan T. Assireu, Enrique V. Mattos and Robson B. Passos
Federal University of Itajuba, Av. BPS, 1303, Pinheirinho, Itajubá MG 37500-903, Brazil

Fernando R. Martins and Madeleine S. G. Casagrande
Federal University of São Paulo, Rua Carvalho de Mendonça, 144, Santos SP 11070-102, Brazil

Yongshi Jie
Aerospace Information Research Institute, Chinese Academy of Sciences, Beijing 100094, China
University of Chinese Academy of Sciences, Beijing 100049, China
National Engineering Laboratory for Integrated Aero-Space-Ground-Ocean Big Data Application Technology, Xi'an 710129, China

Xianhua Ji
Engineering Quality Supervision Center of Logistics Support Department of the Military Commission, Beijing 100142, China

Anzhi Yue
Aerospace Information Research Institute, Chinese Academy of Sciences, Beijing 100094, China
National Engineering Laboratory for Integrated Aero-Space-Ground-Ocean Big Data Application Technology, Xi'an 710129, China
Huizhou Academy of Space Information Technology, Institute of Remote Sensing and Digital Earth, Chinese Academy of Sciences, Huizhou 516006, China

Jingbo Chen
Aerospace Information Research Institute, Chinese Academy of Sciences, Beijing 100094, China
National Engineering Laboratory for Integrated Aero-Space-Ground-Ocean Big Data Application Technology, Xi'an 710129, China

Yupeng Deng
Aerospace Information Research Institute, Chinese Academy of Sciences, Beijing 100094, China

Jing Chen
Aerospace Information Research Institute, Chinese Academy of Sciences, Beijing 100094, China
University of Chinese Academy of Sciences, Beijing 100049, China

Yi Zhang
Aerospace Information Research Institute, Chinese Academy of Sciences, Beijing 100094, China
University of Chinese Academy of Sciences, Beijing 100049, China

Ian R. Young and Ebru Kirezci
Department of Infrastructure Engineering, The University of Melbourne, Melbourne, VIC 3010, Australia

Agustinus Ribal
Department of Infrastructure Engineering, The University of Melbourne, Melbourne, VIC 3010, Australia
Department of Mathematics, Faculty of Mathematics and Natural Sciences, Hasanuddin University, Makassar 90245, Indonesia

Miktha Farid Alkadri, Michela Turrin and Sevil Sariyildiz
Department of Architecture and Engineering Technology, Faculty of Architecture and the Built Environment, Delft University of Technology, Julianalaan 134, 2628 BL Delft, The Netherlands

Francesco De Luca
Department of Civil Engineering and Architecture, Tallinn University of Technology, Ehitajate tee 5, 19086 Tallinn, Estonia

Román Mondragón, David Riveros-Rosas, Mauro Valdés, Héctor Estévez and Adriana E. González-Cabrera
Department of Solar Radiation at the Geophysics Institute of the National Autonomous University of Mexico, Mexico City 07840, Mexico

Wolfgang Stremme
Department of Spectroscopy and Remote Perception at the Geophysics Institute of the National Autonomous University of Mexico, Mexico City 07840, Mexico

Alexandra I. Khalyasmaa and Stanislav A. Eroshenko
Ural Power Engineering Institute, Ural Federal University named after the first President of Russia B.N. Yeltsin, 620002 Ekaterinburg, Russia

Power Plants Department, Novosibirsk State Technical University, 630073 Novosibirsk, Russia

Valeriy A. Tashchilin
Ural Power Engineering Institute, Ural Federal University named after the first President of Russia B.N. Yeltsin, 620002 Ekaterinburg, Russia

Hariprakash Ramachandran
Department of Electrical and Electronics Engineering, Bharath Institute of Higher Education and Research, Chennai 600073, India

Teja Piepur Chakravarthi
Department of Computer Science and Engineering, Bharath Institute of Higher Education and Research, Chennai 600073, India

Denis N. Butusov
Youth Research Institute, Saint Petersburg Electrotechnical University "LETI", 197376 Saint Petersburg, Russia

Anders V. Lindfors, Axel Hertsberg and Aku Riihelä
Finnish Meteorological Institute, 00101 Helsinki, Finland

Thomas Carlund
Swedish Meteorological and Hydrological Institute, 60176 Norrköping, Sweden

Jörg Trentmann and Richard Müller
Deutsche Wetterdienst, 63067 Offenbach, Germany

Jinwoong Park, Jihoon Moon, Seungmin Jung and Eenjun Hwang
School of Electrical Engineering, Korea University, 145 Anam-ro, Seongbuk-gu, Seoul 02841, Korea

Cristina Sáez Blázquez, Pedro Carrasco García, Ignacio Martín Nieto, Miguel Ángel Maté-González, Arturo Farfán Martín and Diego González-Aguilera
Department of Cartographic and Land Engineering, University of Salamanca, Higher Polytechnic School of Avila, Hornos Caleros 50, 05003 Avila, Spain

Index

www.ingramcontent.com/pod-product-compliance
Lightning Source LLC
Chambersburg PA
CBHW080248230326

41458CB00097B/4165